S0-BQM-390

Physiology and Pathology of Interferon System

Contributions to Oncology
Beiträge zur Onkologie

Vol. 20

S. Karger · Basel · München · Paris · London · New York · Tokyo · Sydney

International Symposium on
Physiology and Pathology of Interferon System
Smolenice, May 16—20, 1983

Physiology and Pathology of Interferon System

Volume Editors
L. Borecký, V. Lackovič, Bratislava

85 figures and 89 tables, 1984

S. Karger · Basel · München · Paris · London · New York · Tokyo · Sydney

Contributions to Oncology
Beiträge zur Onkologie

National Library of Medicine, Cataloging in Publication
Physiology and Pathology of Interferon System
Ladislav Borecký, Vladimír Lackovič — Basel; New York; Karger, 1984.
(Contributions to oncology = Beiträge zur Onkologie; v. 20)
I. Title II. Series: Beiträge zur Onkologie; v. 20
ISBN 3-8055-3839-1

Drug Dosage
The authors and the publisher have exerted every effort to ensure that drug selection and dosage set forth in this text are in accord with current recommendations and practice at the time of publication. However, in view of ongoing research, changes in government regulations, and the constant flow of information relating to drug therapy and drug reactions, the reader is urged to check the package insert for each drug for any change in indications and dosage and for added warnings and precautions. This is particularly important when the recommended agent is a new and/or infrequently employed drug.

Printed in Czechoslovakia
ISBN 3—8055—3839—1

Contents

Preface IX

I. Present Status of Interferon System

Dianzani, F.; Dolei, A. (Rome): From Isaacs to Escherichia Coli . . 1
Kawade, Y.; Watanabe, Y. (Kyoto): Molecular Aspects of Interferon: An Overview of Current Studies on Mouse Interferon 15
Zhdanov, V. M.; Yershov, F. I. (Moscow): The Role of Interferon in Homeostasis 25
Bocci, V.; Paulesu, L.; Muscettola, M.; Ricci, M. G.; Grasso, G. (Siena): The Physiological Interferon Response. III. Preliminary Experimental Data Support Its Existence 36
Grossberg, S. E. (Milwaukee): Immunopharmacology of the Interferon System 50
Béládi, I.; Pham, Ngoc Dinh; Rosztóczy, I.; Tóth, M. (Szeged): Relationship between Interferons and other Lymphokines . . . 63
Inglot, A. D. (Wroclaw): Interferons in the Light of the New Theory of Hormones 72
Chany, C.; Cerutti, I. (Paris): Interferon in Antitumor Protection and during Gestation 86
Fantes, K. H.; Finter, N. B.; Toy, J. L. (Beckenham): Production and Purification of Lymphoblastoid (Namalwa). Interferon for Clinical Use 98

II. Effect of Interferon on the Cell and the Consequences of Interferon Induction

Kuwata, T.; Tomita, Y.; Fuse, A.; Sekiya, S. (Chiba): Cellular Response to Varied Action of Interferon 113
Kawade, Y.; Higashi, Y.; Sokawa, Y.; Miyata, T. (Kyoto): Primary Structure of Mouse Interferon — Deduced from cDNA Structure; Considerations on Evolution of Interferon Genes 127

Bodo, G.; Adolf, G. R. (Vienna): Formation of Human IFN-Alpha Subtype Mixtures by Different Human Cells 134
Novák, M.; Borecký, L.; Poláková, K.; Russ, G.; Sekeyová, Z.; Lipková, M.; Fuchsberger, N.; Čiampor, F.; Kontsek, P. (Bratislava): Effect of Interferon on Monoclonal Antibody Producing and Proliferating Capacity of Cell Hybridomas 142
Van Damme, J.; Billiau, A.; De Ley, M.; De Somer, P. (Leuven): A β-Interferon-Inducing Lymphokine? 160
Silverman, R. H.; Krause, D.; Jacobsen, H.; Leisy, S. H.; Barlow, D. P.; Friedman, R. M. (Bethesda, Heidelberg, London): Regulation of 2-5A--Dependent RNase Levels during Interferon-Treatment, Growth Inhibition and Cell Differentiation 169
Votyakov, V. I.; Khmara, M. E., Akhrem, A. A.; Moroz, S. G. (Minsk): Antiviral and Physico-Chemical Properties of a Macromolecular Complex Isolated from the Virus-Infected Cell Extract Treated with Ribonuclease 176
Fritsch, R. S.; Fahlbusch, B.; Schumann, I.; Zippel, D.; Zöpel, P. (Jena): Biochemical and Biological Characterization of Tumor-Associated Lymphokine-Like Factors 185

III. Consequences of Interferon Administration to the Animal Organism

Tovey, M. G.; Gresser, I. (Villejuif): Interferon Induced Disease . . 196
Lipková, M.; Borecký, L.; Novák, M.; Sekeyová, Z.; Fuchsberger, N. (Bratislava): Effect of α, β and γ-Interferons on the Growth of Methylcholantrene Induced Tumor Cells in Mice 205
Molomut, N.; Padnos, M.; Pevear, D. C.; Pfau, Ch. J. (New York): A Role of Interferon in the Oncolytic Activity of Pichinde Virus 216
Mécs, I.; Koltai, M.; Toth, S.; Ágoston, É. (Szeged): The Anti-Inflammatory Effect of Human Interferons in Mice 221
Blach-Olszewska, Z.; Cembrzynska-Nowak, M.; Kwasniewska, E. (Wroclaw): Age Related Synthesis of Spontaneous Interferon in BALB/c Mice 224
Kojima, Y.; Yamaguchi, Y.; Kuramoto, H.; Shibukawa, N.; Tamamura, S. (Tokyo): Effect of Experimental Water-Immersion Stress on the Induction of Interferon and Viral Infection 233
Degré, M.; Rollag, H.; Bukholm, G. (Oslo): Interferon Modifies Host Defence Mechanisms against Bacterial Agents 244

IV. Consequence of Interferon Administration to and Production in the Human Organism

Billiau, A. (Leuven): The Clinical Application of Fibroblast Interferon — An Overview 251
Møller, B.; Berg, K. (Aarhus): A Hydrophylic Interferon Gel in the Treatment of Herpes Virus Infections in Man 270

Galabov, A. S.; Mastikova, M. (Sophia): Dipyridamole As an Interferon Inducer in Man 278
Rentz, E. (Berlin): Influence of Thymosin and Thymopoietin Pentapeptide on the Production of Interferon in Lymphocytes . . . 285
Rovenský, J.; Lackovič, V.; Borecký, L.; Žitňan, D.; Lukáč, J. (Piešťany Spa, Bratislava): Interferon Production by Leukocytes Obtained from Patients with Systemic Lupus Erythematosus . . . 290
Waschke, S. R.; Diezel, W. (Berlin): Interferon in Patients with Psoriasis 298
Vaněček, K.; Lehovcová, A. (Prague): Production of Interferon in Children with Chronic Respiratory Disease 304
Orlova, T. G.; Pavlushina, S. V.; Gavrilova, I. E.; Mentkevich, G. L. (Moscow): Synthesis of Gamma Interferon by Blood and Bone Marrow Cells in Children with Hemoblastoses 310

V. The Perspectives of Interferon and Inducer Therapy

Borecký, L. (Bratislava): Interferon: The Lesson Learned from Endocrine and Immunotherapy 318
Kishida, T.; Imanishi, J.; Nakajima, E.; Okuno, T.; Takino, T.; Matsumura, N.; Yoshikawa, T.; Kondo, M.; Yoshioka, H.; Sawada, T.; Nakagawa, Y.; Ueda, S.; Hirakawa, K.; Hirose, G. (Kyoto): Clinical Testing of Interferon in Japan 333
Stringfellow, D. A. (Syracuse): Interferon Inducers: Application As Antiviral and Antineoplastic Agents 346
Levy, H. B.; Chirigos, M. (Frederick): Studies with the IFN Inducer and Immune Modulator, Poly ICLC 358
De Clercq, E.; Torrence, P. F. (Leuven): Poly (G). Poly (C) As an Inducer of Interferon 375

Preface

In May 1983, in the Castle of Smolenice near Bratislava, the Institute of Virology of the Slovak Academy of Sciences organized its second international conference dealing with interferon. While the first conference in 1964 will be remembered as initiating a stimulating exchange of informations among those who joined the Interferon Exchange Group 6, the second one was convened at a time at which great progress has been achieved in production and understanding of the structure of the interferon molecule but the hopes for its broad exploitation in the therapy received little support. It was the conviction of the organizers of the International Symposium "The Physiology and Pathology of Interferon System" in 1983 that we can move ahead only after learning more about the fate and behavior of interferon in the animal organism, its transport and catabolism and its interactions with other regulatory systems such as the immune or endocrine system. No conference can give an answer to all of these questions. However, during the Symposium and after its closing the participants felt that the meeting was justified and that its work was useful.

Selected papers from this Symposium are now offered to the reader.

L. Borecký, V. Lackovič

The matter-of-fact aspects of articles are guaranteed by the editors; the languages responsibilities lie with the authors, the texts have not gone through any languages revisions of the publishing house. (Publisher's note.)

Present Status of Interferon Concept

Contr. Oncol., vol. 20, pp. 1—14 (Karger, Basel 1984)

From Isaacs to Escherichia Coli

F. Dianzani; A. Dolei

Institute of Virology, University of Rome, Rome, Italy

Premise

In 1957 Isaacs and Lindenmann discovered that fragments of chorion-allantoic membrane infected with virus released a protein capable of reacting with normal cells to render them resistant to infection by a wide variety of homologous or heterologous viruses. Since it was immediately clear that this protein was a major mediator of viral interference, it was called "interferon".

Like almost every other true discovery, the entry of interferon into the sceene of biomedical research was not triumphal and its potential for future clinical application was regarded by the majority of the scientific community with a full variety of opinions, ranging from very cautious optimism to frank skepticism. Even malicious humor was involved and someone proposed for interferon a new name: "misinterpreton" (quoted by Stoker, 70). This type of attitude, however, did not prevent a handfull of investigators from working hard over the next 10 or 15 years, not only to show that interferon was real and effective, but also to elucidate its mechanisms and to explore its physiologic role. Most unfortunately, just in the middle of this exciting race the field lost one of its most imaginative leaders when Alick Isaacs died in 1967. His work not only opened a new field, but also generated many of the concepts which led to its growth from infancy to adulthood.

Today, twenty-five years after its discovery, it is still difficult to predict whether, when, and how interferon will be clinically useful; it is, however, unquestionable that the knowledge originated over these years by the interferon research had, and will have, a tremendous impact on today's virology and cell biology.

A step-by-step description of the various critical acquisitions which led to the present "state of the art" would be impossible, and also probably injust; indeed, in the "romantic era" of interferon research a small company of investigators was working friendly and cooperatively, and therefore almost every new discovery was achieved through a sort of collective effort. Indeed, to facilitate the flow of information before publication of the formal papers, a new priceless tool was devised: the "Interferon Scientific Memoranda", still operating.

Describing interferon as it is today, we mean to pay a tribute to all the "Old Guard", especially to those who left us: Kurt Paucker, Lowell Glasgow, George Svet-Moldavsky, Rudolf Weil, and others.

The Infancy of Interferon

Clearly, much of the early efforts were directed to show that interferon was a broadly active, widely distributed, naturally occurring antiviral mechanism. It was then shown that interferon was produced also by human cell cultures (13), as well as by other mammalian cells (75), that it was present in infected tissues of the whole animal (53), that it could be induced, at protective levels, in mice (5); it was finally found that virus infected children produced interferon at levels which were capable of protecting them against a subsequent viral challenge (63).

The work on the mechanisms of production and action was also growing, especially after the introduction of metabolic inhibitors to modulate the cell response following viral or interferon stimulation of cell cultures. It could be then established that interferon is a newly-produced cellular protein (49, 12), which is directly induced by viruses (21).

Although subsequent studies confirmed the idea, originated by Isaacs et al. (54), that the inducing factor is viral RNA, other substances were shown capable of inducing interferon in vivo and in vitro: endotoxin (69, 50), natural and synthetic RNAs (56, 36), polysaccharides (10), polyanions (19), mitogens (76). The lack of information on the subsequently established heterogeneity of the interferon system made these findings somewhat misleading from the concept of an unifying induction mechanism; however the possibility of inducing endogenous interferon with agents which were non infectious, non antigenic, and non toxic disclosed an alternative approach to clinical use of interferon.

Also the studies on the mechanism of action took advantage of the use of metabolic inhibitors. Indeed inhibition of interferon action by actinomycin D (73) showed that the interferon-induced antiviral state is mediated by newly made effector protein(s) which are induced directly by interferon (20).

But the biochemists were at the door: partial purification of chick interferon was achieved (35) allowing further characterization of the interferon protein. It was additionally identified the first of the many mechanism of interferon action at the molecular level. New activities came to light: anticellular (61), antitumoral (2, 45, 77), immunoregulatory (39, 55), along with some very cautious clinical trials (66, 46).

Finally, large scale production of interferon was made possible by conceptual and technical improvement of the production procedures.

The impact of these observations on the scientific community was such as to attract new bright investigators to the field in exponential numbers and the rate of newly acquired information grew accordingly: the infancy of interferon was over.

Interferon Today

As shown in tab. I, three main types of interferon have been identified and originally designated as fibroblast, leukocyte, and immune interferon. According to the newly adopted nomenclature (68), these interferons are now called beta, alpha, and gamma interferon, respectively. Several subtypes are also being identified for each type of interferon (71, 79, 40, 41).

Fibroblast (or beta) interferon is produced by fibroblast and epithelial cells during viral infections. The induction requires virus internalization and exposure of nucleic acid. Viral RNA, either single or double stranded, appears to be the critical factor in trig-

Table I. Types of interferon

Type	References[1]	Producing cells	Inducers
Fibroblast (beta)	Isaacs & Lindenmann 1957 (52) Taylor 1964 (73)	Fibro-epithelial	Viruses, nucleic acids, bacterial products
Leukocyte (alpha)	Gresser 1961 (44) Duc-Gorian et al. 1971 (33)	Lymphocytes	Viruses, B cell mitogens, foreign cells
Immune (gamma)	Wheelock 1965 (76) Green et al. 1969 (42) Youngner and Salvin 1973 (78)	Lymphocytes	Specific antigens, T cell mitogens

[1] References refer to original observation and/or characterization of molecular properties and mechanism of action.

gering the induction process (22). When inoculated intramuscularly, this type of interferon appears to remain confined in the site of injection since blood levels are hardly detected even after administration of large doses (8). Probably because of its poor diffusion from solid tissues, the interferon produced even in minute amounts by virus infected cells, may very soon reach in the extracellular fluid a concentration sufficiently high to quickly establish antiviral protection in the neighboring cells (23, 25).

It is therefore reasonable to infer that interferon beta may have the function of controlling viral replication at the site of the infection, rather than protecting distant organs.

Leukocyte, or alpha, interferon is produced by lymphoid cells induced not only by viruses, but also by xenogenic or tumor cells, bacteria, and B cell mitogens (42, 74, 9, 6).

It is likely that the activation of the interferon response by non-viral inducers may occur after some kind of surface interaction and without the requirement of internalization of foreign genetic material. Recent findings indicate that induction of high titered interferon occurs also after treatment of lymphoid cells with viral envelopes depleted of genetic material (28) or with cells bearing membrane asociated viral antigen, but fixed with gluteraldehyde (57, 16). This strongly suggests that also viruses may induce alpha interferon through a mechanism occurring at the cell surface rather than intracellularly. This view is substantiated by the finding that optimal IFN production requires cell-to-cell contact among the induced lymphocytes (28).

When injected intramuscularly, alpha interferon is released in the bloodstream giving relatively high plasma levels (15). Being produced by leukocytes, it may protect these cells from viruses spreading through the bloodstream and, perhaps more importantly, may increase viral resistance of endothelial cells and enhance effectiveness of natural barriers, such as blood-brain, blood-lung, etc.

Immune, or gamma, interferon is produced by T lymphocytes stimulated by specific antigens or by T cell mitogens; the presence of macrophages is required for optimal production. The induction occurs following oxidation of membrane bound galactose residues by the enzyme galactose oxidase (27, 31). Since specific depletion of galactose residues prevents also the induction by conventional T cell mitogens, non specific oxidizing agents, and specific antigens (tab. II), it appears that oxidation of membrane linked galactose residues is the sole event capable of initiating the induction process. A second critical event for the induction process is the activation of a calcium flow through the lymphocyte membrane, since

Table II. Effect of depletion of galactose residues in lymphocyte membranes on gamma IFN production

Inducer	Interferon production ($\log_{10}$) by cells pretreated with:	
	None	NaNase + beta Galactosidase[a]
None	<1.0	<1.0
Galactose oxidase[a]	3.0	1.0
ConA[a]	2.5	<1.0
PHA[a]	2.2	<1.0
SEA[a]	2.5	<1.0
Sodium periodate[a]	1.8	<1.0
PPD[b]	2.1	<1.0
MLR[c]	2.4	1.5

[a] Data from: Dianzani et al., Inf. Immunity 1982 (32).
[b] 10 μg/ml/10^7 lymphocytes from tuberculine sensitive donors.
[c] Mixed lymphocyte reaction.

interferon gamma production may be induced also by a calcium ionophore (28). Additionally, interferon production by galactose oxidase, conventional T cell mitogens, and specific antigens is inhibited by the addition of chelating agents (tab. III) and blockers of calcium channels (not shown).

Table III. Effect of calcium depletion on gamma interferon induction by various T cell mitogens.

Inducer	IFN production (units $\log_{10}$) after stimulation in the presence or absence of EGTA[2]	
	EGTA present	EGTA absent
A23187[1]	<1.0	2.0
GO+	1.0	3.3
Con-A[3]	<1.0	2.5
PHA[3]	<1.0	2.2
SEA[3]	<1.0	2.5
PPD[3]	<1.0	1.5

[1] Calcium ionophore A 23187, 2 mM.
[2] 5 μM.
[3] as in table II.

Recent studies from our laboratory (31) have shown that production of moderate amounts of a gamma interferon-like molecule may be triggered in cultures of human lymphocytes by inhibitors of DNA-dependent RNA synthesis, such as actinomycin D, alpha amanitine, and 5,6-dichloro-ribofuranosylbenzimidazole. Since IFN production was prevented by subsequent addition of the protein synthesis inhibitors cycloheximide and fluorophenylalanine after actinomycin D treatment, in could be concluded that this production was not due to release of preformed interferon. Interferon production occurred very early after treatment with the metabolic inhibitors; additionally production was not inhibited by addition of cordycepin, a specific inhibitor of transport of nuclear heterogeneous RNA to the cytoplasm. Taken together these data suggest that lymphocytes carry a preformed mRNA for interferon whose translation is prevented by a rapidly turning over mechanism which requires ongoing RNA synthesis, probably similar to the haemin translational control mechanism described in rabbit reticulocytes.

In conclusion, the data shown here strongly suggest that mitogenic induction of human gamma interferon is activated at the lymphocyte membrane through oxidation of galactose residues which, under the experimental conditions described, are the common trigger for all inducers tested. Activation of a calcium flux across the membrane appears to be equally critical, suggesting that calcium flow may play a role in transmitting the membrane "signal" to intracellular recognition sites. Finally, the presence of untranslated mRNA for interferon in the lymphocyte cytoplasm strongly suggests that a post-transcriptional control operates in the cytoplasm and raises the hypothesis that the membrane associated activating events may be sufficient to trigger the genomic control by acting at cytoplasmic site (s).

The antiviral activity by gamma interferon is somewhat lower as compared with the other types of interferon (37), additionally, the kinetics of cell activation are considerably slower and the derepressional mechanism leading to the cellular antiviral state is more complex (26, 3). On the other hand the antitumor activity is greater (65, 18). Taken together these facts suggest the hypothesis that gamma interferon may not be mainly antiviral but may have some other important function (cell-regulatory, immunomodulatory?). Interestingly, however, gamma interferon is capable of enhancing the antiviral and antitumor activity by alpha and beta interferons (37, 38).

The mechanism of the potentation has not been defined, yet, although a possible explanation could be attempted on the basis of the different mechanism of cell activation (28, 29) which could result in a synergistic rather than simply additive effect. This property may have great practical importance, since a several fold potentiation of the antiviral activity by beta interferon may be obtained with the addition of minimal amounts of gamma interferon.

Most of the above acquisitions have been obtained using rather crude interferon preparations. The availability of pure material will certainly bring new insight in the whole field. Interestingly, several subspecies of IFN alpha have been isolated. Although molecular differences among them appear rather slight, increasing evidence suggests that the various subspecies differ in biological activity; namely, optimal antiviral or growth inhibitory activities are carried by different molecules (62). The implication for future application of these findings is clear.

Mechanisms of Antiviral Action

As already stated, interferon must bind to the cell membrane; interferon alpha and beta share the same receptor, coded for by chromosome 21 (34). IFN gamma binds to a different, still unidentified, receptor.

Cells respond to interferon through activation and expression of specific cell genes (58), that occur at different times for the three types of interferon. Interferon alpha and beta induce directly the effector (antiviral) molecules (3), either by transcription and translation of new products, or by activation of molecules present as inactive precursors. This occurs in a one-step derepressional process which takes no more than 30 minutes for production of specific mRNA and no more than 1 hr to establish the antiviral state (24). Interferon gamma, instead, activates the cell through a multiple-step derepressional process (28, 29). In fact at least one intermediate step of transcription and translation must occur before transcription and translation of the effector molecules. Activation of the antiviral state by interferon gamma, as compared by the other types of interferon, is very slow, taking several hours instead of few minutes (26). Virtually all phases of the virus growth cycle have been shown to be affected by interferon at least to some extent.

Molecular mechanisms which govern inhibition of viral growth are very complex; three main metabolic pathways have been described which, in turn or sequentially, may lead to the establishment of the antiviral state against cytopathic viruses (fig. 1). They act primarily at the level of protein synthesis through a mechanism of cascade activation. In cell-free systems these inhibitors act on both cell and virus messages. The whole cell, however, is capable of distinguishing between own and foreign messages. One possible mechanism of discrimination could be the presence of double-stranded RNA (dsRNA) in infected cells only. Pathway 1 of fig. 1

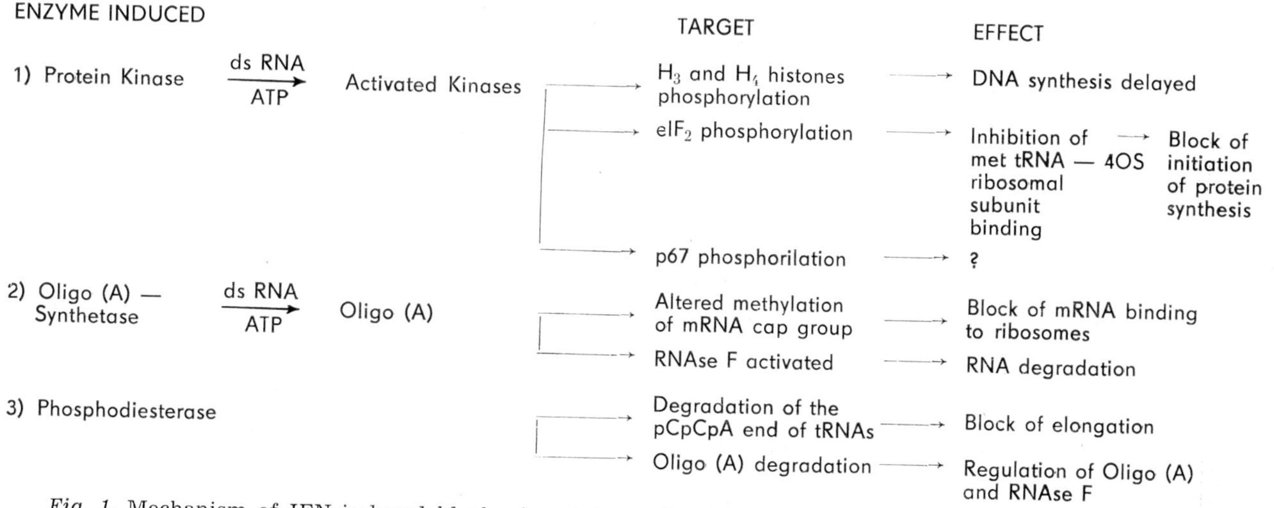

ENZYME INDUCED			TARGET	EFFECT
1) Protein Kinase	ds RNA → ATP	Activated Kinases	→ H_3 and H_4 histones phosphorylation	→ DNA synthesis delayed
			→ eIF_2 phosphorylation	→ Inhibition of met tRNA — 40S ribosomal subunit binding → Block of initiation of protein synthesis
			→ p67 phosphorilation	→ ?
2) Oligo (A) — Synthetase	ds RNA → ATP	Oligo (A)	→ Altered methylation of mRNA cap group	→ Block of mRNA binding to ribosomes
			→ RNAse F activated	→ RNA degradation
3) Phosphodiesterase			→ Degradation of the pCpCpA end of tRNAs	→ Block of elongation
			→ Oligo (A) degradation	→ Regulation of Oligo (A) and RNAse F

Fig. 1. Mechanism of IFN-induced block of protein synthesis.
dsRNA = double-stranded RNA; p67 = protein kinase of M.W. 67.000; eIF_2 = eukaryotic initiation factor 2; cap = a methyl Guanosine residue linked by 5′—5′ condensation to the 5′ end mRNAs. For explanations see text.

is activated by low concentration of dsRNA and is inhibited by high concentration which, in turn, activates pathway 2. Pathway 3, instead, is independent on dsRNA, and can regulate pathway 2.

Moreover, the activation of the three pathways does not occur under the same conditions: for instance, mouse cells treated with low levels of interferon alpha-beta respond with pathway 1 and 2, while activation of pathway 3 requires much higher IFN concentrations.

Beside through direct inhibition of virus growth, the antiviral action of interferon is amplified in vivo through several mechanism, among which activation of macrophages, cytotoxic lymphocytes, and NK cells appear to be the most prominent.

The same is certainly true also for the antitumor action.

Antiviral Action of Interferon in Man

Although small scale treatment of human deseases with interferon and interferon inducers was started in the early 60's, systematic clinical trials were initiated only in the mid 70's and have been greatly expanded over the last few years. Describing the results of these studies would be beyond the purpose of this article. One can only say that properly controlled studies have shown modest, but definitely positive, results in several infections, such as herpes simplex (1), varicella zoster (59, 60), cytomegalovirus (17), rhinovirus (43), juvenile laryngeal papilloma virus (47), and several neoplastic diseases (for a review, see Baron et al., (6)). Many problems, however, are still open and they have to be solved before making interferon available for general use. They include development of large scale production and purification at reasonable cost, more complete definition of pharmacodynamics and pharmacokinetics, establishment of clinical indications for each type, and subtype, of interferon, and possible associations between different types of interferon and other types of treatment.

The Future of Interferon

It is not difficult to predict that the development of new technology, such as recombinant DNA and monoclonal antibody techniques, as well as progress in chemical synthesis, will soon open new areas of research and will clarify some of the many obscure points still existing.

It is not so certain, however; whether we will be soon able to answer a very basic question: what is after all the evolutionary role of the interferon system? The long way from Isaacs to E. coli

Fig. 2. From the tree of biological background, what's in sight? (From S. Steinberg, "The art of living", slightly modified.)

has brought us to a point which is jokingly illustrated by a famous cartoon by S. Steinberg (fig. 2). Apparently we have gotten a solid base but . . . we clearly need another one which may not to be easy to be found. However, if you allow us to express a personal opinion on this matter, it will be found!

References

1 Armstrong, J. A. (1982): Texas Rep. Biol. Med. 41, 571.
2 Atanasiou, P.; Chany, C. (1960): Comp. Rend. Acad. Sci Paris. 251, 1687.
3 Baglioni, C.; Maroney, P. A. (1980): I. Biol. Chem. 255, 8390.
4 Baglioni, G.; Nilsen, T. W.; Maroney, P. A.; de Ferra, F. (1982): Texas Rep. Biol. Med. 41, 471.
5 Baron, S.; Buckler, C. E. (1963): Science, -41, 1061.
6 Baron, S.; Howie, V.; Langford, M.; Mac. Donald, E.; Stanton, J.; Reitmeyer, J.; Weigent, D. (1982): Tex. Rep. Biol. Med. 41, 150.
7 Baron, S.; Dianzani, F.; Stanton, G. J. (1981—1982): In "The Interferon System, a review to 1982", Part I and II Texas Rep. Biol. Med. 141.
8 Billiau, A.; De Somer, P.; Edy, V.; De Clercq, E.; Hermans, H. (1979): Antimicrob. Ag. Chemother. 16, 56.
9 Blalock, J. E.; Langford, M. P.; Georgiades, J. A.; Stanton, G. J. (1979): Cell. Immunol. 43, 197.
10 Borecký, L.; Lackovič, V.; Blaškovič, D.; Masler, L.; Šikl, D. (1969): Acta virol. 11, 264.
11 Braun, W.; Levy, H. B. (1982): Proc. Soc. Exp. Med. 141, 769.
12 Buchan, A.; Burke, D. C. (1966): Biochem. J. 98, 530.
13 Burke, D. C.; Isaacs, A. (1958): Brit. J. Exp. Pathol. 39, 452.
14 Cantell, K.; Strander, H.; Hadhazy, G.; Nevanlinna, (1969): in "The interferons, an international symp," ed. by G. Rita, Acad. Press p. 223.
15 Cantell, K.; Pyhala, I.; Strander, H. (1974): J. Gen. Virol. 25, 453.
16 Capobianchi, M.; Facchini, J.; Di Marco, P.; Dianzani, F. (1983): (In press).
17 Cheeseman, S. H.; Rubin, R. H.; Stewart, J. A.; Tolkoff-Rubin, N. E. (1979): New Engl. J. Med. 300, 1345.
18 Crane, J. L.; Glasgow, I. A.; Kern, E. R.; Youngner, J. S. (1979): J. Nat. Cancer Inst. 61, 871.
19 De Somer, P.; De Clercq, E.; Billiau, A.; Schonne, E.; Claesen, M. (1968): J. Virol. 2, 878.
20 Dianzani, F.; Baron, S.; Buckler, C. E. (1967): in "The interferon, an international symposium", ed. by G. Rita Acad. Press. 147.
21 Dianzani, F.; Buckler, G. E.; Baron, S. (1969): Proc. Soc. Exp. Biol. Med. 130, 519.
22 Dianzani, F.; Pugliese, A.; Baron, S. (1974): Proc. Soc. Exp. Biol. Med. 145, 428.
23 Dianzani, F.; Baron, S. (1975): Nature 257, 682.
24 Dianzani, F.; Levy, H. B.; Berg, S.; Baron, S. (1976): Proc. Soc. Exp. Biol. Med. 152, 593.
25 Dianzani, F.; Viano, I.; Santiano, M.; Zucca, M.; Baron, S. (1977): Proc. Soc. Exp. Biol. Med. 155, 445.
26 Dianzani, F.; Salter, L.; Fleishmann, W. R.; Zucca, M. (1978): Proc. Soc. Exp. Biol. Med. 159, 94.

27 Dianzani, F.; Monahan, T. M.; Scupham, A.; Zucca, N. (1979): Infect. Immunity 26, 879.
28 Dianzani, F.; Monahan, T. M.; Zucca, M.; Scupham, A.; Jordan, C. (1980): in "Interferon Properties and Clinical Uses", ed. by A. Khan and N. O. Hill, Leland Found. Press, Dallas p. 223.
29 Dianzani, F.; Zucca, M.; Scupham, A.; Georgiades, J. A. (1980): Nature 283, 400.
30 Dianzani, F.; Monahan, T. M.; Georgiades, I. A.; Alperin (1980): Infect. Immunity 29, 561.
31 Dianzani, F.; Monahan, T. M.; Jordan, C. A.; Langfod, M. P. (1981): Proc. Soc. Exp. Biol. Med. 167, 338.
32 Dianzani, F.; Monahan, T.; Santiano, M. (1982): Inf. Immunity 36, 915.
33 Duc-Goiran, P.; Galliot, B.; Chany, C. (1971): Arch. Ges. Virusforsh. 34, 232.
34 Epstein, C. J.; Epstein, C. B. (1982): Texas Rep. Biol. Med. 41, 324.
35 Fantes, K. H. (1965): Nature 207, 1298.
36 Field, A. K.; Lampson, G. P.; Tytell, A. A.; Nemes, M. M.; Hilleman, M. R. (1967): Proc. Nat. Acad. Sci. 58, 2102.
37 Fleischmann, W. R. (1977): Texas Rep. Biol. Med. 35, 316.
38 Fleischmann, W. R.; Georgiades, J. A.; Osborne, L. C.; Johnson H. M. (1979): Infect. Immun. 26, 248.
39 Gisler, R. H.; Lindahl, P.; Gresser, I. (1974): I. Immunol. 113, 438.
40 Goeddel, D. V.; Leung, D. W.; Dull, T. J.; Gross, M.; Lawn, R. M.; McCandliss, R.; Seeburg, P. H.; Ulrich, A.; Yelverton, E.; Gray, P. W. (1981): Nature, 290, 20.
41 Goldstein, L. D.; Langford, M. P.; Georgiades, J. A. (1981): Ant. Res. Spec. Iss. ARSRDR 1 (1) 43.
42 Green, J. A.; Cooperband, S. R.; Kibrick, S. (1969): Science 1964, 1415.
43 Greemberg, S. B.; Harmon, M. W.; Tyrell, D. A. J.; Scott, J. M. (1982): Texas Rep. Biol. Med. 41, 549.
44 Gresser, I. (1961): Proc. Soc. Exp. Biol. Med. 108, 799.
45 Gresser, I.; Coppey, J.; Falcoff, E.; Fontaine, D. (1966): Comp. Rend. Seanc. Acad. Sci. Paris 263, 586.
46 Guerra, R.; Frezzotti, R.; Bonanni, R.; Dianzani, F.; Rita, G. (1970): Ann. N. Y. Acad. Sci.
47 Haglund, S.; Lungquist, P.; Cantell, K.; Strander, H. (1981): Arch. Otolaryng. 107, 327.
48 Havell, E. A.; Vilcek, J. (1972): Antimicrobial Ag. Chemother. 2, 476.
49 Heller, E. (1963): Virology 21, 652.
50 Ho, M. (1964): Science 146, 1472.
51 Ho, M.; Tan, Y. H.; Armstrong, J. A. (1972): Proc. Soc. Expl. Biol. Med. 139, 259.
52 Isaacs, A.; Lindnmann, J. (1959): Proc. Roy. Soc. B 147, 258.
53 Isaacs, A.; Hitchcock, G. (1960): Lancet 2, 69.
54 Isaacs, A.; Cox, R. A.; Rotem, Z. (1963): Lancet II, 112.
55 Johnson, H. M.; Baron, S. (1976): CRC Crit. Rev. Biochem. 4, 203.

56 Kleinschimdt, W. J.; Cline, J. C.; Murphy, E. B. (1964): Proc. Nat. Acad. Sci. 53, 741.
57 Lebon, P.; Commoy-Chevalier, M.; Robert-Galliot, B.; Chany, C. (1982): Virology 119, 504.
58 Lengyel, P. (1981): Methods in Enzymology 79, 135.
59 Merigan, T. C.; Rand, K. H.; Pollard, R. B.; Abdallah, P. S.; Jordan, E. W.; Fried, R. P. (1978): New Engl. J. Med. 298, 981.
60 Merigan, T. C.; Gallagher, J. G.; Pollard, R. B.; Arvin, A. M. (1981): Antimicr. Ag. Chempther. 19, 193.
61 Paucker, K.; Cantell, K.; Heule, W. (1962): Virology 17, 324.
62 Pestka, S.; Evinger, M.; Maeda, S.; Rehberg, E.; Familletti, P. C.; Welder, B. (1982): Texas Rep. Biol. Med. 41, 31.
63 Petralli, J. K.; Merigan, T. C.; Willbur, J. R. (1965): New Engl. J. Med. 273, 198.
64 Rubinstein, M.; Levy, W.; Moschera, J.; Lai, C.; Hershberg, R.; Bartlett, R.; Pestka, S. (1981): Arch. Biochem. Biophys. 210, 307.
65 Salvin, S. B.; Younger, J. S.; Nishio, J.; Neta, R. (1975): J. Nat. Cancer Inst. 55, 1233.
66 Scientific Committee on Interferon (1962): Lancet i, 873.
67 Stewart, W. E. (1974): Virology 61, 80.
68 Stewart, W. E. (1980): Nature I 286, 110.
69 Stinebring, W.; Younger, J. S. (1964): Nature, 204, 712.
70 Stoker, M. G. P. (1967): in "Ciba Found symp on interferon", ed. Wolstenholme G. and O'Connor M., Churchill Ltd., London, p. 1.
71 Streuli, M.; Shigekazu, S.; Weissmann C. (1980): Science 209, 1343.
72 Svet-Moldavsky, G. J.; Nemirovskaya, B. M.; Osipova, T. V. S. (1974): Nature 247, 205.
73 Taylor, J. (1964): Biochem., Biophys Res. Com. 14, 447.
74 Trinchieri, G.; Santoli, D.; Knowles, B. B. (1977): Nature, 270, 611.
75 Tyrell, D. A. J. (1959): Nature 184, 452.
76 Wheelock, E. F. (1965): Science 149, 310.
77 Wheelock, E. F. (1966): Proc. Nat. Acad. Sc. U.S.A. 55, 774.
78 Youngner, J. S. and Salvin S. B. (1973): J. Immunol 111, 1914.
79 Seghal, P. B.; Sagar, A. D. (1980): Nature 288, 95.

F. Dianzani, M.D., Institute of Virology, University of Rome, Rome (Italy)

Contr. Oncol., vol. 20, pp. 15—24 (Karger, Basel 1984)

Molecular Aspects of Interferon: An Overview of Current Studies on Mouse Interferon

Y. Kawade; Y. Watanabe

Institute for Virus Research, Kyoto University, Kyoto, Japan.

Introduction

Interferon (IFN) research is now concentrated on human IFN, but mouse IFN is important because many problems related to the physiology and pathology of the IFN system can best or only be studied in the mouse. This article presents a current view of investigations on the nature of mouse IFN molecules, often making reference to the human counterparts.

One of the intriguing problems about the IFN system is the presence of multiple forms of IFN molecules. Its significance in the physiology and defence of the body is at present far from clear, but the nature of the multiple forms of IFN has considerably been clarified. It is now well established that both human and mouse IFNs comprise three antigenically distinct types, α, β and γ. Information on the structure of IFN proteins was difficult to obtain. mainly due to paucity of the materials available, but through application of recombinant DNA technology, human IFN genes were recently cloned (23, 14, 3), and the complete amino acid sequences of human IFN-α, β and γ were determined. Large scale production of human IFNs could also be realized. Cloning of DNAs coding for mouse IFN-α, β and γ was also achieved more recently, and their amino acid sequences were deduced (17, 7, 1). Ways are thus now open for large scale production of mouse IFNs, which will facilitate various mouse model experiments that are relevant to the role of IFN in the body.

Characteristics of Mouse IFN Molecules

Three Types of Mouse IFN and Their Genes

The classification of IFN into the α, β and γ types is based on antigenicity, but they also differ in other properties (19). The cloning of the human IFN genes gave a firm genetic basis for their distinction. It also revealed the presence of multiple IFN-α genes (more than 13) differing slightly in structure from each other (27). For IFN-β and γ, only single genes have been unambiguously identified, although presence of different genes that do not cross-hybridize with those that have already been cloned may not entirely be excluded.

Mouse IFNs are clearly different from human IFNs in antigenicity, target cell spectra and other properties, but the three respective types of human and mouse IFNs resemble each other in various ways (9). At the genetic level, cloned human IFN DNAs were found to be cross-hybridizable to their mouse counterparts. Taking advantage of this, clones of mouse DNA coding for IFN-α (17), β (7) and γ (1) were isolated. The mouse genome was shown to contain, like the human genome, multiple IFN-α genes and single β and γ genes (although, again, presence of more than one β and γ genes may not totally be excluded). These investigations clearly indicated conservation of the IFN system during evolution, and revealed structural homologies between human and mouse IFN proteins.

Molecular Weight

The numbers of amino acids in mouse IFN proteins determined from the base sequences of the isolated clones are 166 and 167 for α (two types), 161 for β, and 146 for γ. These numbers are thus identical or similar to those of human IFNs, as shown in tab. I.

IFN proteins from natural sources have in general higher molecular weights than those of the polypeptides, since they are glycosylated. Human IFN-α is the only exception, for which there has been no clear evidence of glycosylation. Mouse IFN-α, on the

Table I. Multiplicity and molecular weight of human and mouse IFNs

IFN	Number of genes	Number of amino acids[a]	Glycosylation[b]	Molecular weight[c]
Human α	>13	165, 166	No (0)	17,000—24,000
Mouse α	>10	166, 167	Yes (1)	20,000—24,000
Human β	1	166	Yes (1)	22,000
Mouse β	1	161	Yes (3)	26,000—36,000
Human γ	1	146	Yes (2)	20,000, 25,000
Mouse γ	1	136	Yes (2)	22,000, 30,000

[a] Determined from the base sequence of cloned DNA.
[b] Figures in parentheses indicate the number of potential N-glycosylation sites (Asn-X-Ser or Thr) in the amino acid sequence.
[c] Determined by SDS-PAGE for IFN from natural sources.

other hand, is glycosylated, and shows a molecular weight of 20—24K in SDS-polyacrylamide gel electrophoresis (PAGE). IFN-α from L cell (24K) is reduced in molecular size to 18K when produced in the presence of tunicamycin (2). This indicates that most, if not all, of the natural mouse IFN-α molecules are glycosylated. The two mouse IFN-α subtypes that have so far been sequenced have a single N-glycosylation site (Asn-X-Ser or Thr) at a position identical to that in the sequence of human IFN-β (17).

The molecular size of mouse IFN-β (36K for L cell IFN-β) is in general greater than human IFN-β (22K); in some preparations, the size distribution is heterogenous or split into two peaks in SDS-PAGE (24—28K and 30—35K) (9). Since the mouse IFN-β polypeptide has three potential N-glycosylation sites, in contrast to one in the human IFN-β polypeptide, the differences in their molecular sizes may largely be attributable to the carbohydrate moieties (7).

The molecular size of IFN-γ has been controversial. The values estimated by gel filtration were much higher than those of IFN-α and β with both human and mouse IFN-γ. SDS-PAGE, which should give molecular size estimates free of aggregation, was not applied, since IFN-γ is extensively inactivated by SDS, until Yip et al. (29) found that some 10 % of the activity of human IFN-γ survived in SDS, and migrated in two main peaks with 25K and 20K. Our current studies on mouse IFN-γ from a clonal T cell line indicated similarly the presence of two components with 30K and 22K in SDS-PAGE; the IFN produced in the presence of tunicamycin contained 18K species, presumed to be the nonglycosylated form (25). Its size is close to the value (17K) deduced from the structure of cloned cDNA for IFN-γ (there are two N-glycosylation sites in both the human and mouse IFN-γ sequences (3, 1). It seems likely, but has not been proved, that the two electrophoretic components are the products of a single gene, perhaps glycosylated differently or proteolytically processed differently. The problem also remains whether the molecular sizes determined by SDS-PAGE on the SDS-resistant activity truly represent the whole population of the IFN-γ molecules.

Role of Glycosylation

The sugar moiety of IFNs seems not to be essential for the biological activities, since nonglycosylated molecules produced in the presence of glycosylation inhibitiors are active, as are the recombinant IFNs produced by genetically engineered bacteria, which lack carbohydrates. However, in the case of IFN-β, glycosylation seems to have a positive role in the production and secretion from animal cells, since the yields of IFN-β from both human and mouse cells are strongly reduced by the addition of glycosylation inhibitors (6,24). We also found reduction of IFN-γ yields from mouse T cells when tunicamycin was present (24). In contrast, production of mouse IFN-α was not affected by added tunicamycin (2), indicating that the significance of the sugar moiety differs depending on the species of IFN.

In general, the sugar moiety may confer higher heat stability to the molecule. Furthermore, it may have an important role in animal body, affecting pharmacokinetic behavior and tissue interactions of IFN molecules. This question needs to be thoroughly investigated, especially in connection with clinical uses of recombinant IFNs.

Protein Structure

The complete amino acid sequences of mouse α (two subtypes) and β IFNs deduced from the base sequences of cloned DNAs are shown in another article in this volume (11). Human and mouse α IFNs shows high homology in sequence to each other (57 to 62 %) (17), while their β IFNs have less but still strong homology (47 %) (7). The sequence of mouse γ IFN has not yet appeared in print, but was reported to be 40 % homologous to human γ IFN (1).

There is about 25 % homology between human α and β IFNs and also between mouse α and β IFNs, but so far no antigenic cross-reactions have been found, except for a monoclonal antibody which was reported to react with both mouse α and β IFNs (13). Cross-reactions can be demonstrated, although at low relative titers, between human and mouse α IFNs (but not between their β and γ IFNs) using high-titered conventional antibodies (10).

Attempts are being made to deduce the three-dimensional structure of IFN proteins from their primary structure, and to make inferences on structural features responsible for the biological activity (18). Since some recombinant human α proteins were crystallized, it will not be long before their 3-D structure is determined. Various monoclonal antibodies are now becoming available, and they will eventually allow determination of the antigenic sites, cell receptor binding site and so on.

Properties of Some Subtypes of Mouse IFN-α

Mouse IFN-α has weak activities on heterologous cells, in contrast to mouse IFN-β which is virtually inactive (2). Stewart and Havell (20) showed that the activity of mouse IFN-α on human cells was carried by a minor subspecies, which reacted well with antibody to human IFN-α.

Since the genes of human IFN-α subtypes can individually be expressed in bacteria, detailed studies on each of the multiple IFN-α proteins have become possible. The different subtypes of human IFN-α have been shown to differ in their antiviral, cell growth-inhibiting, and cross-species activities (26, 21, 16). Similar studies are now becoming possible with mouse IFN-α. The two subtypes of mouse IFN-α so far cloned showed clear differences in cross-species activities and in neutralizability by antisera (17, 22). To understand the mouse IFN-α system more fully, it will be worthwhile to clone more different mouse α IFN genes and examine the properties of the IFN proteins.

Problems in IFN Induction

In contrast to the progress in our understanding of the nature of IFN molecules and the mechanisms of action, little has been known about the molecular mechanisms of IFN induction. Various cells are presumed to have an identical set of various IFN genes, but it is clear that only some of them are derepressed in a given combination of cell and inducer. For instance, human leukocytes and fibroblasts induced by virus or polynucleotides produce IFN consisting solely or overwhelmingly of α and β IFNs, respectively. With mouse cells, on the other hand, α and β IFNs are concomitantly produced by various cells examined so far, including fibroblasts and leukocytes (although the α type tends to predominate in lymphocyte-derived IFNs and the β type in fibroblast-derived IFNs) (28). The differential expression of a particular gene upon induction may principally be a problem of transcriptional control. One could hope that information on eukaryotic genes and their expression now being accumulated rapidly may shed light on this problem in near future.

IFN-α from natural sources appears in general to be a mixture of many subtypes, but little is known at present of their exact compositions in various preparations. To understand induction mechanisms, and also to be able to precisely evaluate clinical results, it will be necessary to develop methods of determining the subtype compositions.

As to IFN-γ induction, the systems usually employed (blood leukocytes or spleen cells) comprise heterogeneous cell populations, and there are problems still not resolved well, such as what cell types are IFN-producers and what roles accessory cells play. Clonal lymphocytes and hybridomas will therefore be more suitable for molecular analyses of induction mechanisms.

When one comes to in vivo systems, the situation is naturally more complex, and diverse types of inducers that are not effective in tissue culture systems become effective. Lymphocytes and macrophages are important participants, but complex interplay of many factors must take place. For instance, T cell mitogens, that are well known IFN-γ inducers in vitro, do not usually induce IFN when given to mice; mitogens seem to activate an inhibitory mechanism for IFN-γ induction (8). However, when mice are pretreated with certain bacteria, they respond to mitogens with high level production of IFN-γ (15). Interestingly, bacterial lipopolysaccharide, which is known to be an inducer of IFN-(α, β), triggers production of IFN-γ in addition to (α, β) in bacteria-treated mice ("immune" induction of IFN-γ is unlikely here). Physiological significance of this IFN induction is unclear, but could be related to the anti-tumor and immunomodulating actions of certain bacteria.

Production and Purification of Mouse IFN

The commonly used sources of mouse α and β IFNs are L, C243, and Ehrlich ascites tumor cells (9). The IFNs from these three sources were completely purified, and have specific activities of about 10^9 IU/mg protein. The line of L cell we use (denoted L0) produces 0.1–0.2 IU of IFN-(α, β) per cell when induced by Newcastle disease virus, and it is not difficult to prepare, say, 10^9 IU in ordinary laboratories. We can now purify it nearly completely in two steps, using Controlled Pore Glass and antibody columns (24). These IFNs are mixtures of α and β. For their separation, gel filtration or electrophoresis can be used for moderate amounts, but for large-scale separation, other methods must be devised, perhaps utilizing monoclonal antibodies.

IFN-γ is usually produced by mouse spleen or lymph node cells induced by mitogen, and it is laborious to obtain it in large amounts. Clones of normal T cells, propagated with the help of T cell growth factor (TCGF), if capable of producing IFN-γ, may be more suitable, although the supply of large amounts of TCGF may be a problem. Human T cell hybridomas that can grow without TCGF and are capable of IFN-γ production have been obtained (12), and it seems highly desirable to establish similar mouse cell lines.

Purification of mouse IFN-γ has not yet been achieved, and the specific activity of pure IFN-γ is not known. Human IFN-γ has recently been purified well (Yip et al., 1982), and it appears possible that the specific activity of pure IFN-γ is considerably lower (ca. 10^7 U/mg protein) than those of IFN-α and β (10^8–10^9 IU/mg protein) (although the unitage of both human and mouse γ IFNs is poorly defined because at present no international references are available). Purification of mouse IFN-γ produced by a normal T cell clone in serum-free medium is now in progress in our laboratory by a combination of several chromatographic techniques.

Mass production of IFN, however, will eventually rely on genetic engineering techniques, as in the case of human IFNs. One of the advantages is the availability of single α subtype IFNs which cannot readily be obtained from natural sources. Also, comparisons of normal and nonglycosylated IFNs as to their biological activities will be greatly facilitated, if the DNAs are expressed both in bacteria and in animal cells.

Future Prospects

In spite of the great progress in our understanding of IFN proteins and genes and IFN actions in vitro, we know rather little about the role of IFN in normal functioning of the body. For instance, we do not know whether endogeneous IFN really regulates immune responses under normal conditions, nor whether IFN has anything to do in normal development. The effects of exogenously added IFN on immune responses in mice have been examined, but this type of experiments will have to be expanded also to other physiological functions (4). An especially acute problem is the supply of purified IFN-γ which is at present quite inadequate in quality and quantity. Hopefully, the recombinant DNA techniques will solve the problem of supply of various kinds of IFN in near future. Especially, preparations of individual mouse α IFN subtypes are interesting, since they might allow to dissect the biological roles played by the different α subtypes.

In clinical trials of IFN, some patients were found to develop antibodies to IFN. This could be a serious problem for the safety and efficacy of IFN administration. To better understand this unexpected phenomenon, experiments on mouse models appear to be highly desirable.

Administration of IFN-neutralizing antibodies to mice will be another important approach to understand the role of IFN in the body, as has already been done using mixed anti-α-anti-β antibodies in connection with virus infection of mice (5). For this purpose, antisera against individual mouse α, β and γ IFNs with much higher titers than currently available will be needed. A panel of monoclonal antibodies with high neutralizing titers against various

IFN species would be most valuable, because of specificity and the quantity that can be prepared.

Recently, it has become feasible to artificially introduce a particular gene into early embryo cells of mice and grow such mice harboring the extra gene. "IFN mice", which have extra IFN genes chosen by the experimenter, if successfully produced, might tell us much about the action of IFN in the body.

Through these and other approaches based on firm knowledge on the IFN proteins and genes, one may hope that the physiological and pathological significances of the IFN system will be gradually unraveled.

References

1 Derynck, R. et. al. (1983): in "Biology of the Interferon System", ed. by H. Schellekens, Elsevier, Amsterdam (in press).
2 Fujisawa, J.; Y. Kawade (1981): Virology 112, 480.
3 Gray, P. W.; Leung, D. W.; Pennica, D.; Yelverton, E.; Najarian, R.; Simonsen, C. C.; Derynck, R.; Sherwood, P. J.; Wallace, D. M.; Berger, S. L.; Levinson, A. D.; Goeddel, D. V. (1982): Nature 295, 503.
4 Gresser, I. (1982): in "Interferon 4, 1982", ed. by I. Gresser, Academic Press, London.
5 Gresser, I.; Tovey, M. G.; Bandu, M.-T.; Maury, C.; Brouty-Boyé, D. (1976): J. Exp. Med. 144, 1305.
6 Havell, E. A.; Yamazaki, S.; Vilcek, J. (1977): J. Biol. Chem. 252, 4425.
7 Higashi, Y.; Sokawa, S.; Watanabe, Y.; Kawade, Y.; Ohno, S.; Takaoka, C.; Taniguchi, T. (1983): J. Biol. Chem. (in press).
8 Johnson, H. M. (1981): Antiviral Res. 1, 37.
9 Kawade, Y. (1982): Texas Rep. Biol. Med. 41, 219.
10 Kawade, Y.; Watanabe, Y.; Fujisawa, J.; Dalton, B. J.; Paucker, K. (1981): Antiviral Res. 1, 167.
11 Kawade, Y.; Higashi, Y.; Sokawa, Y.; Miyata, T.: this volume.
12 Le, J.; Vilcek, J.; Saxinger, C.; Prensky, W. (1982): Proc. Nat. Acad. Sci. USA 79, 7857.
13 Männel, D.; De Mayer, E.; Kemler, R.; Cachard, A.; De Maeyer-Guignard, J. (1982): Nature 296, 664.
14 Nagata, S.; Mantei, N.; Weissmann, C. (1980): Nature 287, 401.
15 Okamura, H.; Kawaguchi, K.; Shoji, K.; Kawade, Y. (1982): Infect. Immun. 38, 440.
16 Rehberg, E.; Kelder, B.; Hoal, E. G.; Pestka, S. (1982): J. Biol. Chem. 257, 11497.
17 Shaw, G. D.; Boll, W.; Taira, H.; Mantei, N.; Lengyel, P.; Weissmann, C. (1983): Nucl. Acids Res. 11, 555.

18 Sternberg, M. J. E.; Cohen, F. E. (1982): Int. J. Biol. Macromol. 4, 137.
19 Stewart, II W. E. (1979): "The Interferon System". Springer Verlag, Vienna.
20 Stewart, II W. E.; Havell, E. A. (1980): Virology 101, 315.
21 Streuli, M.; Hall, A.; Boll, W.; Stewart, II W. E.; Nagata, S.; Weissmann, C. (1981): Proc. Nat. Acad. Sci. 78, 2848.
22 Taira, H.; Kawakita, M.; Watanabe, Y.; Kawade, Y.; Weissmann, C. (1982): Paper read at the 55th Annu. Meeting Japan. Biochem. Soc., Osaka.
23 Taniguchi, T.; Sakai, M.; Fujii-Kuryiama, Y.; Muramatsu, M.; Kobayashi, S.; Sudo, T. (1979): Proc. Japan. Acad. 55, B, 464.
24 Watanabe, Y.; Kawade, Y. (1983): J. Gen. Virol., in press.
25 Watanabe, Y.; Taguchi, M.; Iwata, A.; Namba, Y.; Kawade, Y.; Hanaoka, M. (1983): in "The Biology of the Interferon System", ed. by H. Schellekens, Elsevier, Amsterdam (in press).
26 Weck, P. K.; Apperson, S.; May, L.; Stebbing, N. (1981): J. Gen. Virol. 57, 233.
27 Weissman, C.; Nagata, S.; Boll, W.; Fountoulakis, M.; Fujisawa, A.; Fujisawa, J.-I.; Haynes, J.; Henco, K.; Mantei, N.; Ragg, H.; Sehein, C.; Schmid, J.; Shaw, G.; Streuli, M.; Taira, H.; Todokoro, K.; Weidle, U. (1982): Phil. Trans. R. Soc. Lond. B299, 7.
28 Yamamoto, Y. (1981): Virology 111, 312.
29 Yip, Y. K.; Barrowclough, B. S.; Urban, C.; Vilcek, J. (1982): Proc. Nat. Acad. Sci. USA 79, 1820.

Y. Kawade, M.D., Institute for Virus Research, Kyoto University, Kyoto (Japan)

Contr. Oncol., vol. 20, pp. 25—35 (Karger, Basel 1984)

The Role of Interferon in Homeostasis

V. M. Zhdanov; F. I. Yershov

D. I. Ivanovsky, Institute of Virology, Academy of Medial Sciences, Moscow, USSR

Within a quarter of century after their discovery, interferons evolved from a peculiar virologic phenomenon to a basic interdisciplinary and integrative division of biology. As a matter of fact, interferons, being perhaps the simplest model of expression of inducible proteins, are of interest for molecular biologists. They attract attention of physicians as an universal antiviral preparation with a broad spectrum of action, and they interest immunologists as an important factor of non-specific resistance.

In my paper I would like to present an approach to interferon, not from the point of view of a specialist in this already vast field of knowledge, but rather from the point of view of an evolutionist who tries to understand how this peculiar system of the biologic homeostasis appeared and developed.

Homeostasis appeared when chemical evolution was replaced by biologic evolution. The first system of homeostasis became the genetic code, which marked the beginning of the biologic era on the Earth. (It is assumed that at first the code was a two-letter one and later became the three-letter code.) At this period of time the final selection of twenty amino acids occured from many thousands of them. These amino acids specify all existing forms of life on the Earth — from the simplest viruses to man or the flower. It is still not known why Nature selected these twenty amino acids; the three-letter genetic code alloved three time as many amino acids as exist now. One may only suggest that combination of only these amino acids resulted in the formation of a biologically active product such as oligopeptide neurohormones etc. The degeneracy of the genetic code became an additional guarantee of its reliability.

With the appearance of the genetic code evolution became inevitable. Life turned from an unsteady equilibrium to a steady disequilibrium.

However, at a certain stage of evolution of Prokaryotes, when their genome reached the size of 10^9 dalton, the genetic code could not guarantee homeostasis of the bacterial cell, as far as reading mismatch errors, e.g. the template DNA synthesis made it impossible to reproduce exactly such large genomes. Therefore a complex system of the template DNA synthesis evolved, which included the synthesis of relatively shorts Okazaki (1) fragments, correction of errors, the repair system etc.

This type of the template DNA synthesis was preserved also in Eukaryotes. Their genomes increased to 10^{12} daltons and were separated (in the nucleus) from the cytoplasm as chromosomes – the DNA threads being associated with histones. It is interesting that such complex template synthesis of DNA did not develop for RNA (both cellular and viral). It would be not justified economically (in the general biologic sense), since errors in RNA synthesis and translation cause no important harm to the cell. It is therefore no accident that the size of RNA never exceeded 10 million dalton – a critical size when reading mismatch errors. However, the Okazaki system that guaranteed the fidelity of replication of the genome had an alternate side of the medal. The genetic apparatus became too rigid, having been unable to adapt quickly to changing conditions of the milieu. Therefore, the evolution of the organic world made the following step (still during the prokaryotic stage): it developed mobile genetic structures – the transposons that were able either to translocate genes or to establish strong promotors before them.

The template synthesis with the correction and repair system solved the problem of heredity and, at the same time, did not close the possibility of mutations that ensured evolution of the organic world on the Earth. It is true that mutations and recombinations, together with natural selection, allowed microevolution, whose maximal result does not exceed species formation. In other words, if only this evolution mechanism existed (mutations and recombinations), nonspecific natural selection would drive evolution of the organic world into a deadlock at a very early stage (from bacteria only various forms of bacteria could evolve).

However, even at the earliest stages of life, progressive evolution or macroevolution took place. This created the diversity of existing forms of the organic world on the Earth. Three mechanisms ensured macroevolution.

Firstly, at the early stages of development of life endosymbiosis played a great role. I cite the concept by L. Sagan, that was brilliantly confirmed by the method of nucleic acid homology. Just endosymbiosis ensured the possibility of transformation of Prokaryotes into Eukaryotes, as far as endosymbiotic Prokaryotes that inhabited the cytoplasm of Protokaryotes gradually transformed to so important organelles like chloroplasts, mitochondria, the mitotic spindle apparatus. We even may speculate which groups of Prokaryotes (ancestors of modern bacteria) gave rise to organelles of Eukaryotic cells. We can say this at least, for two of them: the ancestors of chloroplasts are apparently ancient Cyanobacteria, whereas the ancestors of mitochondria may be ancient Rhodobacteria.

Secondly, at the later stages of evolution duplication of genes and genomes played an important role. As S. Ohno – the author of this concept – states, the beginnings of multicellular organisms needed new genes, and these could originate from abundant cistrons whose mutations created new genes (2).

Thirdly, at all stages of evolution of the organic world viruses played a certain role as carriers of clusters of genetic information – a concept developed by the author of the paper. The genome of bacteria presents indisputably a mosaic, 30 % of which consists of genomes of integrated viruses. Such important genetic information as production of toxins or enzymes that destroy antibiotics were probably transmitted by viruses. But in higher animals including man, the genome contains integrated proviruses, for example endogenous oncoviruses that carry out some important, but still not understood functions. Cancer is only an indication of useful genetic information brought by viruses.

Appearance of multicellular organisms occured in all four branches of Eukaryote evolution, but only in plants and in animals did it become the principal direction of evolution. We will not discuss here the pathways of plant evolution, because it would take us away from the topic of our paper, and therefore we point out some pathways of evolution of animals.

If one considers all the diversity of the animal kingdom, both present and past, we can conclude that the most progressive (from the biologic point of view) evolution is that of Vertebrates. Analysing Invertebrates one may see not only the diversity of evolutionary directions, but also a lack of appearance of intellect. But the absence of intellect is not the only lack in Invertebrates: they have not such perfect systems of homeostasis which Vertebrates possess.

This relates, at first, to the immune system. The system of immunity appeared as a result of the development of multicellular organisms. If one considers the average frequency of mutations of about 10^{-6}, an organism consisting of 10 million cells cannot exist and evolve if it does not possess a system e.g. the immune system of recognition and elimination of mutating somatic cells. Mutations of genes are finally expressed as antigenic variations of proteins coded by them and the immune system has evolved as a system of the protein homeostasis – of recognition and elimination of antigenically changed proteins and, subsequently of cells containing these proteins. As R. V. Petrov (3) states the immunologic supervision over the internal stability of multicellular populations of the organism, e.g. over homeostasis, is the main function of the immune system. Recognition and destruction of foreign cells with alien protein composition, including microorganisms and viruses is the main function of the immune system. Another function is recognition and elimination of neoplastically transformed cells.

However, the formation of the immune system – of the system of protein homeostasis and of genetic stability appeared to be insufficient for providing progressive evolution of Vertebrates. A system was needed that would directly recognize and eliminate the foreign genetic information. Such a system became the system of interferon.

Mutating cells of an organism, including neoplastically transformed ones and diverse microorganisms (bacteria, fungi, protozoa) are sheathed in a protein envelopes, while their naked genetic material is usually non-functional. Quite different are viruses. At certain stages of their intracellular reproduction viruses compete at the genetic level. Hence, in the course of evolution of the organic world a system was formed that insured the nucleic homeostasis not indirectly – through proteins, but directly acting upon nucleic acids through recognition and elimination.

The formation of such system was dictated by the fact that

viruses are not only the causal agents of infectious diseases, but the permanent companions of all forms of the organic world. We have already mentioned the role of viruses in evolution of the organic world, and now we would stress the ubiquity of viruses. According to our definition viruses are autonomous genetic structures capable of functioning and reproducing themselves in the cells of animals, plants, protozoa, fungi and bacteria. This definition fits not only conventional viruses, but also plasmids, viroids and even prions – the agents like scrapie. As everyone now recognizes, viruses origined from genetic elements of cells. We would like to stress that viruses appeared and evolved simultaneously with cellular forms of life and thereafter always accompanied the complete cellular forms of life.

The system of interferon arose at the same time when the system of immunity appeared, e.g., in Vertebrates. But from the very beginning it differed from the immune system in that the target of its action were foreign nucleic acids, and the ways of recognition and elimination of them had no similarity with those of proteins. Proteins are constructed of twenty amino acids that have different configurations thus providing the diversity of the structure of antigenic determinants. Nucleic acids are constructed of four nucleotides that form two conformationally similar pairs (purines, pyrimidines), and therefore antigenic properties of various nucleic acids are similar and there is no diversity of antigenic determinants among them. The immune system possesses specialized cells and organs, and the diversity and specificity of antibodies reflects the diversity of protein antigenic determinants. The interferon system has neither specialized cells nor, moreover, organs. It must exist in every cells, as each cell may be infected by virus and therefore must have a system of recognition and elimination of the foreign nucleic acid. Evolution of the immune system proceeded by increasing the diversity of antibodies and specialization of the immune system cells (B and T lymphocytes, helpers, suppressors, effectors etc.). Evolution of the interferon system developed in the way of species specific recognition.

The diversity of physiologic functions of interferon that is revealed and studied nowadays undoubtedly indicates their controlling and regulatory role in the preservation of homeostasis. The main functions of interferons are summarized. They may be subdivided into antiviral, antimicrobial, antiproliferative (including antitumorogenic), immuno-modulating, radioprotective and other functions.

Even a simple enumeration of interferon effects shows that the importance of this system may be comparable with the system of immunity, and is even more universal than the latter. This universality makes it an important factor of non-specific resistance.

Now, based on these evolutionary considerations, let us analyse some peculiarities of the interferon system, the mechanisms of its action, the main task of the system and its complementary functions, and also the correlation and interrelation of the interferon and the immune systems.

If we proceed from the statement that viruses caused the formation of the interferon system, a natural question's why did this system arise only in Vertebrates? This may be answered by another question: why did the modern immune system appear only in Vertebrates? It is apparent that the evolutionary process proceeded by the method of trial and error. Even intellect was created after two unsuccessful attempts (the Octopus and the social insects). Only when Vertebrates appeared did intellect inevitably evolve. But even in this case evolution marked time for a long period, having lost about 300 million years with dinosaures. Finally the last duplication of the genome occured and Mammalia arose. Therefore I will stop with numerous "whys" that cannot be answered yet and return back to the topic of the paper – the interferon system.

We have noted the main function of the interferon system – the recognition and elimination of foreign nucleic acids. If one considers viruses it should be remembered that their genetic material is extremely diverse: single-stranded and double-stranded RNAs and DNAs, linear, circular and fragmented ones. Nature seems to have tested in viruses all possible forms of genetic material before finally selected its canonical from – the double-stranded DNA – which became universal for all cellular organisms, including all Prokaryotes and Eukaryotes.

Although there is a large diversity in the replication cycles of foreign genetic material, there is one point where it is possible to stop it: initiation of translation. Just here is where the mechanism of the interferon system is working. We may present schematically the mode of action of interferon as follows. Interferon binding to the cell membrane induces the synthesis of a protein kinase which phosphorylates one of initiating factors of translation. The phosphorylated initiation factor becomes unable to participate in the initiation of translation, therefore the formation of the initiating complex is breaked. The selective inhibition of translation of viral templates may be due either to the more pronounced sensitivity of the viral translation system to phosphorylation or to the specific switch off of translation of the infected cell. In cells treated with interferon the synthesis of an enzyme is induced that catalyzes the synthesis of 2,5-oligoadenylic acid. This switches on cellular nucleases which destroy free (viral) messenger RNAs. Thus, those viral messenger RNAs that could not bind to ribosomes, undergo destruction by cellular nucleases. Both effects are connected with the functioning of double-stranded RNAs in interferon-treated cells.

We attempted to reveal both poly-A synthetase and protein-kinase among the products of translation of anti-viral protein (AVP) messenger RNA. For this purpose AVP—mRNA was translated in oocytes of Xenopus laevis. The presence and activity of the first enzyme was established by the synthesis of the oligo-adenylate in the column with poly(I):poly(C)—Sepharose which preliminarily absorbed the translation products of AVP—mRNA. It was shown that among the products, oligo-A-synthetase was found. This catalyzes the synthesis of the (2—5A) translation inhibitor from ATP. The kinetics of inhibition of translation of endogenous (globin mRNA) and exogenous (tobacco mosaic virus mRNA) templates was shown after the addition of the inhibitor to a rabbit reticulocytes cell-free system of protein synthesis.

It was also shown that activation of newly synthesized inhibitors of ^{3}H—RNA hydrolysis occurs after the addition of interferon to the cell sap of. Besides oligo-A-synthetase, the presence of protein-kinase was shown among the products of AVP—mRNA.

Without presenting this important data, which supports our general concept on interferon, we consider it necessary to stress once more that interferon acts through the existing regulation of the nucleic acid synthesis. This is accomplished by activating enzymes and inhibitors, that block translation or degrade the foreign (viral) genetic information.

Thus, the blocking by interferon at the stage of translation initiation and the destruction of messenger RNAs is the universal mechanism for blocking viral infections.

It is not clear how the interferon system recognizes foreign nucleic acids, or, more specifically, alien messenger RNAs. The price for this recognition is the stoppage of the general translation in the cell – hence the antiproliferative action of interferon. In general, this negative effect of interferon has a positive side – inhibition of neoplastic growth. Here the immune system joins with the interferon system.

But this fact is far from exhausting the interaction of the immune and interferon systems. Interferon possesses immunomodulating action. It stimulates phagocytosis, the activity of natural killer cells and the expression of the major histocompatibility antigen complex. From the other side, it inhibits the formation of antibodies, the development of anaphylactic shock, delayed-type hypersensitivity, lymphocyte proliferation, transplantation reactions and the complement fixation-reaction.

Many of these effects of interferon are easy to understand, if one remembers the main function of interferon and the mechanism of its action. Since interferon action stops translation, it is easy to understand the inhibition of antibody formation, of lymphocyte proliferation, and of cell reactions to a transplant. But we need to study and explane such effects as the stimulation of phagocytosis the enhancement of killer activity or the expression of the main histocompatibility complex.

It was interesting to study the influence of interferon inducers upon the vaccinal response. The prophylactic effect of the combined application of interferon with vaccines was determined on the models of the experimental influenza, herpetic encephalitis and rabies and tick-borne encephalitis.

The vaccines and interferon inducers, being employed separately, did not possess a sufficiently high protective action. However, with combined application a marked additive (herpes) or stimulatory effect (influenza, rabies, tick-borne encephalitis) was observed (in a range of 3 to 8 fold). Thus, these experiments show additionally, that the system controlling nucleic homeostasis (interferon) reliably influences the system controlling protein homeostasis (immunity).

Les studied is feedback inhibition. That is how does the immune response influence interferon. There are many reasons to assume, that this influence does exist and that interferons may be considered as specific kinds of cytokines. For example, we know that there is a sharp decrease of interferon production activity of leukocytes obtained from patients with the decreased immunoreactivity, e.g. in oncologic, autoimmune, and chronic viral infectious diseases.

As it is true for the formation of antibodies, the production of interferon needs induction. There are various inducers of interferon synthesis: the most active of them are double-stranded RNAs (replicating or transcribing RNA viral genomes) or imitators of them — copolymers of poly G:C or poly I:C type, and, similar polyanions. That's quite understandable. It's more difficult to understand why interferon inducers are also bacterial endotoxins, or quite simple chemical compounds. The problem of interferon inducers needs further study.

For those studying interferon a well known phenomenon is hyporeactivity after the repeated interferon induction. It develops very soon (1—2 days) after the first induction and lasts for 5—6 days. We found that such hyporeactivity could be overcome by the change of interferon inducers. In other words, the use of inducers of different nature, particularly when a successful pair of inducers is found, could support the production of interferon at a sufficiently high level.

Interferon does not act directly, it acts through the cell membrane. This is also a problem that needs further study. To schematize, one may say that interferon does not rescue the cell infected by virus, but it protects the neighbour cells from viral infection.

Since interferon is the main factor in nucleic homeostasis it must be connected with the whole preexisting system of DNA synthesis. This problem is almost completely open for study, however there is at least one fact showing the interrelation of the interferon system with the system of DNA synthesis. This is the radioprotective action of interferon that is undoubtedly connected with the system of DNA repair.

From this point of view one more problem needs study. Interferon has apparently no influence upon integration processes that occur in cells infected with oncogenic viruses and particularly with oncoviruses and other retroviruses. This phenomenon is not consistent with the close relation of the interferon system with the system of DNA synthesis.

Undoubtedly interferon is an important or even the main factor in recovery from viral infection. Let us take as an example influenza. In favourable conditions an influenza patient recovers within 3 days, yet immunity (even IgM antibodies) appears no earlier than in 4—5 days. Of course, immunologists will criticize me for this heresy, mentioning the existence of non-specific killers that also appear within 2—3 days after onset of infection. Therefore I don't want to argue, I consider that both sides are right.

And, finally, the last question concerns the plurality of interferon genes which are located in different chromosomes. A detailed analysis of complementary DNAs obtained from different interferon messenger RNAs and cloned in recombinant plasmids allowed determination of amino acid sequences of human interferons. This work established the existence in human chromosomes of at least 12 genes of α-interferons, 5 genes of β-interferons and 4 genes of γ-interferons. New interferon genes may still be found. To explain this fact, it is worthwhile to recall the existence of a multitude of immunoglobulins and their classes. Different classes of immunoglobulins — IgM, IgG, IgA, IgD, IgE — are not identical structurally and functionally, although the source of their origin is the same. It is apparent that something similar occured with interferons. As evolution proceeded due to the duplication of interferon genes, their functional specialization also developed. Therefore the initial function of interferons was modified and nowadays we found about two dozen interferons subdivided into three classes that have different inducers and different functions. But, nevertheless, all of them have preserved the main function — ensuring the nucleic acid homeostasis — not by a detour, but directly by putting a shield against the expression of foreign genetic information.

Concluding our report, we would like to formulate some tasks for further study of the interferon system. Though some have been mentioned, we shall try to summarize as follows:

1. The study of the control by interferon of cell metabolism through the cytoplasmic membrane.
2. The interrelation of the interferon system with the system of DNA synthesis and transcription.
3. Establishing the recognition mechanism by interferon of the foreign genetic information.
4. The specification of enzymes and cofactors that protect cells

against foreign genetic information at the level of the initiation of translation.

5. Study of the influence of interferon on integration of foreign genetic information into the genetic apparatus of the cell (chromosomes).
6. The study of the influence of interferon and factors induced by it upon the process of DNA replication, correction and repair.
7. The role of interferon in formation of persistent infection in cells and in organisms.
8. The possibilities and the limitations of the use of interferons and their inducers for the treatment of acute and chronic infections.
9. The uses and abuses of the application of interferons for the treatment of neoplastic processes.
10. The preparation and the use of various classes of interferons obtained by methods of genetic engeneering.

And finally, I would like to stress once more, that the interferon system that appeared in vertebrates, almost simultaneously with the immune system, developed with it in a close inter—relationship, however, in spite of a very complex tie with the latter, it preserved one main function – the direct nucleic acid homeostasis.

References:

1 Ohno, S.; Taniguchi, T. (1982): Nucleic Acids Res. 10, 967.
2 Okazaki, R. T.; Okazaki, K.; Sakabe, K.; Sugimoto, K.; Sugino, A. (1968): in "Imunologiya", Moskva, Medicina.
3 Petrov, R. V. (1983): in "Imunologiya", Moskva, Medicina.

V. M. Zhdanov, M.D., Institute of Virology, Academy of Medical Sciences, Moscow (USSR)

Contr. Oncol., vol. 20, pp. 36—49 (Karger, Basel 1984)

The Physiological Interferon Response. III. Preliminary Experimental Data Support Its Existence

V. Bocci; L. Paulesu; Muscettola; M. G. Ricci[a]; *G. Grasso*[b]

Institutes of General Physiology, of Obstetrics and Gynecology[a] and of Human Anatomy[b] of the University of Siena, Italy.

Interferons (IFNs) have caught people's imagination so much as a panacea that some health food stores in the USA are making money by selling tablets each one containing 500 U of IFN in combination with vitamin A. The idea is simply preposterous and called for some justified comments by Annabel Hecht (15.), a member of the Public Health Service (FDA). However her last one: "The body produces IFN only when it is attacked by a disease. A healthy person does NOT need IFN" attracted our attention and the present paper is in fact a challenge to the statement that a healthy person does not need IFN.

There is no doubt that during a viral disease one of the first important responses is the production of IFNs which, by inhibiting viral replication, tends to abort the disease. Nonetheless there are a number of reasons suggesting that IFNs, as well as other lymphomonokines, must be released in trace amounts also in conditions of health. Even though it is not readily apparent, animals and plants are continuously fighting against an array of microorganisms each one trying to obtain the best nutritional and reproductive conditions.

The equation: health ⇌ disease → death indicates that health or disease are reversible situations and that health is a dynamic state due to a temporary prevalence of the body's defence system against any agent trying to alter homeostasis. Any environment, even one in the most virgin region of the earth, is not germ-free and therefore it is unavoidable that immediately after birth a number of foreign, and possibly endogenous agents, become in contact with the gut (GALT) – bronchial (BALT) – skin and possibly the "internal" — associated lymphoid tissue (IALT) and are able to induce synthesis of IFN as well as of other cytokines.

The concept of the physiological IFN response (as opposed to the acute IFN response occuring during emergency situations) can help to explain how the immune system remains efficient for long periods in the majority of individuals (2). There are a few points that need to be emphasized: one is that during the physiological response, a state of generalized refractoriness to IFN induction is not likely to occur because, at any time, only a limited number of cells are responsive and therefore inducible. Another relevant assumption is that secretion of IFN during the physiological response is paracrine[1] and little, if any, IFN reaches the general circulation, where it is practically undetectable owing to dilution and rapid turnover (3). A corollary to this is that, although the physiological response is localized (even though dispersed in several anatomical areas), it can deeply influence the immune system. In fact in normal conditions there is no need for circulating IFN, so far undetectable in plasma (20), which could have deleterious "chalone-like" effects on rapidly turning-over tissues (3). A fundamental characteristic of the lymphocytes is their migration pattern which allows them to pick up IFN molecules, thus rendering the circulation of IFN superfluous. If this reasoning is correct, the "spontaneous" activity of NK cells can be understood by considering that inactive precursor NK cells, during their circulation, have a chance to capture IFN molecules released in the pericellular environment and undergo activation. It is worth noting that no other factor than IFN seems involved in the enhancement of NK cell cytotoxicity, a two stage-process as described by Senik et al., (23). In line with this thinking, Wigzell (28) has presented evidence that endogenous IFN is probably the major, if not unique, regulator of normal NK cells: newborn mice, or mice reared under pathogen-free conditions, have practically no background NK activity but they acquire rapidly normal reactivity when transferred to conventional conditions. These findings do not contradict the hypothesis (2) that even germ-free animals, fed with a synthetic diet, have an IFN response runn-

[1] Paracrine secretion is defined as the process by which IFN is delivered from the producing cell to the neighbouring targets through the extracellular fluid.

ing at a minimum level owing to the scarcity of inducers. Encouraging data already show the physiological presence of enhanced NK cytotoxicity in the lung (21) and in the GALT (25) of normal mice. The production of human IFN gamma and alpha (acid-labile) by normal large granular lymphocytes (undifferentiated NK cells?) in vitro without apparent induction, is unexpected although the complex fractionation procedure may have acted as a derepressive stimulus (11).

There are also other indications, although indirect, that the physiological IFN response is a reality: enzymes such as (2′—5′) oligo A synthetase (22), protein kinase (16) and indoleamine 2,3-dioxygenase (29) induced in response to IFN, can be either measured in cells or in plasma, where apparently they leak and remain for longer periods than IFN because of their slower turnover. The findings of a constant level of protein kinase activity, analogous to that found in IFN-treated HeLa cells, in the plasma of healthy individuals and of (2′—5′) oligo A synthetase activity in normal mononuclear cells are exciting because they suggest that, even if IFN is not measurable in plasma, it must have been present in some extracellular fluids in order to induce enzymatic activities. In fact the physiological response has been characterized by: (a) trace amounts of inducers acting in different anatomical areas, (b) production of IFN released by a few cells at any time with the formation of a micro-environmental concentration gradient, (c) short-range effects exerted at the site of IFN gradient on fixed or circulating cells, which, by virtue of their mobility, can generalize defensive functions and (d) little, if any, spill over of IFN into the general circulation where IFN is not detectable with the current biological assay but indirectly revealed by enzymatic markers.

The placenta and its annexes as another site of the physiological IFN response (4) is becoming a controversial subject: Fowler et al., (13) and more recently Weislow et al., (27) found IFN-like activity in the murine placenta and Lebon et al., (19) have detected little but significant IFN-alpha activity in human amniotic fluids (AF). On the contrary, more recently, Cesario et al., (10) excluded the presence of substances with IFN-like effects but actually found antiviral antibodies and inactivators against IFN beta. Thus the significance of all these findings requires further study and we wish to report preliminary data collected during the last two years.

Although the search was at times fruitless, most of the following results support the existence of the physiological IFN response.

To begin with, among the five sites potentially producing a basic IFN level, the GALT is the most likely to yield IFN. Although it was assumed that the physiological IFN response would be strictly localized without an important leakage into the general circulation this may not be necessarily so, and therefore it was considered worth while to investigate the eventual presence of antiviral activity (AA) in the abdominal and thoracic lymph. Comparative controls could be represented by the evaluation of AA in hind leg lymph (presumably formed in the absence of IFN inducers) and in the peripheral plasma where normally AA is not detectable. To spare the reader all the methodology, which is being reported elsewhere (6), we can say that abdominal and thoracic lymph collected from healthy (New Zealand, Californian and Hungarian) rabbits contained consistent level of AA (about 40 IU and 5 IU/ml of lymph, when titrated in primary rabbit cell cultures and rabbit cell line, respectively, using VSV as the challenge virus). In contrast plasma, hind leg lymph and urine, obtained simultaneously from the same animals, did not have AA (≤ 3 IU/ml). Because AA could be due to many other factors other than IFN, we checked whether it was due to the presence of lipoproteins (particularly low-density lipoproteins), to anti-VSV antibodies, to bacterial endotoxins, or to non-specific, non-IFN, virus inhibitors (CVI) as reviewed by Kumar and Baron (17). It suffices here to say that none of these factors, although present (except endotoxins) in variable amounts in lymph, can account for the observed AA. We then characterized the lymph AA according to classical criteria and it appeared that AA could be due really to a heat-acid-labile IFN. Unexpectedly, lymph IFN displayed a marked cross species activity particularly on human cell lines (Wish, HEp2); because rabbit IFN induced by PPD in BCG sensitized rabbits showed a similar cross species activity, we are tentatively considering the lymph IFN to be preponderantly of gamma type.

Also the abdominal lymph collected from healthy rats tested on L929 cells (VSV), in comparison to plasma (<3 IU/ml), contained some AA (12 ± 2). The AA was slightly higher in L929 cells than in RATEC cells (a rat cell line kindly provided by Dr. H. Schellekens). The rat lymph AA, after characterization, appeared to be a heat-acid labile IFN with a broad species activity. We are currently testing whether the rat lymph IFN can be neutralized by anti-rat IFN (Sindbis induced) antibodies. In fact these findings needs to be checked by using the appropriate anti-IFN antibodies before clarifying the type(s) of IFNs during the physiological response and it will not be too surprising to find that, depending upon the different nature of inducers hitting different lymphoid cell types, variable mixtures of IFNs are produced. Anyway, the presence of AA in abdominal lymph already indicates that the GALT of healthy animals, after becoming in contact with alimentary proteins, bacterial flora and their products, may indeed produce a small amount of proteins closely resembling IFN. However IFN was never detectable in the peripheral circulation because, as it was postulated (2), the lymph IFN dripping into the venous pool would undergo very rapid catabolism and elimination.

Secondly, we would like to describe results obtained after incubation of organs belonging to GALT, BALT and spleen (IALT) aiming to test whether or not there exists any IFN release. Although we undertook this project in February 1981, results have been so unreliable that we have had to try several experimental approaches. The last one is still probably not the best but at least it has allowed us to obtain reproducible results: the rationale behind the investigation was that if organs (from healthy rabbits kept in conventional conditions) such as the appendix, the terminal ileum (usually containing a well visible Peyer's patch), the liver, the lungs (including the macrophages of the first tracheal-bronchial-alveolar lavage) and the spleen had been minimally stimulated in vivo by exogenous or endogenous inducers that ought to release traces of IFN-like activity immediately after and/or for a few hours during the successive incubation in vitro (in RPMI 1640 medium additioned with 5% FCS and piperacillin: 100 μg/ml). A liver segment and the spleen collected in aseptic conditions were weighed and immediately dispersed through a stainless steel mesh in medium (1 g wet weight tissue/5 ml medium) and several aliquots were incubated at 37 °C in air/CO_2 (95/5%) for different periods of time up to 6 hrs. After cannulation of the trachea, the lungs were initially inflated with 20 ml of medium and, after a gentle massage, about 60 % of the lavage medium was recovered containing an average of 13×10^6 cells of with 90–95 % were macrophages. These were divided into

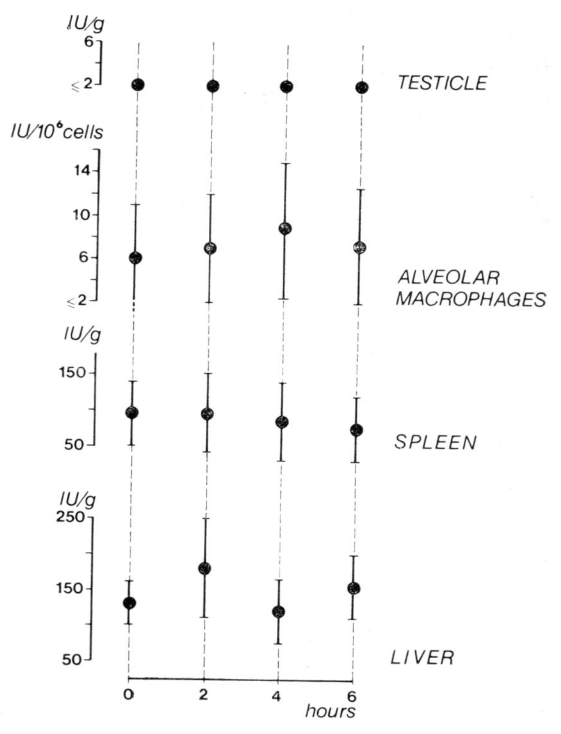

Fig. 1. Kinetics of release of IFN-like activity from dispersed rabbid tissues and alveolar macrophages incubated at 37° C. Bars indicate mean of 5 experiments ± SD.

5 aliquots of which the first was the zero time sample and the others were incubated for varying periods up to 6 hrs. The lungs were then partially refilled with 15 ml of medium and, immersed in medium, were maintained at 37° C in air/CO_2. Every two hrs, some (4—8 ml) of the bronchial-alveolar fluid was drained off the tracheal catheter and each time the lungs were replenished with fresh medium. The same method was adopted for the incubation of the appendix (an organ particularly developed in the rabbit) and of the terminal ileum. This was tied at the valve end and, first of all, both were everted in order to wash out all the luminal content and then they were cannulated and either incubated in the everted position or normalized. In both cases an adequate volume of medium was added and, at the end of any incubation period the

medium could be rapidly drained off for collection and substituted with a fresh one. All samples were centrifuged at 3200x g for 30 min at +2° C and the supernatants were filtrated through 0.22 μ Millipore filters and stored frozen until titration of AA was performed on RL cells, using VSV as a challenge virus according to the method of Langford et al. (1981).

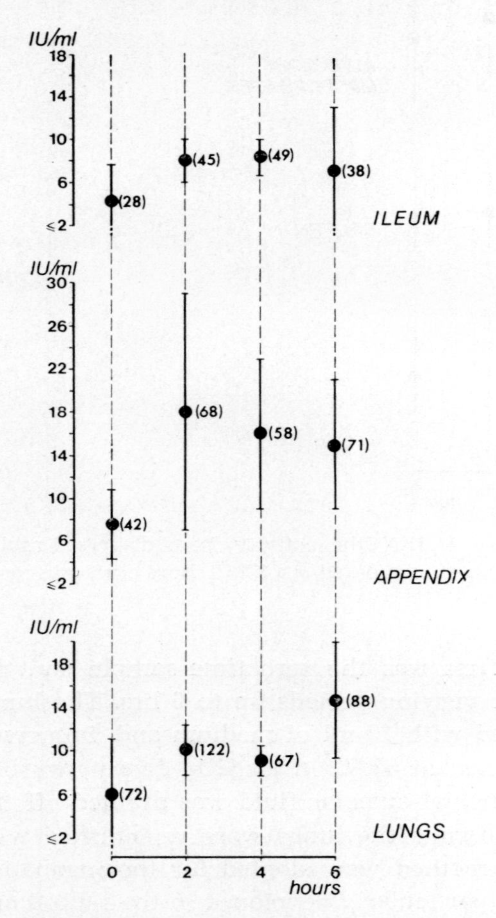

Fig. 2. Kinetics of release of IFN-like activity from rabbit organs incubated at 37° C.
Organs were filled with medium at 0 (twice), 2 and 4 hrs. At each time the recovered medium was assayed for antiviral activity. Bars indicate mean of 5 experiments ± SD. Numbers (within brackets) indicate total Units recovered in the medium during 1 min (0 time) and 2 hrs interval.

Results are reported in figures 1 and 2 and it must be noted that while hepatic, splenic, testicular cells and alveolar macrophages (fig. 1) were incubated in the same medium for different periods, the lungs, appendix and ileum (fig. 2) underwent an almost total change of medium at each prefixed interval. This may in part explain the very low AA levels (referred as International Units: IU/ml when compared to the IRP of RaIFN-NDV induced) displayed by the lung, appendix and ileum. There is no doubt that the latter organs are the most representative ones of the BALT and GALT system and being potentially rich in IFN inducers ought to release into the medium considerable AA. However, four important factors have to be taken into account: the first is that the prevalent movement of the IFN molecules released by the lymphoid cells present in the intestinal mucosa is conditioned by the lymph flow and therefore is directed towards the submucosa rather than to the lumen. This is most probable also when using everted intestinal segments. Secondly, there is a possible proteolytic component which may rapidly destroy the AA. Addition of a single protease inhibitor (pepstatin, leupeptin and aprotinin) or their mixture, does not appear to be effective. Thirdly, an inhibitor blocking the AA of IFNs could be produced at the same time (12). Fourthly, when studying pulmonary catabolism of IFN, we have clearly demonstrated that IFN instilled in the bronchial-alveolar space is rapidly absorbed by the mucosa (8).

In spite of all this, these mucosae appear to release some AA that is probably a small fraction of the activity produced in the GALT, BALT microenvironments and we are still improving the technique in order to increase the yield. Alveolar macrophages release a little AA either with or without addition of FCS. Thus it seems possible that macrophages activated and induced while floating in the alveolar fluid, release traces of IFN later in the medium. If this point of view is accepted, it is clear that, in physiological conditions, most of the IFN production is induced rather than being spontaneous. Hepatic and splenic cells release an AA corresponding to 148 IU/g of liver and to 90 IU/g of spleen. Bocci (2) has discussed at length the possible role of traces of alimentary lectins, endotoxins, bacteria normally present in portal blood and of endogenous lectins and/or proteases as IFN inducers in liver and spleen respectively. It is not, therefore, surprising to find AA shortly after tissue dispersion, a procedure implying the dilution into the medium of the IFN already present in the pericellular space. It is worth noting that the AA levels measured in the liver and spleen cell cultures (fig. 1) versus time are stationary or tend to decrease (catabolism prevails over synthesis?) whereas the IFN levels began to increase shortly after addition of 5 μg/ml of lipopolysaccharide endotoxins (LPS W E. Coli 0111:B4, Difco). By using the Limulus amebocyte lysate (LAL) test, Dr. F. Lattanzi (Istituto SVT "A. Sclavo" Siena) kindly tested a number of our samples and while liver, spleen, lung, macrophages were free of detectable endotoxin this was not the case for the ileum and appendix media. Moreover, on the basis of Blach-Olszewska and Cembrzyńska-Nowak (1) findings, we carried out a whole set of experiments at two different temperatures, namely +26° C and +37° C. The former temperature, according to Borecky et al. (9), is the most favourable for production of LPS-induced IFN but at +26° C insignificant AA was detected in all cases. Thus, on the whole, we believe that AA released during the incubation at +37° C is more probably due to a preceding in vivo than an in vitro induction. Finally we would like to emphasize that testicles, (Fig. 1) used as a possible control, yielded no AA when the organs were dispersed and incubated as described for liver and spleen. Last, but not least important, was the fact that supernatants from homogenized organs did not show AA.

Obviously the crucial question is: does AA correspond to IFN? Previous experiments (6) exclude that, by repeatedly washing the cell monolayer before VSV challenge, lipoproteins, antibodies and CVI could contribute significantly to the observed AA. The physicochemical and biological properties of the AA released from liver and spleen are reported in tab. I. The AA of the other organs has not been characterized because of too low activity. The AA appears to be an undialysable protein heat-and almost acid-stable with a remarkable activity on MDBK cells. Unfortunately, because of the present lack of antirabbit-IFN antibodies, we can only suggest the presence of an IFN-like component resembling more type I-IFN than type II.

As the placenta is one of the five sites of physiological IFN response and having acquired experience with perfused organs (3), about two years ago we began to investigate if the isolated and

Table I. Characterization of rabbit interferon (IU/ml) in liver and spleen using homologous and heterologous cells and VSV as challenge virus.

Treatment	Liver	Spleen
None	30	16
pH 2.0	25	16
Heating (56° C for 1 hr)	30	15
Trypsin	<3	<3
Dialysis (1 day at +2° C)	30	14
DNase	28	16
Ultracentrifugation (105,000 g for 1 hr)	30	14
Cycloheximide (10 μg/ml)	<3	<3
RL	30	16
MDBK	50	60
HEp 2	10	10
RATEC	4	3

perfused human placentas at term (exclussively after Caesarean section) produce and release any AA. After considerable preliminary work, we have recently perfused five normal placentas which, judged from biochemical and morphological criteria, remain viable for up to 13 hrs and probably longer. Both the umbilical arteries and the vein were rapidly cannulated and perfusion started with RPMI medium without phenol red and with 5 U/ml heparin to prevent intravascular coagulation. The actual perfusion was carried out with the same medium (2.2 l) containing heparin (1 U/ml), insulin (120 U/l) and piperacillin (100 μg/ml). After a first period of 7 hrs, the medium was drained off and fresh medium containing also human serum albumin (5 mg/ml) was added for the second perfusion period. Temperature (37° C) and pH (7.3) were continuously monitored. The gas mixture was O_2/CO_2 (95/5 %), but CO_2 content was progressively reduced as the pH tended to shift below 7.3. We would like to anticipate that, with some variability ("good" or "bad" producers!), all placentas released into the medium detectable AA (Wish and HEP 2 cells with VSV). The AA rose progressively from below 3 up to a maximum of 25 IU (average 13 $\pm$ 7 IU, calculated on the basis of an IRP of HuIFN alpha). After changing the medium, the AA fell again below 3 but slowly increased once more as during the first perfusion period. Although the AA was low, it could become measurable even in the necessarily large perfusion volume (total units were $2.8 \pm 1.4 \times 10^4$). AA reported as IU/g dry tissue was $3.9 \pm 2.4 \times 10^2$ IU/g. Because of the complex work entailed in each experiment, we carefully selected the patients most likely to undergo Caesarean operation: only those free of viral or other infections and without anti-Rhesus agglutinin would be placenta donors. After birth, newborns had normal karyotype and none had signs of congenital virus infections. Levels of antibodies in the mother's, cord plasma and AF (when available) against CMV, rubella, EBV, toxoplasma were in the low range of detection. All samples collected during the perfusion were sterile and endotoxin-free (or below 0.05 μg/ml according to the LAL standard and measurements performed by Dr. F. Lattanzi). Thus the perfused placentas could be classified as normal organs.

The presence of IFN in the human AF is still a controversial subject: while Lebon et al., (19) detected alpha-IFN in 50/52 AF samples collected between the 16th and 20th week and in 10 samples near term, Tan and Inoue (26) detected IFN only in 26/101 samples collected between the 16th—18th week of gestation and Cesario et al. (10) could not measure any AA. Owing to transplacental transfer of protein molecules, IFN present in AF might also derive from the mother's blood as not infrequently AA become detectable in peripheral plasma during pregnancy. Thus, our finding is the first unequivocal report showing that human placentas at term to produce and release AA. After appropriate concentration of the perfusion medium, the activity has been extensively characterized (also with IFN antisera) and mainly correspond to beta IFN (7). This is not surprising as induced human amnniotic cells produce beta IFN (24) but it remains to clarify why AF contains alpha IFN only. Work in progress aims to define to cell type(s) involved in the production and the IFN physicochemical and antigenic characteristics.

Finally, we would like to recall that cells attached to intrauterine devices release IFN in the medium during incubation in vitro (14). Although limited to the uterine cavity, this is another example of the mechanism underlying the response.

Conclusions

After having laid down the theoretical background of the physiological IFN response, we have summarized the anatomical, microbiological and immunological data which indirectly support its existence. More recent data based on the detection of typical enzymes elicited by IFN and on the presence of IFN in AF have provided further indirect evidence.

We have reported here three different experimental approaches sought to provide a direct demonstration of the production of IFN in physiological conditions. Presence of IFN has been shown in lymph of two animal species and in the medium perfusing the human placenta. IFN-like components appear to be released also by the liver, spleen, lungs, ileum and appendix. These are the first direct proofs that the BALT, GALT, IALT and placenta are indeed the sites of the IFN response in normal conditions. Although almost nobody can believe that "the body produces IFN only when it is attacked by a disease. A healthy person does NOT need IFN", till recently, the role of the physiological IFN response has been completely overlooked. We believe, instead, that our "well being" depends at least in part upon its existence and well tuned function. We would like to emphasize that one of the main advantages of the response lies in the limited and localized production of IFN: its effects are exerted near the synthesizing site on cells in traffic which, by virtue of their mobility, can generalize defensive functions essential for the survival of the host. The concept of a paracrine secretion of IFN limits the need of circulating IFN in plasma and implies the absence of a generalized antiproliferative and toxic effects. It looks probable that in the microenvironments where IFN is produced, it displays a number of effects which, most likely, vary among different organs. The excitement of understanding another important natural mechanism is for us the most encouraging stimulus for carrying out further work.

Acknowledgement

This work was supported by contract no. 82.02387.52 from CNR, Rome (Progetto finalizzato: Controllo delle Malattie da Infezione).

References

1 Blach-Olszewska, Z.; Cembrzýńska-Nowak, M. (1979): Acta Biol. Med. Ger. Band 38, 765.
2 Bocci, V. (1981a): Biol. Rev. 56, 49.
3 Bocci, V. (1981b): Pharm. Ther. 13, 421.
4 Bocci, V. (1982a): in Human Lymphokines ed. by A. Khan & N. O. Hill, Academic Press, New York, p. 241.
5 Bocci, V. (1982b): Drugs Exptl. Clin. Res. VIII, 683.
6 Bocci, V.; Muscettola, M.; Paulesu, L.; Grasso, G. (1983a): J. gen. Virol. (submitted for publication).
7 Bocci, V.; Paulesu, L.; Ricci, M. G. (1983b): Placenta (submitted for publication).
8 Bocci, V.; Muscettola, M.; Paulesu, L.; Pessina, G. P.; Pacini, A.; Mogensen, K. E. (1983c): J. Lab. Clin. Med. (submitted for publication).
9 Borecký, L.; Lackovič, V.; Fuchsberger, V.; Hajnická, V.; Žemla, J. (1973): in Non specific factors influencing host resistence, ed. by W. Braun & J. Ungar, Karger, Basel p. 391.
10 Cesario, T.; Goldstein, A.; Lindsey, M.; Dumars, K.; Tilles, J. (1981): Proc. Soc. Exp. Biol. Med. 168, 403.
11 Fischer, D. G.; Koren, H. S.; Rubinstein, M. (1982): The 3rd Annual International Congress for Interferon Research, Miami, Florida.
12 Fleischmann, W. R.; Georgiades, J. A.; Osborne, L. C.; Dianzani, F.; Johnson, H. M. (1979): Infect. Immun. 26, 949.
13 Fowler, A. K.; Reed, C. D.; Giron, D. J. (1980): Nature 286, 266.
14 Grasso, G.; Muscettola, M.; Bocci, V. (1983): Proc. Soc. Exp. Biol. Med. 173 (in press).
15 Hecht, A. (1981): FDA Consumer HHS Publication No (FDA) 82—1090.
16 Hovanessian, A. G.; Rollin, P.; Riviere, Y.; Pouillart, P.; Sureau, P.; Montagnier, L. (1981): Bioch. Biophys. Res. Comm. 103, 1371.
17 Kumar, S.; Baron, S. (1981—1982): in "The Interferon System: A review to 1982 Part 1" ed. by S. Baron, F. Dianzani & G. J. Stanton, Texas Reports on Biology and Medicine 41, 395.
18 Langford, M. P.; Weigent, D. A.; Stanton, G. J.; Baron, S. (1981): in Interferons Part A, ed. by S. Pestka, Academic Press, New York, p. 339.
19 Lebon, P.; Girard, S.; Thepot, F.; Chany, C. (1982): J. gen. Virol. 59, 393.
20 Levin, S.; Hahn, T. (1981): Clin. Exp. Immun. 46, 475.
21 Puccetti, P.; Santoni, A.; Riccardi, C.; Herberman, R. B. (1980): Int. J. Cancer 25, 153.
22 Schattner, A.; Merlin, G.; Levin, S.; Wallach, D.; Hahn, T.; Revel, M. (1981): The Lancet II, 497.
23 Senik, A.; Kolb, J. P.; Orn, A.; Gidlund, M. (1980): Scand. J. Immunol. 12, 51.
24 Sorrentino, V.; Di Francesco, P.; Soria, M.; Rossi, G. B. (1982): J. gen. Virol. 63, 509.
25 Tagliabue, A.; Luini, W.; Soldateschi, D.; Boraschi, D. (1981): Eur. J. Immunol. 11, 919.

26 Tan, Y. H.; Inoue, M. (1982): in Interferons, ed. by T. C. Merigan & R. M. Friedman, Academic Press, New York, p. 249.
27 Weislow, O. S.; Allen, P. T.; Fowler, A. K. (1982): in The 3rd Annual International Congress for Interferon Research, Miami, Florida.
28 Wigzell, H. (1981): in Cellular Responses to Molecular Modulators, p. 403.
29 Yoshida, R.; Imanishi, J.; Oku, T.; Kishida, T.; Hayaishi, O. (1981): Proc. Natl. Acad. Sci. 78, 129.

V. Bocci, M.D., Istituto di Fisiologia Generale, Via del Laterino 8, Universita' degli Studi di Siena — 531 00 Siena (Italy)

Contr. Oncol., vol. 20, pp. 50—62 (Karger, Basel 1984)

Immunopharmacology of the Interferon System

S. Grossberg

Department of Microbiology, The Medical College of Wisconsin, Milwaukee, U.S.A.

Introduction

Immunopharmacology, a relatively new discipline, involves the study of (i) the effects of drugs on the immune system, (ii) the mechanisms of action of the endogenous substances capable of modulating the immune response, and (iii) the possible pharmacodynamic and pharmacokinetic interactions between immunotherapeutic agents and other drugs, including the inhibitory effects of immune stimulants on drug biotransformation (20). A comprehensive review of the immunopharmacology of the interferon system, which we have been asked to address, should touch on virtually all aspects of the interferon system: not only do different cellular elements of the immune system respond to a variety of interferon inducers by producing interferons and lymphokines which can then act on all manner of somatic cells, but also the actions of interferon on nonimmune somatic cells may resemble or relate to those on cells of the immune system. Within this broad context, we shall attempt to deal in a more limited way with the pharmacology and toxicology of interferons, including their interactions with cell receptors, their production by cells of the immune system, and their immunomodulatory effects.

Pharmacology and Toxicology

The pharmacological fates of exogenous or endogenous interferon are schematically presented in fig. 1. Interferon is not detectable in the circulation of normal individuals. After steady-state equilibration by intravenous infusion into patients, the half-life of α interferons is 2 to 4 hours. Maximum levels in blood following intramuscular injection are achieved in 4 to 8 hours and in about 6 to 10 hours following subcutaneous inoculation. Doses given every 12 hours can provide relatively steady serum levels. In adults a single daily intramuscular dose of $80-100 \times 10^6$ units will provide circulating levels of 100—1000 units per milliliter of blood.

Hypothetical pharmacological fates of interferon

IFN	(endogenous or exogenous)
↓	
Plasma	Free IFN ⇌ Bound IFN (albumin, gangliosides, RBC)
⇅	Clearance
Tissues	Free IFN ⇌ Bound IFN (receptors, binding proteins)
	Metabolic actions
	Degradation (proteolysis)
	Biotransformation (desialylation)

Fig. 1. Hypothetical pharmacological fates of interferon.

Doses of approximately $3-6 \times 10^6$ units intramuscularly will give about 5- to 10-fold lower levels. The rate of disappearance of interferon from the bloodstream is initially high but slows later (2, 13, 3, 6, 19, 21, 23). In general, interferon in normal individuals equilibrates slowly and reaches low levels in extravascular spaces and interstitial fluids; for example, 3 % or less of the level in serum is detectable in cerebrospinal fluid (3). Interferon injected into the cerebrospinal fluid was subsequently detectable in blood; and intrathecal administration showed rapid diffusion through the CSF of monkeys but no interferon was recoverable in white matter or cortex. Most investigators have found that intramuscular injection of HuIFN-β (of the same potency as that of HuIFN-α (Le)) gave lower or undetectable levels of serum interferon; a minimum of $10-30 \times 10^6$ units had to be injected intramuscularly for interferon to become detectable in serum. HuIFN-β given intramuscularly may not enter the circulation perhaps because it can be inactivated by muscle extract (14). Intravenous infusion of HuIFN-β is necessary to obtain appreciable levels in blood. Low levels in serum may be due to its removal by its binding to erythrocytes (18) or vascular endothelium, or inactivation by circulating gangliosides. However, contradictory findings by different investigators may relate to the molecular heterogeneity observed within partially purified HuIFN-β preparations detectable by isoelectric focusing; a HuIFN-β subpopulation with a pI of 5.9 got into the circulation after intramuscular injection as well as HuIFN-α (Le), whereas another molecular subpopulation having a pI of 5.2 gave 30-fold lower blood titers (3). Further, different molecular forms of HuIFN-α may be cleared at different rates (28).

HuIFN-α (Le) appears to be filtered through the glomerulus. Very little interferon is found in the urine. Renal artery ligation (in rats) can result in as much as a 10-fold increase in serum interferon. The kidney appears to be a major site of IFN-α degradation in the tubules, apparently by lysosomal proteases. HuIFN-α2 was rapidly removed by the kidney, HuIFN-β, either as the native product or partially desialylated, is taken up rapidly by the liver, perhaps by hepatic binding protein (34, 7, 4).

Interferon given orally does not appear subsequently in the blood of older animals. In suckling mice, however, about 1 % of interferon given by mouth can be detected in blood. Interferon levels in milk of lactating mice are about 10—20 % of levels in serum, and suckling mice from such mothers were protected against virus challenge.

In contrast to most drugs that are active only when adequate concentrations are maintained in target tissues, interferon causes tissues to resist virus infection for substantial periods of time after the interferon disappears, perhaps because of its induction of a long-lasting antiviral state in cells or its effects on immune system components.

A number of reversible side effects have been associated with HuIFN-α and -β administration, even with increasingly purified preparations. Fever is the most common reaction, being most prominent after the first injections and tending to become less severe after prolonged treatment. Chills and malaise often accompany the fever, and occasionally transient glycosuria (22). Lassitude is a common complaint. A degree of depression of bone marrow elements, including leukocytes, platelets and reticulocytes, is usually noted in most patients and is readily reversible when the interferon is stopped; this depression is never so severe as that seen with antimetabolites or other potent cytostatic drugs. In infants, transient fever, rises in serum transaminase, decreased feeding and weight loss have occurred. The tender erythematous reactions that sometimes occur at injection sites appear to be more related to the purity of the interferon. With high doses of HuIFN-α ($>10 \times \times 10^6$ IU), anorexia and general fatigue can sometimes be so pronounced as to be incompatible with normal ambulation (21). Weight loss and mild to moderate abnormalities in biochemical liver function tests can be observed with continued administration but without recognized histopathological alterations. Other symptoms include headache, hair shedding (most often in patients with a familial history), and peripheral sensory neuropathy. When doses greater than 150×10^6 IU are given daily by continuous intravenous infusion, central nervous system toxicity manifest as severe drowsiness and even seizures has been observed (26). HuIFN-α2 purified to about 4×10^8 IU per milligram of protein has produced side effects similar to those observed with less pure preparations (15).

The suggestion that HuIFN-α could have endorphin or adrenocorticotropin-like activity (5) has been disputed by the failure to find either functional or structural homology of HuIFN-α with human α-ACTH or β-endorphin (11). Recently, Scott et al. (23) have observed rises in serum 11-hydroxycorticosteroids (and falls in plasma zinc levels) after HuIFN-α2 administration. One important pharmacological effect of IFN-α, β, and γ is the depression of drug biotransformation, apparently mediated by the hepatic cytochrome P-450 monooxygenase system (27).

It is now appreciated that interferons can be weakly antigenic in homologous species. Although no interferon antibody was detected in 20 patients receiving HuIFN-α (Le) for 6–18 months (16), antibody to HuIFN-α has been found in at least two patients never treated with interferon. Further, three out of 16 patients receiving HuIFN-α2 developed neutralizing IgG antibody (33). Similar information is not yet available on HuIFN-γ. It is not known whether resistance to IFN therapy relates to the presence of such antibody.

Receptor Binding

Like most pharmacologically active substances, all three types of interferon must first bind to specific cell surface receptors, the human genes for which are located on chromosome 21 (11). However, α and β interferons appear to bind to the same receptor which is different from the receptor for γ interferon. Much circumstancial evidence indicates that interferon need not enter cells to act. Binding occurs at 0° C but demonstration of the antiviral state requires subsequent incubation at physiological temperatures. Gangliosides interfere with interferon binding, and thyrotropin and cholera toxin appear to compete for the same receptors. From studies of somatic cells hybrids, it is clear that the receptors are species-specific in nature (10). Antibodies to cell surface components coded for by human chromosome 21 inhibit interferon action. The degree of interferon sensitivity appears to relate to the number of duplications of chromosome 21 such that cells from children with Down's syndrome (trisomy 21) have much greater sensitivity to the action of the interferon. The number of receptor sites per cell has been estimated to be only between a few hundred and a few thousand (12, 9). The receptors are located apparently at random on cell surfaces (17), or in coated pits (18) to which some bacterial toxins and peptide hormones are known to bind. There is conflicting evidence as to whether interferon is internalized. Although one study indicates that radiolabeled interferon did not enter cells (1), another (9) showed that about 50 % entered cells, whereas cell surface binding alone caused an increase in 2′–5′A synthetase. IFN binding can be enhanced greatly by treatment of interferons with lanthanide salts (J. J. Sedmak, personal communication), which also results in enhanced antiviral activity (24). Blocking experiments with polyclonal and monoclonal antibodies to interferon suggest that different domains may be involved in binding some α/β interferons. Thus, different interactions of these interferons with membrane receptors may selectivily trigger different IFN effects (J. Taylor-Papadimitriou, personal communication).

Immunocyte Interferon Production, Pharmacological Effects, and Immunomodulation

Every principal cell type of the immune system can produce interferons, primarily either IFN-α or IFN-γ types, depending upon the kinds of inducers utilized (see tab. I). Viruses, polyribonucleotides, heterologous species cells, and B-cell mitogens induce predominantly IFN-α types; whereas T-cell mitogens, microbial antigens, tumor cells, and antibody to the OKT3 T-cell antigen induce

Table I. Interferon production by cells of the immune system in relation to inducer

Inducer	Principal cell types	Principal interferon produced
Viruses	B, T, M, null	α
Poly(I) . poly(C)	M, null (T, B)	α
Foreign cells (allogeneic)	B, null	α
B mitogens[1], bacteria	B, null	α
T mitogens[1]	T	γ
Antigens (viruses, bacteria, etc.)	T (sensitized)	γ
Tumor cells	null	γ
Antibody (vs. T-cell OKT3 antigen)[2]	T	γ
Carboxymethylacridanone	M	α/β

Abbreviations: B = bone marrow-(or bursa-) derived lymphocytes
T = thymocyte-derived lymphocytes
M = monocytes/macrophages

[1] SEA-, PHA-, and PWM-induced HuIFN-γ can be contaminated with HuIFN-α; conA-induced HuIFN-γ can be contaminated with HuIFN-β; Lentil lectin- and PPD-induced HuIFN-γ in sensitized lymphocytes can be contaminated with HuIFN-α; C. parvum-induced HuIFN-α can be contaminated with HuIFN-γ.

[2] Monoclonal antibodies vs. OKT4 (on T helper cells) or OKT8 (on T suppressor cells) do not induce interferon.

principally IFN-γ (2). Carboxymethylacridanone induces IFN-α/β in vivo and in macrophages (30, 29). It should be appreciated that lymphokines, such as interleukin-2, lymphotoxin, and macrophage activating factor, can be found in supernatants of peripheral blood leukocyte cultures in which IFN-γ is produced (J. Vilcek, personal communication).

The stimulatory or inhibitory pharmacological actions of interferon are summarized in tab. II through IV, be they of enzymatic or biochemical nature (tab. II), or alterations of cell surfaces (tab. III), or somatic cell and toxicological activities (tab. IV) (32, 8, 2, 31). Despite a great deal of work, it remains unclear how the two major antiviral, translation inhibitory pathways induced by interferon, the 2′—5′oligoadenylate-mediated ribonuclease system and the protein kinase phosphorylation of initiation factor 2, are involved in interferon's cellular effects (31). Of interest is the recent demonstration of an apparent identity between the interferon-induced protein kinase activated by polyamines and the one activated by dsRNA, both of which phosphorylate a nuclear 67K dalton polypeptide (25). Ornithine decarboxylase (ODC), which is important in DNA synthesis, has about the same molecular weight and can be phosphorylated by the polyamine-dependent kinase. Since interferon treatment of cells can inhibit ODC induction, leading to reduced cell growth, it is possible that cellular activities dependent on polyamine synthesis can be altered.

The major immunomodulatory effects, both inhibitory and stimulatory, of the interferons are summarized in tab. V, in relation to the types of immunocompetent cells affected (reviewed in Bor-

Table II. Pharmacology of the interferons: enzymatic and biochemical Changes

Stimulation or increase	Inhibition or decrease
Primary enzymatic changes	Enzymatic changes
2—5An synthetase induction	Ornithine decarboxylase induction
Latent endonuclease activation by 2—5A_n	Hepatic cytochrome P-450-linked monooxygenases
Protein phosphokinase induction	Induction of glucocorticoid-inducible enzymes:
Other enzymatic changes	glutamine synthetase,
Aryl hydrocarbon hydroxylase activity	glycerol-3-phosphate dehydrogenase, and.
Cyclooxygenase	tyrosine amino transferase
Prostaglandins E_2, F_2	
Guanylate cyclase	Nucleotide and protein changes
Nucleotide and protein changes	Basal and stimulated DNA synthesis
cGMP elevated	Calmodulin
2—5A_n synthesis	
Phosphorylated protein	Lipids
P1 (67K daltons)	High density lipoprotein cholesterol
P2 (37K daltons) — subunit of peptide chain initiation factor eIF-2	Total cholesterol
	Phospholipid methylation

Table III. Pharmacology of the interferons: cell surface alterations

Stimulation or increase	Inhibition or decrease
Ligand binding	Protein hormone or toxin binding
Concanavalin A	Thyrotropin
Estrogen and progesterone	Cholera toxin
	Diphtheria toxin
Compositional changes	Compositional changes
Intramembranous granules	Cell and membrane charge more negative
β_2 microglobulin release	Natural killer cell susceptibility
Carcinoembryonic antigen release	Unsaturated fatty acid content
	Plasminogen activator release
Antigenic site expression	Membrane transport
Alloantigens	Thymidine uptake
Histocompatibility antigens	Uridine release
Cytoskeletal changes	Cytoskeletal changes
Bundles of actin filaments	Cell motility
Long filaments of fibronectin	
Membrane rigidity	

Table IV. Pharmacology of the interferons: Somatic cell and toxicological activities

Stimulation or increase	Inhibition or decrease
Phagocytosis	Growth (normal, tumor, embryonic)
Differentiation	Differentiation
Tumor cells	Normal cells
HL60 leukemia (Hu)	myoblast → muscle
Ml leukemia (Mu)	fibroblast → adipocyte
Friend erythroleukemia	Tumor cells
	HL60 leukemia
	Friend erythroleukemia
Neuron excitability	Myeloid maturation
Myocardial beat	Myocardial beat

den and Ball (8), and in Baron et al., (2) to which the reader is referred for the many specific references, especially before 1979). The proliferative response of primarily T-helper lymphocytes, indicated primarily by radiolabeled thymidine incorporation after exposure to lectins (phytohemagglutinin, concanavalin A, or pokeweed mitogen), antigens (e.g., mumps or herpes viruses or tetanus toxoid), and allogeneic lymphocytes is inhibited by IFN-α or IFN-β. The greatest inhibition occurs when interferon is administered at the time of the proliferative stimulant, whereas interferon pretreatment appears to have little or no effect. IFN-γ causes marked inhibition of the mixed lymphocyte culture response. Interferon also inhibits the production and effect of T-lymphokine, or leukocyte migration factor. IFN-α and IFN-β increase the cytotoxicity of T cells against allogeneic lymphocytes or leukemia L1210 cells, despite depression of T-cell proliferative response. Possible mechanisms that might be involved include an inhibition of T-suppressor cell activity, an increase in target cell antigens, especially of the histocompatibility types (e.g., β-microglobulins), or an increase in the maturation of cytotoxic T cells manifest as an increase in Fc γ receptors but a decrease in Fc μ receptors, thereby altering

Table V. Pharmacology of the interferons: immunomodulation

Stimulation or increase	Inhibition or decrease
T cells	T cells
cytotoxicity	proliferation
suppression	leukocyte migration inhibition
B cells	B cells
after proliferative phase	proliferation
antibody response (after antigen)	antibody response (before antigen)
NK cells	NK cells
cytotoxicity	target cell sensitivity
K cells	Macrophages
cytotoxicity	monocyte differentiation
Macrophages	Antigen expression
phagocytosis	Fcμ receptors
tumoricidal activity	
Basophils	
histamine release	
Antigen expression	
histocompatibility	
tumor-associated antigens	
Fcγ receptors	

receptor-mediated immunologic functions. The activity of K cells, which are Fc-receptor positive, is increased by interferon, manifest as an increase in antibody-dependent cell-mediated cytotoxicity.

Macrophages are stimulated by exposure to IFN-α and IFN-β in two major ways. First, phagocytosis is increased, as demonstrated by an increased uptake of colloidal particles or sheep erythrocytes by peritoneal macrophages, for which a one-hour pretreatment or as little as 5 units of interferon appear to be the minimum requirements. Such effects can be demonstrated in vivo by administration of interferon inducers or interferon itself. The second major effect of interferon is enhancement of macrophage cytotoxicity, e.g. against herpesvirus- or cytomegalovirus-infected cells and leukemia cells. Such cytotoxicity can be enhanced by interferon inducers and is blocked by anti-interferon antibody. Finally, it should be noted that interferon inhibits the maturation of monocytes into macrophages.

In B cells, the primary or secondary antibody response manifest by, for example, the Jerne plaque technique, can be inhibited 80–99 % by IFN-α, IFN-β or IFN-γ. IFN-γ can be as much as 250-fold more potent than IFN-α or IFN-β. The maximum effect is seen when interferon is added along with antigen. In mice, treatment by interferon in very high doses will diminish the levels of circulating IgM and IgG. In man, no effect has been seen on antibody production, but the B-cell number has been noted to diminish. Interferon can also enhance antibody production in vitro if the IFN is added 2–4 days after the antigen (sheep erythrocytes), or in vivo by low doses of interferon or when the addition is made 2–3 days after the sheep red blood cells; in this case as much as a 2- to 6-fold increase in antibody production is observed.

Interferon inhibits cell-mediated hypersensitivity (delayed-type immunity). There is inhibition of the efferent response, where sensitized individuals are given interferon before or with the challenge antigen, which may be viral or hapten. Interferon can also inhibit the afferent or sensitizing response if the interferon is given before the primary injection of antigen. Sensitization, however, can be enhanced by giving interferon after the injection of antigen.

Allograft survival can be altered by interferon. Survival can be increased across the mouse major histocompatibility complex barrier, but this requires high doses (50,000 units per mouse per day) right after grafting. Allograft survival can be diminished by low doses of interferon (200–3,000 units), which will accelerate the rejection. Interferon diminishes the graft versus host reaction with large doses.

Other important immunomodulatory effects of interferon include inhibition of hemopoietic (bone marrow) cell proliferation, which has been a consistent clinical manifestation in man, in which reticulocytes, lymphocytes and platelets circulating in blood are diminished. Further, interferon can increase the IgE-mediated release of histamine from basophils. Finally, interferon stimulates expression of cell-surface antigens, including those of the histocompatibility complex.

Conclusion

Interferons are produced by many types of somatic cells, perhaps best of all by those of the immune system. The pharmacological effects of the interferons are extremely varied, depending at least in part on the level of differentiation of cells or the specialized cell function being studied. Virtually every cell type of the immune system can be affected by interferon treatment: more primitive stem cells are susceptible to arrest of differentiation or maturation, proliferative responses to mitogens are inhibited, cytotoxic activities of various mononuclear cells are enhanced, macrophages are activated, and B lymphocytes produce less antibody or more, depending upon when they are exposed to interferon in relation to antigen. Gamma interferon seems to exert more potent cellular effects, especially on immune functions, than alpha or beta interferons. The interferon system can be regarded as a complex network which alters other seemingly unrelated networks, such as growth factors, or hormones in the adrenal-pituitary axis, or drug detoxification systems such as the cytochrome P-450 monooxygenases. Like many pharmacologically active compounds, interferons are able to act only after binding to specific cellular receptors, but the underlying bases for the variegated effects and actions that interferons achieve remain to be elucidated.

References

1 Aguet, M.; Blanchard, B. (1981): Virol. 115, 249.
2 Baron, S.; Grossberg, S. E.; Brunell, P. A.; Klimpel, G. R. (1983): in "Antiviral Agents and Viral Diseases of Man (Second Edition", ed. by G. J. Galasso; T. C. Merigan, R. A. Buchanan, Raven Press, New York (in press).
3 Billiau, A. (1981): Arch. Virol. 67, 121.
4 Bino, T.; Edery, H.; Gertler, A.; Rosenberg, H. (1982): J. gen. Virol. 59, 39.
5 Blalock, J. E.; Smith, E. M. (1980): Proc. Nat'l Acad. Sci USA 77, 5972.
6 Bocci, V. (1981): Biol. Rev. 56, 49.
7 Bocci, V.; Pacini, A.; Bendinelli, L.; Pessina, G. P.; Muscettola, M.; Paulesu, L. (1982): J. gen. Virol. 60, 397.
8 Borden, E. C.; Ball, L. A. (1981): in "Progress in Hematology XII", ed. by E. B. Brown, Grune and Stratton, New York, pp. 299—339.
9 Branca, A. A.; Faltynek, C. R.; D'Alessandro, S. B.; Baglioni, C. (1982): J. Biol. Chem. 257, 13291.
10 Chany, C.; Rousset, S.; Chany-Fournier, F. (1982): Tex. Rep. Biol. Med. 41, 307.
11 Epstein, L. B.; Epstein, C. J. (1982): Tex. Rep. Biol. Med. 41, 324.
12 Epstein, L. B.; Rose, M. E.; McManus, N. H.; Li, C. H. (1982): Biochem. Biophys. Res. Commun. 104, 341.
13 Greenberg, S. B.; Harmon, M. W.; Couch, R. B. (1980): in "Interferon and Interferon Inducers, Clinical Applications", ed. by D. A. Stringfellow, Marcel Dekker, New York, p. 57.
14 Hanley, D. F.; Wiranowska-Stewart, M.; Stewart, W. E. II (1979): Int. J. Immunopharmac. 1, 219.
15 Horning, S.; Levine, J.; Miller, R.; Rosenberg, S.; Merigan, T. (1982): JAMA 247, 1718.
16 Ingimarsson, S.; Bergström, K.; Broström, L. A.; Cantell, K.; Strander, H. (1980): Acta Med. Scand. 208, 155.
17 Kushnaryov, V. M.; Sedmak, J. J.; Bendler, J. W. III; Grossberg, S. E. (1982): Infect. Immun. 36, 811.
18 Kushnaryov, V. M.; MacDonald, H. S.; Sedmak, J. J.; Grossberg, S. E. (1983): Infect. Immun. 40, 320.
19 Lucero, M. A.; Magdelenat, H.; Friedman, W. H.; Pouillart, P.; Billardon, C.; Billiau, A.; Cantell, K.; Falcoff, E. (1982): Eur. J. Cancer clin, Oncol. 18, 243.
20 Mullen, P. W. (1981): in "Advances in Immunopharmacology", ed. by J. Hadden, I. Chedid, P. Mullen and F. Spreafico, Pergamon Press, New York, pp. 3—36.
21 Priestman, T. J.; Johnston, M.; Whiteman, P. D. (1982): Clin. Oncol. 8, 265.
22 Scott, G. M.; Secher, D. S.; Flowers, D.; Bate, J.; Cantell, K.; Tyrrell, D. A. J. (1981): Br. Med. J. 282, 1345.

23 Scott, G. M.; Ward, R. J.; Wright, D. J.; Robinson, J. A.; Onwubalili, J. K.; Gauci, C. L. (1983): Antimicrob. Ag. Chemother. 23, 589.
24 Sedmak, J. J.; Grossberg, S. E. (1981): J. gen. Virol. 52, 195.
25 Sekar, V.; Atmar, V. J.; Krim, M.; Kuehn, G. D. (1982): Biochem. Biophys. Res. Commun. 106, 305.
26 Smedley, H.; Katrak, M.; Sikora, K.; Wheeler, T. (1983): Br. Med. J. 286, 262.
27 Sonnenfeld, G.; Harned, C. L.; Nerland, D. D. (1982): Tex. Rep. Biol. Med. 41, 363.
28 Stewart, W. E. II; Hanley, D. F.; Nason, K. C.; Wiranowska-Stewart, M. (1981): in "Advances in Immunopharmacology", ed. by J. Hadden, L. Chedid, P. Mullen and F. Spreafico, Pergamon Press, New York, pp. 25—28.
29 Storch, E.; Kirchner, H. (1982): Eur. J. Immunol. 12, 793.
30 Taylor, J. L.; Schoenherr, C. K.; Grossberg, S. E. (1980): Antimicrob. Ag. Chemother. 18, 20.
31 Taylor, J. L.; Sabran, J. L.; Grossberg, S. E. (1983): in "Handbook of Experimental Pharmacology", ed. by P. Came and W. Carter, Springer-Verlag, Berlin (in Press).
32 Taylor-Papadimitriou, J. (1980): in "Interferon 2", ed by I. Gresser, Acad. Press, New York.
33 Trown, P. W.; Kramer, M. J.; Dennin, R. A.; Connell, E. V.; Palleroni, A. V.; Quesada, J.; Gutterman, J. U. (1983): Lancet 1, 81.
34 Vilcek, J.; Sulea, I. T.; Zerebeckyj, I. L.; Yip, Y. K. (1980): J. Clin. Microbiol. 11, 102.

S. Grossberg, M.D., Department of Microbiology, Medical College of Wisconsin, 8701 Watertown Plank Road, Milwaukee, Wisconsin 53226 (U.S.A.)

Contr. Oncol., vol. 20, pp. 63—71 (Karger, Basel 1984)

Relationship between Interferons and Other Lymphokines

I. Béládi; Pham Ngoc Dinh; I. Rosztóczy; M. Tóth

Institute of Microbiology, University Medical School, Szeged, Hungary

Introduction

Interferon (IFN) systems, the host's most rapid defence against viruses, are proteins secreted by body cells when they are stimulated by viruses, bacteria or numerous other compounds. The secreted IFNs then stimulate surrounding cells to produce other proteins, which in turn may regulate the immune response, cell growth and other cell functions. At least three distinct types of IFN may be produced, depending on the type of stimulus and the type of cell stimulated.

Lymphokines are produced when sensitized lymphoid cells (lymph node, spleen, peritoneal exudate or peripheral blood cells) come into contact with specific antigen, or when non-sensitized lymphoid cells are cultured with mitogens or allogeneic cells. The term "lymphokine" was first used to describe biological activity resulting from the antigen activation of sensitized lymphocytes (4).

Although IFN can be produced by many cells (e.g. leukocytes (IFN-α) and fibroblasts (IFN-β) stimulated by virus or RNA, immune IFN (IFN-γ) is produced by lymphoid cells in response to antigen or mitogen, and it is therefore by definition a lymphokine. As lymphoid cells stimulated by viruses produce IFN-α, this type of IFN can also be considered a lymphokine in respect of its origin.

One of the chemically best-characterized lymphokines, LIF, selectively inhibits the random migration of polymorphonuclear (PMN) leukocytes in vitro. It acts as a serine protease and its molecular biology has recently been reviewed (1). The lymphocyte population(s) responsible for the production of this mediator has not been established. Antigen-stimulated enriched T cells obtained by passing human blood mononuclear cells through Sephadex anti-human Fab columns produce LIF equally as well as the enriched B cells obtained by eluting the same columns with immunoglobulin (2).

Since in some cases the same stimulus can trigger the cells to produce both IFNs and lymphokines and IFN can regulate its own synthesis (priming and blocking), it seemed of interest to study the LIF content of IFN-α and IFN-γ preparations, and the possible effect of IFN on the production of LIF by leukocytes.

Materials and Methods

Chemicals

Concanavalin A (Con A) was purchased from Calbiochem, Ficoll 400 from Pharmacia Fine Chemicals, and Uromiro from Bracco.

Interferons

Partially purified human IFN-α from Sendai virus-induced human leukocytes (spec. act. 3.72×10^5 U/mg protein) were used. The partially purified human IFN-α preparation was devoid of LIF and MIF activities. Crude human IFN-α (spec. act. 260 U/mg protein) and partially purified chicken leukocyte derived IFN (spec. act. 4.3×10^5 U/mg protein) induced by adenovirus type 12 were included in other experiments. IFNs were assayed in microtitre plates as described previously (6). One unit of IFN used represents one unit of reference standard preparation G-023-901-527 for human IFN-α and one unit of MRC Research Standard A, 62/4 for chick IFN, respectively. Leukocytes were incubated with IFN at 37 °C for different periods. At the end of the incubation period, IFN was removed by centrifugation before stimulation of leukocytes with Con A.

Stimulation of the Production of Lymphokines by Con A

Human leukocyte cultures were prepared by the method described earlier (6). Cells were distributed in test tubes at a concentration of 1×10^7 cells/ml. Con A (5 μg/ml) was added and cultures were incubated at 37 °C for 1 hr in a Girotory incubator (New Brunswick Sci. Co.) at 80 rpm shaking frequency. At the end of the incubation period, cells were pelleted by centrifugation at $800 \times g$ for 5 min and washed twice with MEM to remove residual Con A. Following this, cells were resuspended in MEM and incubated further for 48 hr. At this time cells were removed by centrifugation and the supernatant was stored at –70 °C until the determination of LIF activity.

LIF Assay

The conventional capillary tube method was used (David and David, 1971; Rocklin, 1974).
Migration inhibition (MI) was expressed as follows:

$$\text{percent MI} = 100 - \frac{\text{100 x migration area in the presence of experimental supernatant}}{\text{migration area in the presence of supernatant from control cultures}}$$

Results

The LIF contents of crude preparations of Sendai virus-induced IFN were studied. The preparations tested contained LIF, with different activities.

The study of the species-specificity of LIF has revealed similar activities on homologous and heterologous leukocytes. Thus, in contrast to the IFN present in the same preparation, the LIF had no species-specificity when chicken or rabbit leukocytes were used as indicator cells (fig. 1).

The LIF and IFN-α activities could be separated by treatment of IFN-α preparations at pH 2. LIF proved to be sensitive to acid pH, while the IFN activity was retained. Partially purified IFN-α preparations were devoid of LIF activity, became of their sensitivity to pH 2.

Next, the effect of IFN on the production of LIF by Sendai virus-induced human leukocytes was studied. The leukocytes were

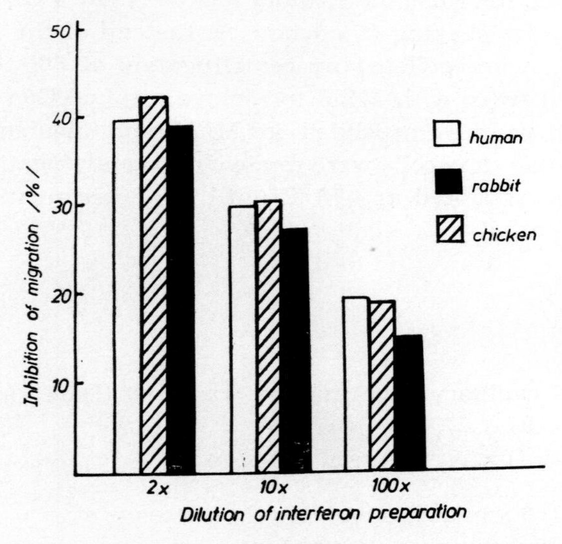

Fig. 1. The effect of human LIF on the migration of leukocytes of different species.

Table I. Effect of duration of pretreatment with IFN-α on LIF production by human leukocytes

Duration of pretreatment (hr)	IFN prethreatment dose (IU) ml 0	100	10000
	Migration inhibition %		
0	23	—	—
1	—	23	23
1.5	—	27	36
2	—	30	30
4	—	53	5
6	—	50	3
8	—	48	5
24	—	17	10

treated with different quantities of IFN-α for 4 hr before Sendai virus infection. No effect was found following 100 or 1000 IU/ml IFN-α pretreatment. In contrast, when leukocytes were treated in the same way before Con A induction, a biphasic effect of IFN-α

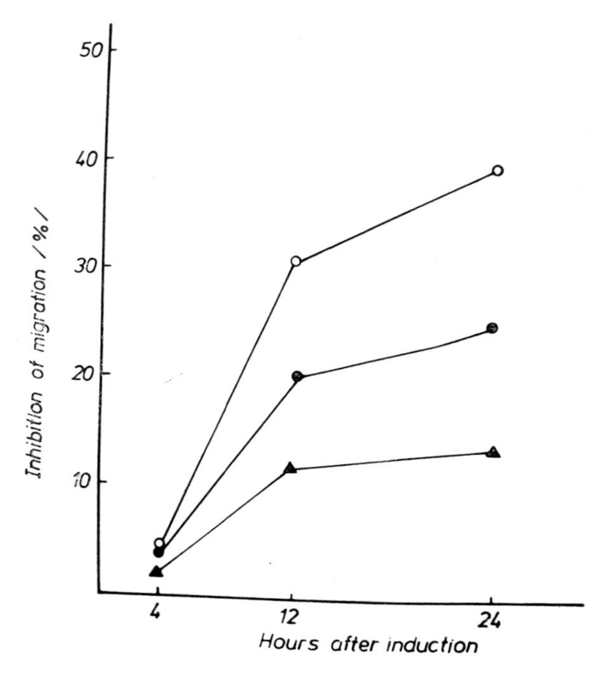

Fig. 2. The effect of interferon pretreatment on Con A induced LIF production:
(●—●) untreated
(▲—▲) 10 000 IU of interferon
(○—○) 100 IU of interferon

could be observed. The lower dose of IFN-α enhanced, while the higher dose impaired LIF production (fig. 2). The effect of the duration of pretreatment was next investigated. It can be seen (tab. I) that a 4—8 hr pretreatment with 100 IU/ml IFN-α resulted in stimulation of the production of LIF by leukocytes; at the same time, 1000 IU/ml IFN-α inhibited the formation of LIF.

It seemed of interest to learn whether the IFN-pretreatment of leukocytes before Con A induction affected the quantity of LIF formed, or resulted in the earlier production of LIF. The data obtained clearly showed that the kinetics of the production run parallel in IFN-α-treated and untreated leukocytes.

The question arose of whether the presence of monocytes was needed for LIF production and therefore monocyte-depleted leukocytes were also tested. It can be seen that the LIF production does not differ significantly in the presence or absence of monocytes when different doses of IFN-α are applied for pretreatment (fig. 3).

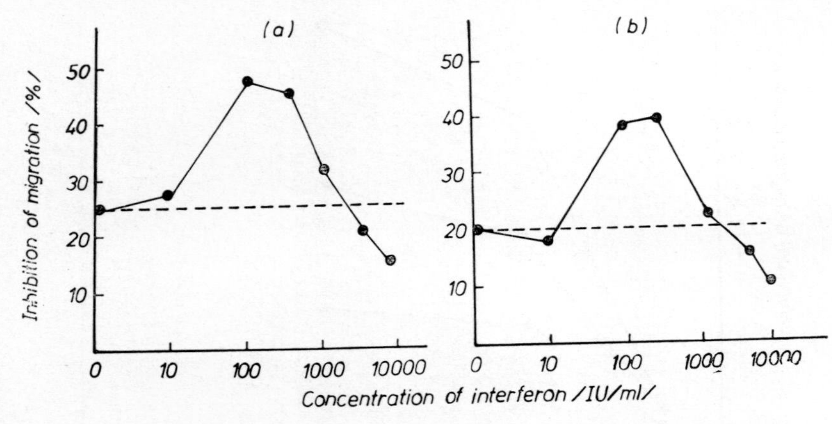

Fig. 3. The effect of interferon on the production of LIF in the presence (a) and absence of monocytes (b).

The stability of LIF activity against trypsin, heat and pH 2 treatment was studied. LIF proved to be resistant to heat treatment at 56 °C for 30 min, but sensitive to trypsin and to pH 2.

It might be assumed that the impurities in the IFN-α preparations used for pretreatment of the leukocytes caused the biphasic effect on LIF production. The effect of a partially purified IFN-α preparation was therefore compared with that of the crude IFN-α. The resulting data strongly indicated that IFN-α itself is responsible for both the stimulation and the inhibition of the production of LIF by leukocytes.

Attempts were made to clarify whether heterologous (chicken) IFN-α can influence the LIF production of human leukocytes, similarly to the homologous IFN-α. We observed that chicken IFN-α had no effect on the LIF production by human leukocytes.

It was our aim to see how general a phenomenon is the priming effect of IFN on LIF production. For this purpose the effect of chicken IFN-α on the LIF production of chicken leukocytes was tested. It can be seen (fig. 4) that pretreatment of chicken leukocytes with chicken IFN-α influenced the production of chicken LIF, similarly to the results obtained with the human systems.

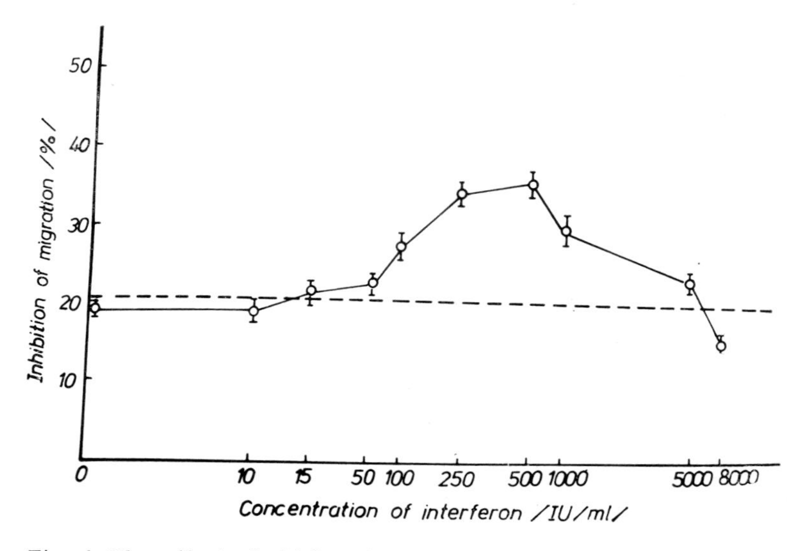

Fig. 4. The effect of chicken interferon on LIF production of chicken leukocytes.

Discussion

Besides their antiviral effect, the IFNs exhibit numerous non-antiviral activities. Among them, the priming effect of IFN on its own production has been known for decades. The IFNs can enhance the production of various substances by different cells. It is of interest that IFN-α, which may also be considered a lymphokine, can influence the production of another lymphokine. We observed that not only the LIF, but also the MIF production of human leukocytes can be modulated by IFN-α pretreatment (9). Vilcek et al. (12) reported that, depending on the dose used, IFN-γ can enhance or inhibit the production of IL-1 and IL-2. They found that IL-2 also enhances IFN-γ production. We did not observe that LIF influences

IFN-α production (data not shown). It is possible that the dose of LIF applied was not sufficient for this purpose, as the preparation used gave rise to a significant inhibition of migration of PMN only up to a 1:64 dilution.

In the preparation of IFN-γ, use was made of the fact that priming of human leukocytes with IFN-α before challenge with mitogen causes an increase in IFN-γ production (Wiranowska-Stewart et al., 1980). The production of phytohaemagglutinin (PHA)-induced IFN-γ by human lymphocytes was significantly enhanced by the phorbol ester, 12—0—tetradecanoyl phorbol—13—acetate (11, 8). Similarly, production of IL-2 by murine splenic lymphocytes induced by Con A was enhanced by phorbol myristate acetate (5).

Mundy and Yoneda (7) found that the presence of prostaglandins was essential for the generation of osteoclast activating factor. Thus, the same mitogens and other substances can stimulate the parallel production of different lymphokines and these can exert a mutual enhancing effect on their production.

Once an antigen is recognized and a response initiated, there is some mechanism to amplify this response via the recruitment and activation of lymphocytes which lack the capacity to respond to that particular antigen (i.e. uncommitted lymphocytes). One functional role of lymphokines may be to fulfil this need for amplification.

Summary

Pretreatment with 25—250 IU/ml human IFN-α enhanced, while 5000—10 000 IU/ml IFN-α inhibited the production of LIF by Con A-stimulated human leukocytes.

Modulation of lymphokine production depended not only upon the amount of IFN-α used, but also on the duration of IFN-α pretreatment.

IFN treatment of human leukocytes before stimulation with Sendai virus enhanced IFN synthesis, but did not affect their LIF production.

References

1 Bendtzen, K. (1980): Lymphokine Reports 1, 41.

2 Chess, L.; Rocklin, R. E.; MacDermott, R. P.; David, J. R.; Schlossmann, S. F. (1975): J. Immunol. 115, 315.

3 David, J. R.; David, R. A. (1971): in "In vitro methods in cell-mediated immunity", ed. by B. R. Bloom and R. P. Glade, Academic Press, New York, 249.

4 Dumonde, D. C.; Wolstencroft, R. A.; Panayi, G. S.; Matthew, M.; Morley, J.; Howson, W. T. (1969): Nature, New Biol. 224, 38.

5 Farrar, W. L.; Johnson, H. M.; Farrar, J. J. (1981): J. Immunol. 126, 1120.

6 Mécs, I.; Béládi, I. (1977): in "Proc. Symposium on Preparation, Standardization and Clinical Use of Interferon", ed. by D. Ikic, Zagreb, 23.

7 Mundy, G. R.; Yoneda, T. (1980): in "Biochemical Characterization of Lymphokines", ed. by A. L. de Weck, F. Kristensen, and M. Landy, Academic Press, New York, 123.

8 Nathan, I.; Groopman, J. E.; Quan, S. G.; Bersch, N.; Golde, D. W. (1981): Nature 292, 842.

9 Pham Ngoc Dinh; Béládi, I.; Rosztóczy, I.; Tóth, M. (1980): J. of Interferon Research 1, 23.

10 Rocklin, R. E. (1974): J. Immunol. 112, 1461.

11 Vilcek, J.; Sulea, I. T.; Volvovitz, F.; Yip, Y. K. (1980): in "Biochemical Characterization of Lymphokines", ed. by A. L. de Weck, F. Kristensen, and M. Landy, Academic Press, New York, 323.

12 Vilcek, J.; Le, J.; Donna, S.; Stone-Wolf, D.; Pearlstein, K.; Siegel, D.; Yip, Y. K. (1983): Antiviral Research, Special Abstract Issue 1, 21.

13 Wiranowska-Stewart, M.; Lin, L. S.; Braude, I. A.; Stewart II, W. E. (1980): in "Biochemical Characterization of Lymphokines", ed. by A. L. de Weck, F. Kristensen, and M. Landy, Academic Press, New York, 331.

I. Béládi, M.D., Institute of Microbiology, University Medical School, H-6720 Szeged, Dóm tér 10 (Hungary)

Contr. Oncol., vol. 20, pp. 72—85 (Karger, Basel 1984)

Interferons in the Light of the New Theory of Hormones

Anna D. Inglot

Institute of Immunology and Experimental Therapy, Polish Academy of Sciences, Wrocław, Poland

Introduction

Roth et al. (35) recently proposed a new theory of the evolutionary origins of hormones and other extracellular chemical messengers. According to this idea the hormones are the ancient form of cell-to-cell communication. They began in the unicellular organisms as factors that stimulate cells to grow or come together or otherwise react biochemically. The biochemical elements of intercellular communication (receptors, messengers, effectors) can be detected even in the most primitive organisms. With evolution, the anatomical elements of the system have become increasingly complex, specialized and diverse. However, there is a much greater unity and scope to hormonal-like processes than was previously recognized.

Essentially all cells are secretory cells that release into the extracellular fluid messenger molecules which act on other cells (and on themselves) to regulate biological functions. The morphological changes in the complex multicellular organisms lead to highly differentiated secretory and target cells. However, "the biochemical vocabulary through which the gland cell talks to the target cell is ancient and remains highly conserved" (35, 36). There is a similarity in vertebrate between tissue factors and classic hormones. At the level of unicellular organisms all of the messengers are like tissue factors; with evolution only some of the messenger molecules were recruited as endocrine hormones, while the majority remained tissue factors (36).

I have recently published the rewievs and new formulation of the hormonal concept of interferon (20, 21) that fits very well to the theory of Roth et al. I shall summarize and extend its main proposals.

Firstly, interferons (IFNs) are the "non-classical" hormones. This group encompasses diverse hormone-like tissue factors which are elaborated not by specialized glands but by various other tissues (fig. 1, tab. I).

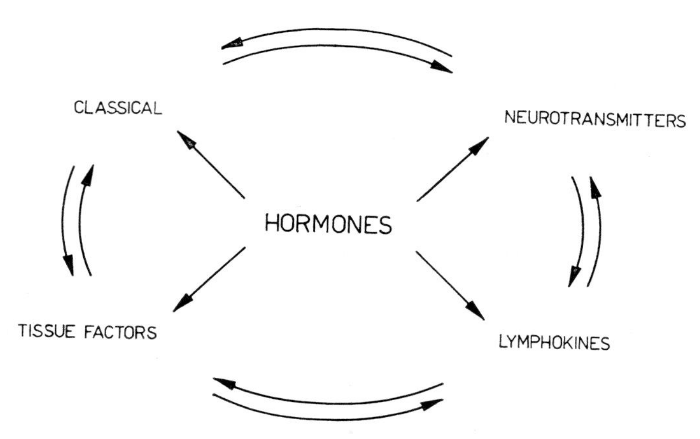

Fig. 1. Interconnexion among hormones.

Secondly, IFNs and growth factors can be grouped together as two families of hormones with opposite actions.

Thirdly, IFNs together with several classical hormones and lymphokines, may be classified as the resistance modulatory hormones.

From such viewpoints, processes involved in the IFN induction and actions may be described using the terminology of endocrinologists. Furthermore, the hormonal concept of IFN suggests new lines of investigations.

Processes of induction of interferons

Induction of synthesis of a hormone by another hormone or by a factor that mimics the hormone action may occur by one of three mechanisms: via receptors at the cell membrane, via cytoplasmic receptors or via nuclear receptors (8, 23).

Table I. Non-classical hormones

Tissue factors	Examples
Growth factors	EGR-URO, FGF, PDGF, NGF, TGF, insulin-like growth factors
Interferons	IFN α, β, γ
Chalones	Antimitogenic factors from skin, blood cells, hypothalamus
Lymphokines	IL-1, IL-2, IFNs, MAF, LIF, MIF, lymphotoxins, suppressor factors
Prostaglandins	PGE, $PGF_{2}\alpha$, tromboxane, prostacyclin, leukotrienes

EGF-URO — epidermal growth factor — urogastrine, FGF — fibroblast growth factor, PDGF — platelet derived growth factor, NGF — nerve growth factor, TGF — transforming growth factor, IL — interleukin, MAF — macrophage activating factor, MIF — macrophage migration inhibition factor, LIF — leukocyte migration inhibition factor, PG — prostaglandin.

Regarding that the large group of known IFN inducers may simulate the action of some hormones regulating IFN synthesis one can tentatively divide the IFN inducers into two categories depending on the localization of putative receptors (tab. II).

The existence of the cell membrane receptors for such IFN inducers as viruses, microorganisms (bacteria, protozoa, moulds, fungi) and their products, plant mitogens, antigen-antibody complexes and similar substances is almost obvious and does not seem to require argumentation. In view of the hormonal concept of IFN one can also suggest the existence of the cytoplastic and/or nuclear high affinity receptors for IFN inducers.

Interestingly, Greene et al. (13), De Clercq and Torrence (10) and Marcus (27) in their working models of IFN induction either by polynucleotides or by viruses proposed the involvement of cellular proteins with high affinity binding to dsRNA which are postulated to be IFN-inducer receptors. More specifically, it has been suggested (but not yet proved) that "the interaction of poly(I).po-

Table II. The tentative classification of interferon inducers

Localization of the putative receptors	Inducers	Hormone produced
Cell membrane and/or cytoplasm	Viruses, microorganisms, lectins, LPS, SEA and other microbial products, anionic polymers, nucleic acids, antigen-antibody complexes, "foreign" cells, TPA and other tumor promoters	IFN α, β, γ interleukins, several classical hormones
Cytoplasm and/or nucleus	Low molecular weight inducers (fluorenones, acridanones, pyrimidinones, anthraquinones, alkyl diamines, levamisole), polynucleotides	IFN α, β, γ interleukins, possibly some classical hormones

LPS — lipopolysaccharide, SEA — staphylococcal enterotoxin A, TPA – 12-0 – tetradecanoylphorbol-13-acetate.

ly(C) with the cell may occur in two stages . . .: the first stage would involve the recognition of a large region of the dsRNA to permit binding of the nucleic acid to the cellular receptors; the second stage would require recognition of a much smaller region of the dsRNA corresponding to roughly 6—12 basepairs in length, which would be responsible for triggering the interferon induction process" (10, 13).

Many low molecular weight IFN inducers have been discovered but their mode of action remains obscure. The most potent among them are: tilorone, acridanones, substituted pirimidinones and some others (28, 41—43). The investigators of the inducers learned long ago that the compounds strongly acting in vivo are usually weakly active or inactive in cultured cells with a notable exception of some lymphoid tissues (thymus, spleen) (28, 41—43).

As postulated by the hormonal concept of IFN, the tentative mode of action of the low molecular weight IFN inducers would be analogous to the action of steroid or thyroid hormones.

On entry into target cells the ligands associate with the receptors and in an "activated" form are bound to the cell genome. Then, by yet undefined process, the cell responds to the hormone by increased RNA and protein synthesis (8). For example Evans et al. (11) showed that both glucocorticoids and thyroid hormone regulate the growth hormone gene by rapidly increasing its transcription rate.

The induction process of new proteins by the sex steroids was found to be inhibited by Actinomycin D and cycloheximide. The inhibitors used in the appropriate way could also enhance the transcription and/or translation of the induced proteins. The superproduction of the hormone-dependent proteins clearly resembled the superinduction of IFN by the same inhibitors (8, 41).

To summarize, it appears that to become effective IFN inducers some if not all, substances have to fulfill two criteria: they have to bind with high affinity receptors (cell membrane, cytoplasmic or nuclear) and they have to interact (directly or indirectly) with DNA of genome. Although the putative receptors for IFN inducers are unknown there are several reports showing the interactions of IFN inducers with DNA (9, 12).

Mode of Action of Interferon

IFNs were found to bind specifically to cell membrane receptors and further processing of the complex seems to be similar to that of several classic protein hormones. There are several reports describing the details of binding of labeled pure IFNs to various target cells (1, 2, 5). Thus, it appears that IFN-receptor complexes diffuse and cluster in invaginations of the plasma membrane designated as "coated pits". IFN is subsequently detected first in intracellular vesicles and then in lysosomes, where the ligand-receptor complexes are degraded.

The internalization of IFN, as well as some other messengers, may not be required for some of its activities but it may play a role in the turnover of receptors (5). The homospecific receptor regulation ("down-regulation") has also been described for IFN (5).

By analogy to the classic protein hormone receptors one can suggest that the IFN receptor could be a rather protein or glyco-lipo-protein, possibly composed of subunits, partially embedded in the membrane lipid environment. The receptors may function in membranes as complex mobile enzyme-like oligomers (16).

The ligand-receptor induced series of events: a "cascade" of enzymatic steps, interaction with genome and gene activation, mRNA and protein syntheses, lead to an altered cellular function. The altered function depends on the target cell (fig. 2).

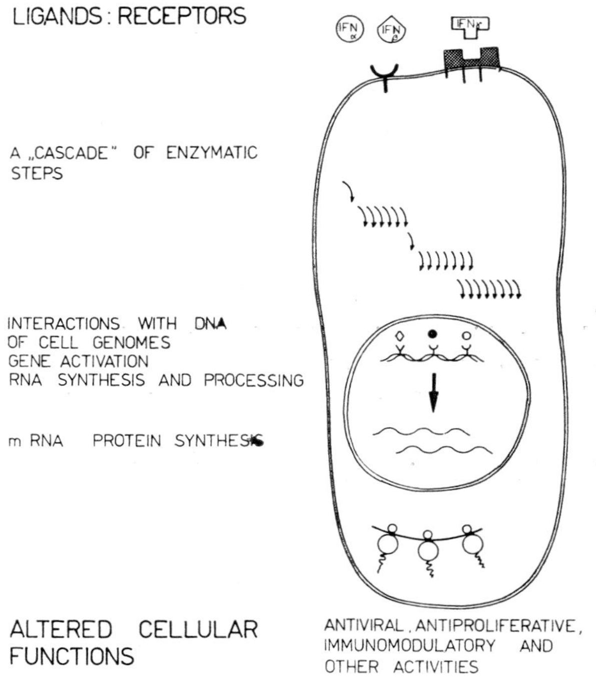

Fig. 2. Model of interferon (hormone) action.

The IFN treated cell infected with a virus will develop a resistance against the invader. The cell cycling will be inhibited. The process of the autostimulation of a tumor cell for abnormal proliferation may be disturbed. The immunocompetent cells will be suppressed or stimulated in their differentiated functions such as immunoglobulin production or cell-killing or cell-activation. Many cells will enlarge, stop moving and will have their membrane and cytoskeleton changed. Cells will start the synthesis of some macromolecules and suppress the production of several other macromolecules (20, 21, 41).

There are many gaps in our knowledge of the post-receptor

mode of action of IFN as well as other tissue factors and classic hormones. However, the correctly asked questions may help to suggest the proper answers. For example, a number of phosphorylation reactions mediated by the classic hormone-receptor complexes have been described. Some of them are cyclic AMP dependent whereas other are cyclic AMP independent (17). Because the phosphorylation reactions were also found to play an important role in the IFN action (41, 44) it would be interesting to find out how much the reactions are reminiscent of the reactions induced by classic protein hormones. And converesely, the processes characteristic for IFN action may be shared by some but not all of the classic hormones.

Smith and Blalock (40) showed that "IFN could cause multiple hormonal responses and hormones could cause cell specific antiviral responses".

The various forms of interaction of IFN with cyclic nucleotides implicated in a number of hormonal processes of cell regulation have been described. The problem was recently reviewed in details by Tovey (44).

The suggestion that there is a common outline of the action for all of the intercellular communicators by no means have to imply that all of the biochemical pathways induced by the hormones are the same or similar (40). On the contrary, there seem to exist a number of diverse and often opposing hormonal pathways within one common scheme of the response to messenger (16, 20, 21).

Plants and Interferons

Roth et al. (35) suggests that hormones appeared in the evolution before plants and animals diverged. Thus, animals and plants could make similar hormones.

In the literature on IFN there are reports in accord with this idea. Sela et al. (39) described the antiviral protein (AVP) produced by tobacco mosaic virus (TMV) infected plants and protecting the plants against the infection. AVP resembles animal IFN because it is induced by virus or poly(I). Poly(C), it exhibits antiviral activity at picomolar level and it induces the synthesis of 2—5 A (intracellular mediator of IFN action).

More recently also Loebenstein and Gera (25) showed that an IFN-like substance inhibiting virus replication is released into the medium from TMV-infected protoplasts of a cultivar in which the infection in the intact plant is localized.

Orchansky et al. (31) reported that human IFNs can protect plants from virus infection. Human alpha IFNs either natural or recombinant (bacterially synthesized) were found to be active in suppressing multiplication of TMV in tobacco leaf discs. Human IFN beta becomes as active as alpha IFN upon incubation with glycosidases. The effect of both IFNs is reversible.

It remains to be determined whether animal IFN and plant IFN-like factors are using similar mechanisms to alter the cellular functions.

Numerous experiments were performed showing that several substances of plant origin react with the animal cells and induce IFNs as well as some other tissue factors. For example, lectins known as T cell mitogens (Con A, PHA, PWM, WGA etc.) bind specifically to the animal cell membrane receptors and initiate the production of new proteins including IFN. On the other hand, the lectins may also modify the IFN actions (41).

According to the new theory of hormones such specificities as described above may not be accidental. They could develop because plants as well as animals have hormones, receptors and efectors. The elements of plant and animal cells could be similar in the general outline but different in details.

Interferon and Nervous System

The new hormonal theory postulates the overlaps of endocrine and nervous systems. The nervous system does not give rise to the endocrine system either embryologically or phylogenetically. Instead both systems are derived independently from a common ancestral system. However, they are interrelated.

It appears interesting to discuss main observations concerning the interrelationship between IFN and the central nervous system (CNS).

It is well established that CNS infected with the neurotropic viruses can produce even large amounts of IFN (41). In the beginning of the studies on the antitumor action of IFN the extracts of the West Nile virus-infected mouse brain were even used as a source of IFN (14).

The blood-brain barrier prevents the unlimited diffusibility of IFN from the brain or to the brain (41).

Neural and glial cells are responsive to the action of IFN (41). An enhanced excitability of neurons after the contact with IFN was observed (6).

Noradrenaline, a hormone of the adrenal gland which is also made by nerves for neurotransmission, was reported to induce IFN-like antiviral activity in the responsive cells (40).

Human leukocyte alpha IFN but not beta or gamma IFNs, were found to bind to opiate receptors in vitro. When injected intracerebrally into mice Hu IFN alpha was found to cause endorphin-like opioid effects including analgesia, lack of spontaneous locomotion and catalepsy. All of these actions of IFN alpha were preventable and reversible by the opiate antagonist naxolone (40).

These findings require confirmation by experiments with pure IFN, preferably of bacterial origin, because the same authors reported that human lymphocytes can produce ACTH and endorphins together with IFN (40). Thus, the preparations of natural IFN may be contaminated with the classic hormones.

When 30 to 100 million units of human IFN alpha of bacterial or natural origin were administered to patients for several days, a severe but reversible neurotoxicity was observed. The symptoms of the toxicity were: lethargy, restlessness, drowsiness, slowing of the EEG activity and of conduction velocities of EMG (7, 26). It remains to be seen which of the effects are due to the direct action of IFN on CNS and which are related to other hormones mobilized by IFN.

Although the neurotoxicity induced by high doses of IFN administered for a short time is reversible (7, 26) the prolonged exposure of CNS to IFN may be noxious.

One could infer that the progressive brain damage observed during the slow virus infections may be due to an unbalanced interactions between small amouts of IFN and certain classes of growth factors responsible for the recycling and renewal of components of neurons.

Interestingly, Libíková et al. (24) found IFN in 29 to 57 % of the cerebrospinal fluids of psychiatric or neurological patients. The authors suggested that IFN might cause disturbances in the neurons leading to altered psychic activity.

The neurologic side effects of IFN administration to patients does not exclude a possibility that IFN used in cautious and proper way may be useful for the treatment of brain tumors. There are already reports describing the antitumor effects of IFN applied interacerebrally to patients with gliomas (29).

The quoted studies clearly suggest that IFNs like other hormones exert various effects on CNS. However the biological significance of the interactions remains to be discovered.

Interferons As the Resistance Modulatory Hormones

The hormonal systems with evolution at the level of multicellular organism have become increasingly complex and diverse. Several primitive intercellular messengers probably evolved to resistance modulatory hormones. It seems that the group of the hormones is large and multifunctional. It may include such classic hormones as corticosteroids and messenger polypeptides produced by the thymus as well as the nonclassical hormones such as prostaglandins, lymphokines, cytokines and IFNs.

For example the lymphokines with toxicity for yeasts or for Treponema pallidum (33) have been described. Another class of lymphokines is apparently associated with the inhibition of the intracellular multiplication of viruses, bacteria and parasites of which only IFNs have been identified as definite protein entities (3, 4, 30, 38). The relationship of the other lymphokines to IFNs and interleukins with known hormonal-like structure and actions remains to be established.

Sam Baron, co-worker of Alick Isaacs and one of the pioneers of interferon research wrote (3): "The interferon system is the earliest appearing of the known host defences. It is operative within hours of infection ... The available evidence strongly supports a casual relationship between the interferon system and natural recovery from many established viral infections ... The interferon system appears to play its defensive role during viral infections of body surfaces, solid tissues and also during virus spread through cells of the vascular system to target organs". IFN was found to be an important factor in resistance to various tumor viruses (14, 22, 41) and to orthomyxoviruses governed by the Mx gene (15).

The possibility that IFN may additionally serve to protect the host against certain bacteria and protozoa has also been suggested although the evidence for this is not compelling (4).

From the viewpoint of the hormonal concept of interferon all of the above statements remain valid. However, IFN should not be regarded as a separate defensive system but as one of the resistance modulatory hormones. Such interpretation not only rearranges the contemporary views but also indicates possible new lines of studies.

Firstly, a few defects in the production of IFN in man were found to be associated with complex immune deficiency syndroms (18). It remains to be determined whether there exist defects in the IFN function accompanying or imitating the classic endocrinological disorders. Because many manifestations of disease in man are associated with autoantibodies against hormonal receptors, eg. for insulin, TSH, acetylocholin, gastrin etc., it seems also worth while to look for such autoantibodies for IFN receptors.

Recently, at least in one case, the autoantibodies hyperstimulating the thyroid gland were found to be associated with antibodies against a gram negative bacterium, Yersinia enterocolitica. This lead to an interesting hypothesis that antibodies produced to some bacterial components may cross react with elements of self TSH receptors on thyroid cells (36).

In the literature on IFN one can find many facts apparently bearing to the problem. Systemic lupus erythematosus, a complex organ unspecific autoimmune disease is associated with the frequent presence of IFN in the serum (18). Thus, it is tempting to suggest that the stimulation for the hyperproduction of IFN may be due to autoantibodies reacting with receptors for IFN inducers. It has been long known that the synthesis of IFN can be triggered by the antilymphocytic sera (41). Besides antibodies against cell surface components may inhibit the action of IFN (34). The antibodies against the cell-membrane components might react as well as ligands with the IFN receptor or IFN-inducer-receptor. The "activation" of the receptors may lead to a hormonal chain reaction.

The same can be said about the role of IFN in the pathogenesis of rheumathoid arthritis and other allergic diseases, especially when virus etiology is suspected (18).

The lymphocytic choriomeningitis virus (LCM) induced disease is considered to be due to the development of sensitized T lymphocytes which are cytotoxic for virus-infected host cells. Because anti-mouse interferon globulin protects mice against mortality due to LCM virus it was suggested that IFN is one of the offenders (37). Regarding IFN as a hormone one can suggest that hyperproduction of IFN may disturb the hormonal balance. Details of the putative disturbance remain to be discovered and may be unusually interesting because LCMV-induced disease is a good model to study infections of central nervous system associated with autoimmunity.

Secondly, because cooperativity among hormones is almost a general rule one can expect various forms of interactions between IFN and other resistance modulatory hormones.

There is an increasing number of reports showing interactions between IFN and other lymphokins during the immune response (30). Corticosteroids were usually found to inhibit the production of IFN in vivo but under some condition the hormones may potentiate IFN synthesis and action (41). A single dose of thymosin fraction 5 approximately 40 hr before induction of IFN with Newcastle disease virus was found to stimulate significantly the synthesis of IFN in mice. On the other hand, two other purified thymosin polypeptides, alpha 1 which induces maturation of helper T cells and alpha 7 which induces maturation of suppressor T cells, had opposite effects (19). Various, often conflicting effects of prostaglandins on the production and action of IFN have been described (42, 44).

All of the above mentioned facts support the view that the immune status of the organism is under control of a family of different hormones including IFN.

Biological Role of Interferons

The physiological role of IFNs is known only to a limited extent. I have briefly discussed the possible significance of IFNs as the resistance modulatory hormones but it does not seem to be their only function. On the contrary, the new theory of hormones strongly suggests that also IFNs are ancient messengers which may be present in the unicellular organisms. At present (1983) there is no experimental evidence for the suggestion. However, the relative ease with which the connected with plasmids IFNs genes are transcribed and further processed by bacteria and yeasts (32) may mean that the genes are not very "foreign" to the microorganisms.

In the primitive microorganisms the messenger molecules are used mainly to regulate feeding and mating behavior. However, with the evolutionary development, IFNs as well as other messengers, were probably recruited to control the growth and differentiation. It remains yet to be determined how IFNs function within the hormonal system.

References

1 Aguet, M. (1980): Nature, 284, 459.
2 Anderson, P.; Yip, Y. K.; Vilček, J. (1982): in Abstracts, The Third Annual International Congress for Interferon Research, Miami, Fl. USA.
3 Baron, S.; Dianzani, F.; Stanton, G. J. (1982): Texas Rep. Biol. Med. 41, 1.
4 Baron, S.; Howie, V.; Langford, M.; Macdonald, E. M.; Stanton, G. J.; Reitmeyer, J.; Weigent, D. A. (1982): Texas Rep. Biol. Med. 41, 150.
5 Branca, A. A.; Faltynek, C. R.; D'Alessandro, S.; Baglioni, C. (1982): J. Biol. Chem. 257, 13291.
6 Calvet, M. C.; Gresser, I. (1979): Nature 278, 558.
7 Cantell, K.; Mattson, K.; Nüranen, A.; Kaupinnen, H. L.; Jivanainen, M.; Bergström, L.; Farhilla, M.; Holsti, L. R. (1982): In Abstracts, The Third Annual International Congress for Interferon Research, Miami, Fl. USA.
8 Chan, L.; O'Malley, B. W. (1976): New Engl. J. Med. 294, 1322 and 1372.
9 Chandra, P.; Woltersdorf, M. (1976): Biochem. Pharmacol. 25, 877.
10 De Clercq, E.; Torrence, P. F. (1982): Texas Reps. Biol. Med. 41, 76.
11 Evans, R. M.; Birnberg, N. C.; Rosenfeld, M. G. (1982): Proc. Natl. Acad. Sci. USA 79, 7659.
12 Fikus, M. (1983): Personal communication.
13 Greene, J. J.; Alderfer, J. L.; Tazawa, I.; Tazawa, S.; Ts'O, P. O. P.; O'Malley, J. A.; Carter, W. A. (1978): Biochemistry, 17, 4214.
14 Gresser, I. (1972): Adv. Cancer Res. 16, 97.
15 Haller, O. (1981): Current Top. Microbiol. Immunol. 92, 25.
16 Hollenberg, M. D. (1982): Trends Pharmacol. Sci. 3, 25.
17 Hollenberg, M. D. (1982): Trends Pharmacol. Sci. 3, 271.
18 Hooks, J. J.; Moutsopoulos, H. M.; Notkins, A. L. (1982): Texas Rep. Biol. Med. 41, 164.
19 Huang, K. Y.; Kind, P. D.; Jagoda, E. M.; Goldstein, A. L. (1981): J. Interferon Res. 1, 411.
20 Inglot, A. D. (1982): Texas Rep. Biol. Med. 41, 402.

21 Inglot, A. D. (1983): Arch. Virol. (in press).
22 Inglot, A. D.; Oleszak, E. (1978): Arch. Immunol. ther. Exp. 26, 529.
23 Jänne, O. A.; Kontula, K. K. (1980): Annals Clin. Res. 12, 174.
24 Libíková, H.; Breier, S.; Kočišová, M.; Pogády, J.; Stünzner, D.; Ujházyová, D. (1979): Acta Biol. med. Germ. 38, 879.
25 Loebenstein, G.; Gera, A. (1982): Texas Rep. Biol. Med. 41, 213.
26 Madejewicz, S.; Creaven, P.; Ozer, H.; O'Malley, J.; Grossmayer, B.; Pontes, E.; Miltelman, A.; Solomon, J.; Ferraresi, R. (1982): In Abstracts, The Third Annual International Congress for Interferon Research, Miami, Fl. USA.
27 Marcus, P. (1982): in Abstracts of the Third International Congress for Interferon Research, Miami, Fl. USA.
28 Mayer, G. D.; Kruger, R. F. (1980): in "Interferon and Interferon Inducers. Clinical Applications" ed. by D. A. Stringfellow. Marcel Decker Inc. N. York, Basel, p. 187.
29 Nagai, M.; Arai, T.; Kohno, S.; Kohase, M. (1982): Texas Rep. Biol. Med. 41, 693.
30 Neta, R.; Salvin, S. B. (1982): Texas Rep. Biol. Med. 41, 435.
31 Orchansky, P.; Rubinstein, M.; Sela, I. (1982): Proc. Natl. Acad. Sci. USA. 79, 2278.
32 Pestka, S.; Maeda, S.; Chiang, T. R.; Costello, L.; Rehberg, E.; Levy, W. P.; M. Candliss, R. (1982): Texas Rep. Biol. Med. 41, 198.
33 Podwińska, J.; Metzger, M. (1981): Arch. Immunol. Ther. Exp. 29, 671.
34 Revel, M.; Bash, R.; Ruddle, F. H. (1976): Nature 260, 139.
35 Roth, J.; Le Roith, D.; Shiloach, J.; Rosenzweig, J. L.; Lesniak, M.; Havrankova, J. (1982): New Engl. J. Med. 306, 523.
36 Roth, J.; Le Roith, D. (1983): (personal communication).
37 Saron, M. F.; Riviere, Y.; Gresser, I.; Guillon, J. C. (1982): Ann. Virol. 133E, 241.
38 Schuman, I.; Fahlbusch, B.; Vasiljeva, I. G. (1979): Acta biol. med. germ. 38, 807.
39 Sela, I. (1981): Adv. Virus Res. 26, 201.
40 Smith, E. M.; Blalock, J. E. (1982): Texas Rep. Biol. Med. 41, 350.
41 Stewart II, W. E. (1979): The Interferon System. Springer Verlag, Wien, N. York.
42 Stringfellow, D. A. (1982): Texas Rep. Biol. Med. 41, 116.
43 Taylor, J. L.; Grossberg, S. E. (1982): Texas Rep. Biol. Med. 41, 158.
44 Tovey, M. G. (1982): in "Interferon" ed. by I. Gresser, Acad. Press Inc. London 4, 23.

A. D. Inglot, M.D., Institute of Immunology and Experimental Therapy, Polish Academy of Sciences, Wroclaw (Poland)

Contr. Oncol., vol. 20, pp. 86—97 (Karger, Basel 1984)

Interferon in Antitumor Protection and during Gestation

C. Chany; I. Cerutti

Institut National de la Santé et de la Recherche Médicale, Hôpital Saint Vincent-de-Paul, Paris, France

Introduction

In this presentation we focus on interferon's (IFN's) possible role in antitumor protection and during gestation. Our hypothesis is based on the observation that long-term treatment with relatively low IFN concentrations results in a reconversion of malignant cells to a normal phenotype (and selection in some cases of a normal genotype). The morphological changes are easily detectable and will be analyzed here. The constant presence of IFN (consisting sometimes in part of unusual molecular species) in the placenta and amniotic fluid of humans and rodents suggests furthermore the role of this substance in embryonic and fetal development. The mechanism involved could be comparable to that observed in transformed cells.

Effect of IFN on Tumor Development

In association with Dr. Atanasiu, we reported that IFN blocks the oncogenic expression of polyoma viruses in newborn hamsters (3). It is well-known that in this system, transformation is obtained in the absence of virus replication. In vitro studies by Allison (1) using polyoma virus and Oxman et al. (24) with SV-40 clearly demonstrate the blocking effect of IFN on cell transformation. However, at that time, it was considered that IFN treatment of the cells did not modify oncogenic functions once they were integrated into the genome and expressed.

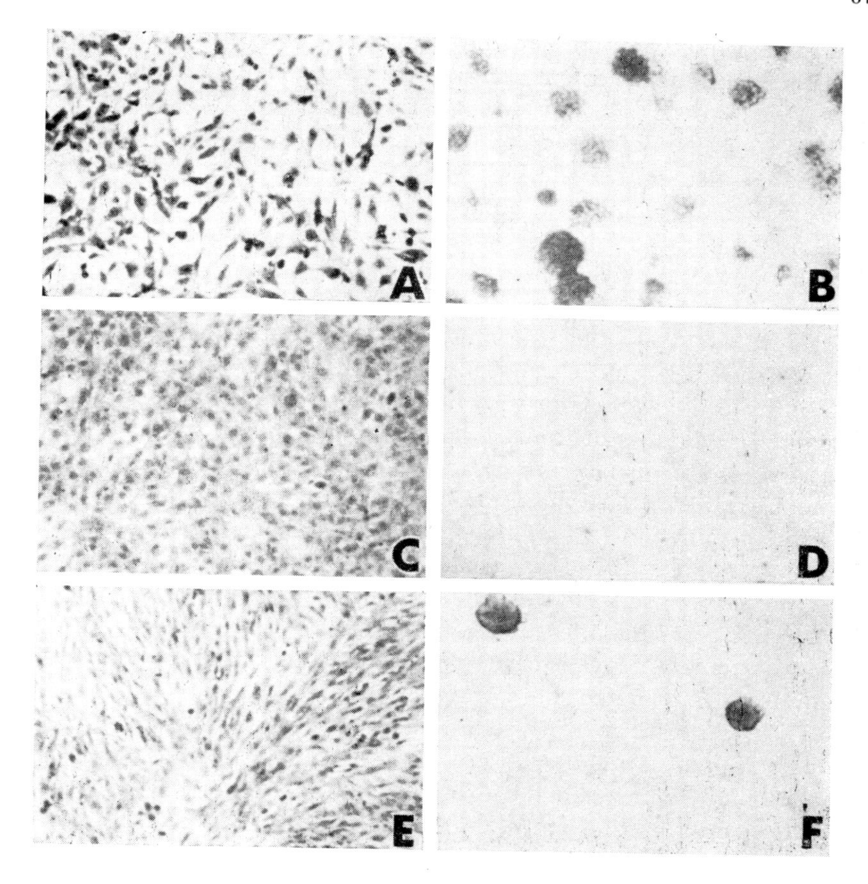

Fig. 1. Long-term effects of IFN on MSV-transformed cells: phenotypic reversion and selection of "normal" revertants.

A. Murine sarcoma virus (MSV)-transformed mouse embryonic fibroblasts (MEF).

B. Same cells producing colonies in soft agar.

C. Same cells grown in presence of IFN for 200 passages (MSV-IF+).

D. MSV-IF+ cells grown in soft agar. Practically no colony development.

E. Normal MEF used for transformation by MSV.

F. Reappearance of colonies in MSV-IF+ cells when IFN is omitted from the tissue culture medium for 30 passages.

Reprinted with modification from the J. gen. Virol. (1970; Chany & Vignal; 7, 203—210).

In preliminary studies using IFN in the treatment of myelogenous human leukemias, an increase in the survival of patients (4/11 cases) was shown (20). These results suggested that IFN could affect the development of the disease even in the absence of detectable associated viruses. On the basis of present knowledge, this implies that IFN could change to some extent the cellular expression of already integrated oncogenes as well. A subsequent study was undertaken to explore this IFN effect after prolonged treatment of mouse cells transformed by the Moloney strain of murine sarcoma virus (MSV) (14). In these studies, IFN treatment was applied to the transformed cells and carried for two years. The surprising results are that the cells become resistant to IFN's antiviral effect, but produce large amounts of the substance even after repeated induction, thus losing the usual refractoriness. The malignant cells recover contact inhibition, cell adhesion and are unable to produce colonies in agar despite the fact that they grow normally (at a slower rate) in the presence of IFN. IFN treatment for long periods results only in a phenotypic reversion of the cells to normal. Indeed, when IFN is omitted from the medium for another 200 subsequent passages, the cells never recover malignancy completely and only a few of them produce colonies again in agar (fig. 1).

It is interesting to correlate these observations with those of Paucker et al. (25) who reported that suspended murine cells replicate at a slower rate in the presence of IFN. The decrease cannot be interpreted as the expression of an anticellular effect but indicates, on the contrary, a reduction of cell growth to normal or even to arrest. This is obviously a physiological and not a pathological process since most of the cells, when differentiated, become constitutive components of tissues and do not replicate (or do so very slowly).

IFN Effect on Cytoskeletal and Extracellular Matrix Components

Further developments in our knowledge on membrane structure, motility and adhesion of the cells focus our attention on the cytoskeletal system and extracellular matrix. Preliminary studies using inhibitors of the development of microfilaments and microtubules show that not only the expression of IFN's antiviral effect but also its maintenance depend on the integrity of these constituents (6, 10). For example, drugs such as cytochalasine B, vinblastine, colchicine, griseofulvine or chlorpromazine, block the development of antiviral protection and decay it even after its establishment (tab. I).

Table I. Effect of cytoskeletal disrupting drugs on the antiviral action of IFN

Drug employed	Inhibition of EMC replication[1]			
	Exp. 1	Exp. 2	Exp. 3	Exp. 4
Control	2300	230	7000	220
Cytochalasin B	150	28	—	—
Vinblastine	65	—	—	—
Cytochalasin B [2]Vinblastine	11	—	—	—
Colchicine	—	27	—	—
Colchicine [2]Cytochalasin B	—	8	—	—
Chlorpromazine	—	—	41	—
Griseofulvine	—	—	—	2

[1] Mouse L929 cells were treated with IFN and the different drugs for 4 h and then challenged with encephalomyocarditis (EMC) virus. The virus yield was estimated by plaque-forming units/ml. Inhibition represents the ratio of virus yield in control/IFN-treated cells.

[2] Reprinted with modification from Lymphokines (1981; C. Chany, 4, 418).

During the course of these studies, a number of reports identified the Rous sarc (15, 16) and, thereafter, other oncogenes (28). The transformation process is generally attributed to a gene product (mostly a phosphoprotein) which disorganizes the cytoskeletal system and consequently the extracellular matrix. The cells become round and lose their capacity to adhere to matrix and to respond to extracellular signals which are responsible for contact inhibition.

On the other hand, previous studies in our laboratory have shown that after thransformation of cells with oncogenic DNA or RNA viruses, their IFN sensitivity decreases rapidly, sometimes in a few hours (9, 13). It was therefore tempting to explore whether this oncogene-induced degradation of the cytoskeletal system could explain or not the loss of IFN sensitivity of these transformed cells. Since other laboratories have shown that butyrate salts reverse the transformed phenotype to normal in a manner somewhat reminiscent of the IFN effect (2), we explored the possibility of using butyrate salts to reconstitute oncogene-induced decay of the cytoskeletal elements (8). The results clearly confirm our hypothesis, since 18-h pretreatment with sodium butyrate is enough to recover full IFN sensitivity in transformed but not in normal cells (fig. 2). These in vitro observations can be extended to in vivo studies, as shown further on.

It is of interest to explore the parallelism between the effects of butyrate salts and IFN. In sarc gene transformed cells, IFN de-

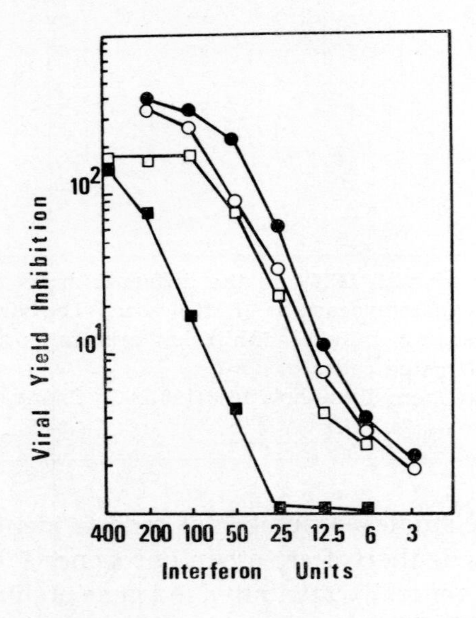

Fig. 2. Effect of sodium butyrate on the antiviral action of IFN in normal and MSV-transformed murine cells.
(●) Normal murine cells treated with IFN alone; (○) Normal murine cells treated with butyrate and IFN; (■) MSV-transformed murine cells treated with IFN alone; (□) MSV-transformed murine cells treated with butyrate and IFN. The results are expressed as the ratio of the virus yield in control cells/IFN-treated cells. Reprinted from Int. J. Cancer (1979; Bourgeade & Chany; 24, 314—318).

velops a phenotypic repair very slowly, while butyrate does it rapidly. The effect of IFN is more stable, especially when treatment is prolonged for longer periods, whereas that of butyrate disappears as soon as the drug is removed from the medium.

Role of IFN in Embryonic Development

The effect of IFN on sarc gene products in transformed cells can probably shed light on the possible role of IFN during embryonic or fetal life. It is known that during early embryonic development some cells migrate and accumulate in areas where they settle and differentiate into tissues. When the cells are mobile, the cytoskeleton and extracellular matrix elements are absent. When they become anchored, the genes responsible for their synthesis are expressed at varying time intervals and in different areas.

It is tempting to attribute to IFN a role in these events after having explored its connection with the cytoskeleton during the transformation process. Therefore, the finding that alpha-IFN is constitutively secreted in human amniotic fluid at very low titers is of great interest (28). Similar observations have been made in mouse (22) and rat (Chany & Cerutti, unpublished experiments) embryos in which the detectable IFN titer in the placenta has been shown to increase up to the 15th day of gestation and level off thereafter. It is obvious that IFN secreted under these conditions, especially in the human species, has no relationship with accidental viral infections.

Furthermore, some of these IFNs have somewhat unusual molecular and antigenic compositions, at least those induced in amniotic membranes. Since early studies, it is known that the infected amniotic membrane produces 160,000 molecular weight (MW) IFNs when estimated by gel filtration. They are thus significantly heavier than those released by leukocytes or fibroblasts using the same techniques (21, 17). More recent studies have indicated that about half of the IFN molecules are of the 21 K-dalton (KD) α type (18). The other half contains 15–17 KD β-IFN and three other IFNs migrating at 26 KD and at high MW ranges of 43 KD and 80 KD. These latter are neutralized to the same extent by antibodies raised to α and β-IFN. It is presently unknown whether these unusual IFNs are products of unknown genes or the result of unusual transcription or other processes. Their physiological role could also be somewhat different from other known IFNs.

Effect of IFN on the Immune System

The relationship of IFNs with the immune system, especially the α and β types ,is well-documented. Somewhat less is known concerning γ-IFN, considered as the "premier lymphokine" by Vilcek et al. (27). Both α and β-IFNs enhance NK cell production and augment T cell-mediated cytotoxicity. These effects appear rapidly but are also limited in time since, in about 18 h, IFN treatment results in target cell anticytolytic resistance. The cells are protected, regardless of the activity of the effector cells which bind poorly to the targets and separate without delivering the lethal hit (4, 5).

IFN slows down or inhibits cell proliferation in almost every somatic cell, thus all the immune functions depending on cell proliferation are impeded. It seems likely therefore that the overall effect of IFN on the immune system is depressive. This point is of importance in the interpretation of IFN's possible role in embryonic development and antitumor therapy.

The almost only immune enhancing activity of IFN is that on homed macrophages which, after infiltration in the inflammatory area, become immobilized, adhere to the target cells, and kill because of increased cytotoxicity (26).

Enhancement of the Antitumor Effect of IFN by Butyrate in Mice Pretreated with Corynebacterium Parvum

1. Effect on Grafted Tumors: 180 TG Sarcoma Cells in Swiss Mice

It has been previously shown using the grafted ascitic 180 TG Crocker sarcoma that sodium butyrate increases the antitumor effect of IFN (7). Since this sodium salt is difficult to employ in vivo, it has been replaced by arginine butyrate which increases antitumor protection even better (11). The depressive effect of IFN on the immune system is counterbalanced by the associated use of an immune stimulant, such as Corynebacterium parvum (CP) (12).

The best therapeutic regimen in this animal model is a single CP injection, followed by 9 alternating shots of arginine butyrate and IFN. This treatment increases the mean survival time (MST) and final survival rate of the mice and significantly delays the appearance of tumors. It is noteworthy that IFN and CP, used separately, are ineffective under our study conditions; however, their association considerably enhances antitumor protection.

These favorable therapeutic results using grafted tumors prompted us to extend these assays to another animal model where lymphomas are developed spontaneously.

2. Effect on Spontaneous Malignancies: AKR Mice

The AKR disease is characterized by a vertical transmission of MuLV, the etiological agent of the lymphoma. Female AKR mice develop spontaneously lymphoid leukemia in about 95–100 % of

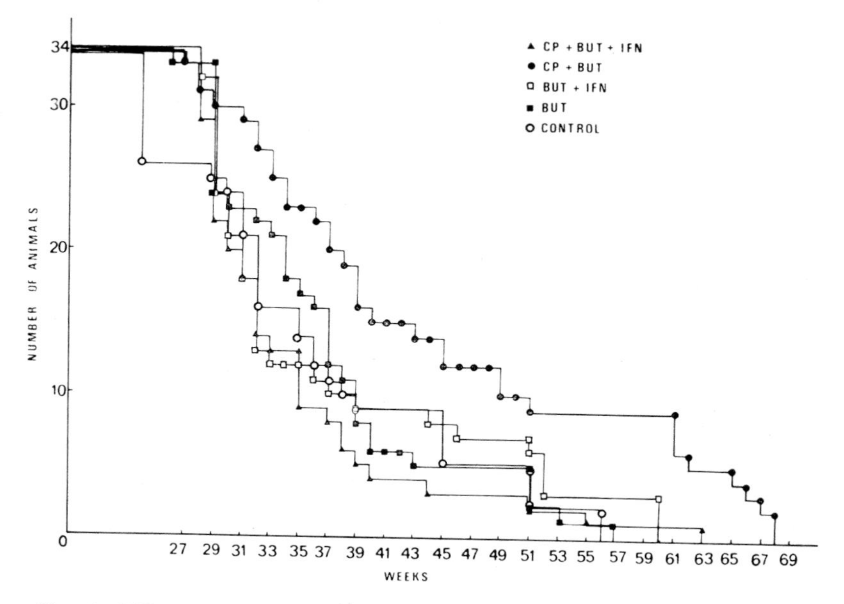

Fig. 3. Effect of associated treatments on the life-span of AKR mice.

the cases. This disease seems to be an excellent experimental model since it bears certain similarities to human lymphoma.

The different treatments are initiated at the 15th week of life when the mice are no longer able to respond to IFN by activating splenic NK cells. After stimulation of the animals with a single injection of CP, 9 alternating shots of both IFN and arginine butyrate are administered. This treatment is stopped for one month and then renewed three times.

In first series using 200 μg of CP, 20,000 international units (IU) of IFN, and 50 mM of butyrate, only CP associated to butyrate augments significantly the MST and final survival rate of the mice (fig. 3). In further experiments, the concentration of IFN is reduced to 10,000 IU and 5,000 IU and that of butyrate to 6 mM. The results are signnificantly improved since the combinations of CP + butyrate + IFN, CP + IFN and IFN + butyrate, protect the animals to a great extent against the spontaneous disease and enhance substantially the MST and final survival rates. It is important that CP and IFN alone are ineffective (tab. II and III).

These promising results using experimentally grafted tumors and naturally occurring leukemias stress the importance of immune stimulation prior to IFN and/or butyrate treatment and the use of both of these substances at low concentrations.

Table II. Comparison of the mean survival time (in weeks) of AKR mice treated or not with C. parvum and different concentrations of arginine butyrate and/or IFN

Drug	BUT 50 mM; IFN 20,000 IU	BUT 6 mM; IFN 10,000 IU
CP + BUT + IFN	35 ± 2.27	41.4 ± 1.86[1]
CP + BUT	42.5 ± 3.66[1]	40.2 ± 1.94[1]
CP + IFN	ND	42.73 ± 3.29[1]
CP	ND	38.8 ± 2.17
BUT + IFN	35.3 ± 2.48	41 ± 2.34[1]
BUT	36.05 ± 2.25	39.2 ± 2.7[2]
	ND	35.73 ± 2.4
Control	35.17 ± 1.67	36 ± 1.87

[1] $P < 0.001$;
[2] $P < 0.02$, compared to control and IFN.

Table III. Comparison of final survival rates of AKR mice treated or not with C. parvum and different concentrations of arginine butyrate and/or IFN

Drug	BUT 50 mM; IFN 20,000 IU	BUT 6 mM; IFN 10,000 IU
CP + BUT + IFN	1/35 (2.8 %)	8/15[1] (53.33 %)
CP + BUT	9/35[2] (25.7 %)	4/15[2] (26.6 %)
CP + IFN	ND	10/15[2] (66.7 %)
BUT + IFN	3/35 (8.5 %)	4/15[2] (26.6 %)
BUT	0/35	6/15[2] (40 %)
IFN	ND	1/15 (6.6 %)
Control	0/35	0/35

[1] $P < 0.01$;
[2] $P < 0.02$, compared to control and IFN

General Conclusions and Practical Applications

Much can be learned from a parallel study of IFN's physiological role during pregnancy and malignant transformation. IFN could play a role in transforming mobile cells to an anchored state and simultaneously slow down their capacity to replicate. Under the condition that membrane-bound receptors are available and functional, the presence of IFN could result in the appearance of cytoskeletal constituents, such as microfilaments, intermediate filaments and microtubules.

These same physiological functions of IFN could be involved both during gestation and malignant transformation. If this interpretation is correct, it would be necessary when studying the physiology of IFN to distinguish between differentiation and cell growth, malignant or not. In most instances during transformation the cells lose their tissue specificity. However, some tumors maintain their differentiated functions and others acquire new ones which are unexpressed in normal circumstances. This is the case of some lung carcinomas which secrete placental gonadotropin hormones (19). During embryonic development, IFN could also play a role in migration and stabilization of cells in the process of differentiating. Furthermore, it could limit at adequate moments the growth of already differentiated tissues.

By its inhibitory effect against effector cell-induced cytolysis on target cells, IFN could protect in utero the fetal tissues against maternal lymphocytes which could occasionally cross the placental barrier, or vice-versa, protect the mother against a graft versus host reaction. The multiplicity of the different IFN genes and the occurrence of unusual molecular forms in the placenta point to the possibility that different IFNs could have specialized functions or at least be more active in some metabolic areas than in others.

In conclusion, the analysis of IFN's likely physiological role during pregnancy and antitumor protection can be employed in developing a new strategy in antitumor therapy. Indeed, pretreatment of the patients with a potent immune stimulation compensates to some extent IFN's immune repressive effects. Through the modulation of the cytoskeleton, IFN further enhances macrophage activity. Moreover, butyrate by its own effect on malignant phenotype increases IFN sensibility in a number of malignant cells. All of these substances should be used at the lowest possible concentration delivered as closely as possible to the target area.

References

1 Allison, A. C. (1961): Virology 15, 47.
2 Altenburg, B. C.; Via, D. P.; Steiner, S. H. (1976): Exp. Cell Res. 102, 223.
3 Atanasiu, P.; Chany, C. (1960): C. R. Acad. Sci. (Paris) 251, 1687.
4 Bergeret, M.; Grégoire, A.; Chany, C. (1980): Immunology 40, 637.
5 Bergeret, M.; Fouchard, M.; Grégoire, A.; Chany, C.; Zagury, D. (1983): Immunology 48, 101.
6 Bourgeade, M. F.; Chany, C. (1976): Proc. Soc. Exp. Biol. Med. 153, 501.
7 Bourgeade, M. F., Chany, C. (1979): Int. J. Cancer 24, 314.
8 Bourgeade, M. F.; Cerutti, I.; Chany, C. (1979): Cancer Res. 39, 4720.
9 Chany, C. (1961): Virology 13, 485.
10 Chany, C. (1981): in "Lymphokines" ed. by E. Pick, New York Acad. Press Inc. 4, 409.
11 Chany, C.; Cerutti, I. (1981): C. R. Acad. Sci. (Paris) 293, 367.
12 Chany, C.; Cerutti, I. (1982): Int. J. Cancer 30, 489.
13 Chany, C.; Robbe-Maridor, F. (1969): Proc. Soc. Exp. Biol. Med. 131, 30.
14 Chany, C.; Vignal, M. (1970): J. Gen. Virol. 7, 203.
15 Collett, M. S.; Erikson, R. L. (1978): Proc. Nat. Acad. Sci. (USA) 75, 2021.
16 Collett, M. S.; Purchio, A. F.; Erikson, R. L. (1980): Nature 285, 167.

17 Duc-Goiran, P.; Galliot, B.; Chany, C. (1971): Arch. Gesamte. Virusforch. 34, 232.
18 Duc-Goiran, P.; Robert-Gaillot, B.; Chudzio, T.; Chany, C. (1983): Proc. Natl. Acad. Sci. (USA) 80, 2628.
19 Dulbecco, R. (1982): La Recherche 13, 1426.
20 Falcoff, E.; Falcoff, R.; Fournier, F.; Chany, C. (1966): Ann. Inst. Pasteur 111, 562.
21 Fournier, F.; Falcoff, E.; Chany, C. (1967): J. Immunol. 99, 1036.
22 Fowler, A. K.; Reed, C. D.; Giron, D. J. (1980): Nature 286, 266.
23 Lebon, P.; Girard, S.; Thépot, F.; Chany, C. (1982): J. gen. Virol. 59, 393.
24 Oxman, M. N.; Baron, S.; Black, P. H.; Takemoto, K. K.; Habel, K.; Rowe, W. (1967): Virology 32, 122.
25 Paucker, K.; Cantell, K.; Henle, W. (1962): Virology 17, 324.
26 Tamm, I.; Wang, E.; Landsberger, F. R.; Pfeffer, L. M. (1982): in "11th Annual UCLA Symposia", Acad. Press New York, 6, 81.
27 Vilcek, J.; Le, J.; Stone-Wolff, D.; Pearlstein, K.; Siegel, D.; Yip, Y. K. (1983): in "Antiviral research. The biology of the interferon system". Second International TNO Meeting, Elsevier Biomed., Abstract 1, 21.
28 Weinberg, R. A. (1982): Cell 30, 3.

C. Chany, M.D., I.N.S.E.R.M., Unité 43, Hôpital Saint Vincent-de-Paul, 74 Avenue Donfert-Rochereau, F-75014 Paris (France)

Contr. Oncol., vol. 20, pp. 98—112 (Karger, Basel 1984)

Production and Purification of Lymphoblastoid (Namalwa) Interferon for Clinical Use

K. H. Fantes; N. B. Finter; L. L. Toy

Wellcome Research Laboratories, Beckenham, Kent, England

Introduction

Looking at the theme of this conference and at the titles of its five workshops, it would seem highly desirable to use only pure interferons (IFNs) throughout, in order to forestall ambiguities and possible arguments, whether any of the observed effects were indeed due to IFN, or to some of the many unknown impurities that might accompany it. The object in our laboratories was to prepare pure IFN-α on a large scale for clinical use. When we decided in 1974 to embark on this project, a number of encouraging clinical results with IFN-α had been published, but all of them had been achieved with IFN produced in primary leukocytes (29), a method not adaptable to the scale we had in mind. At that time the only suitable way to produce large enough amounts of IFN-α seemed to us to grow an IFN-yielding cell line, such as the lymphoblastoid cell Namalwa (30) in suspension culture in large tanks. However, one immediately apparent obstacle to such a scheme was the fact that Namalwa cells originated from a child with Burkitt lymphoma. We argued nevertheless that it should be possible to devise processes and tests which would show beyond any reasonable doubt that a safe product could be obtained from such cells, and we decided to go ahead in spite of many warnings and criticisms (10, 11).

Production

Our early cell growth experiments were carried out in small stirred vessels, but we were soon able to use 50 and 100 l stirred and aerated tanks, and by 1978 we had graduated to a 1,000 l tank in our pilot plant. The cell growth medium we use is RPMI 1640 supplemented with γ-irradiated bovine serum. IFN-induction experiments showed that of a number of viruses tested a strain of Sendai virus gave the highest titres especially when prepared under certain defined conditions (16). Our colleague, Dr. Michael Johnston (15), found that preincubation of the cells for 2–3 days with low concentrations (1–3 mM) of sodium butyrate not only considerably increased IFN titres, but also led to much more reproducible results, and butyrate treatment is now used routinely in our large-scale production. We have also progressively increased the size of our tanks and an 8,000 l tank is now under construction. IFN titres obtained in these large cultures usually range between 4.0 and 4.5 log international units per ml.

Purification

Purification presented a real challenge. Not only did we have to try and isolate some 50–100 mg of IFN protein from 2 to 5 kg of total protein (as contained in 1,000 l of crude IFN), i.e. achieve a purification factor of 20,000 to 100,000, but we also had to use a final process which could be shown to reduce any real or hypothetical contaminants to levels, which could no longer be detected by the most sensitive known methods.

When we started work on purification we examined many of the methods known to protein chemists, but eventually narrowed these down to four, which we investigated in some detail: These were precipitation with trichloroacetic acid (TCA) and extraction of the resulting precipitate with acid ethanol, gel chromatography on Ultrogel AcA54, ion-exchange chromatography on SP-sephadex and antibody affinity chromatography. Every one of these steps provided some but never complete purification. As an example fig. 1 shows the further purification of partly purified interferon on AcA54. Complete purification could only be achieved by a combination of methods which included amongst others TCA-precipitation, acid ethanol extraction and antibody affinity chromatography (7). The finally chosen series of steps could be used on a very large scale and has consistently led to IFN of a specific activity of greater than 10^8 international units per mg protein.

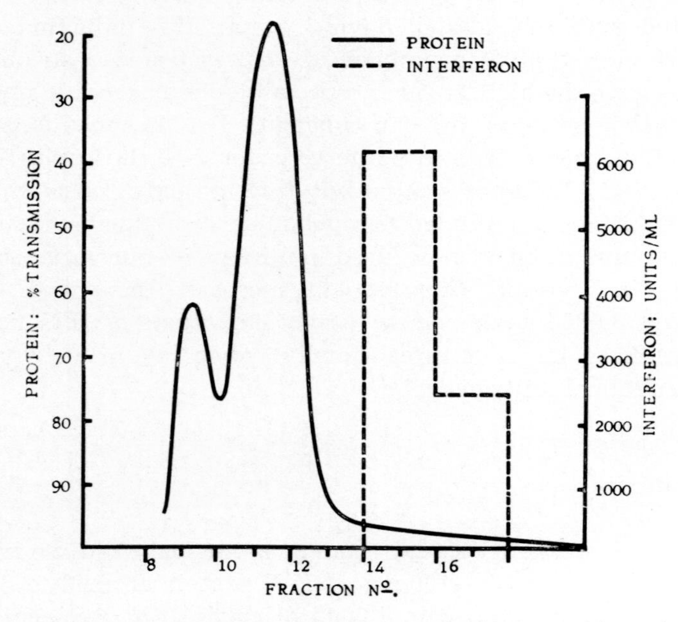

Fig. 1. Chromatography of Namalwa interferon on Ultrogel AcA 54.

Heterogeneity of Purified Namalwa IFN

Samples purified in the above way were analysed by chromatography on Sephadex G 75 (fig. 2) and by polyacrylamide gel electrophoresis (fig. 3) (8). Both methods revealed the presence of seemingly two active materials (peaks A and B in fig. 2, two bands in fig. 3). However, when individual fractions from the Sephadex G75 column were subjected to polyacrylamide gel electrophoresis (PAGE) we found that the purified IFN actually consisted of eight major components, not just of two. This multiplicity of components was confirmed by sequencing studies (1, 2) and had also been observed by others in other leukocyte IFN preparations (21, 13).

When we cut out the bands that were obtained from subjecting Sephadex G75 fractions to PAGE and assayed them for antiviral activity we found that only those bands that had originated from fractions within the double peak area of fig. 2 were active, any bands found in the early stages outside this area were inactive. This allowed us for the first time to express purity and IFN content of

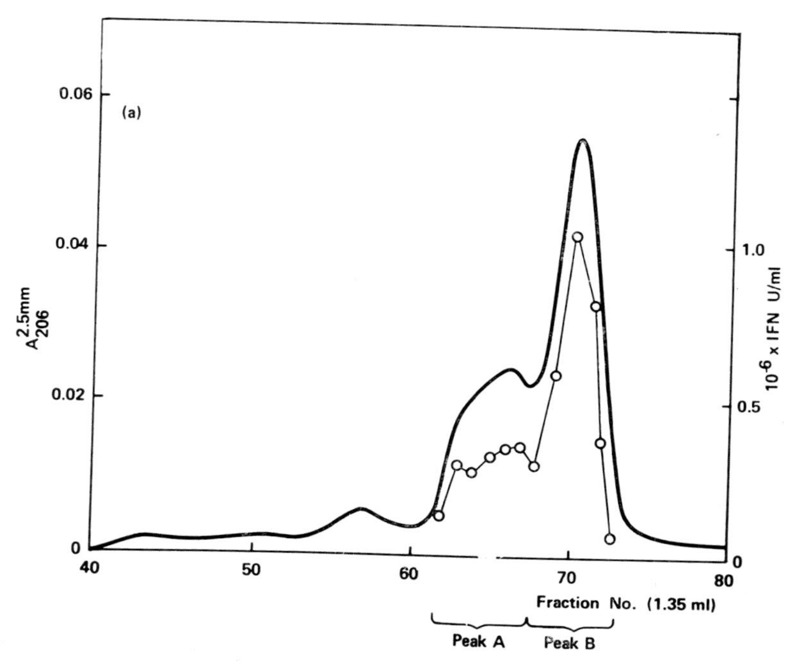

Fig. 2. Chromatography of purified Namalwa-IFN on Sephadex G75.
——— Absorption at 206 nm
—o—o— Antiviral activity of chromatographic fractions.

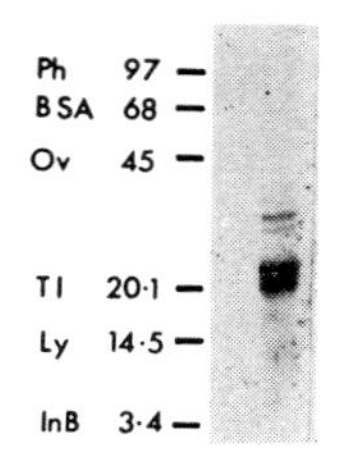

Fig. 3. P A G E of purified Namalwa interferon.

Table I. Specific activity and % purity of eight consecutive, routinely purified batches of lymphoblastoid interferon, determined by two methods

Batch No.	Biological		Sephadex G75 profile			Bio-titre p[4]
	U/ml ± S.D.[1] (×10⁶)	Spec. activity[2] (U/mg protein × 10⁶)	U/ml[3] (×10⁶)	Spec. activity[3] (U/mg protein × 10⁶	% Purity[3]	Profile-titre
CIN — 8	10.96 ± 1.20	122	11.93	145	92	1.06
CIN — 9	10.47 ± 1.29	105	12.62	134	85	1.17
CIN — 10	14.13 ± 1.20	118	12.51	126	80	0.86
			13.75	119	75	0.95
CIN — 11	8.13 ± 1.12	151	8.24	125	79	0.98
CIN — 12	9.12 ± 1.25	114	8.59	134	85	0.91
CIN — 13	7.59 ± 1.17	141	6.53	123	78	0.84
CIN — 14	9.12 ± 1.20	138	8.18	119	75	0.87
CIN — 15	7.24 ± 1.13	134	7.22	118	76	0.97

[1] Antiviral titre obtained by routine bioassay (mean of 12 determinations).
[2] Protein assayed by a Coomassie blue method.
[3] Values obtained from U.V. absorption trace at 206 nm.
[4] Allowing for 3% dilution of "bio-titre".

a highly purified sample in non-biological terms: from the ratio of the area outside and inside the double peak we could calculate the percentage purity, and by using protein standards and by having determined the specific activity of pure Namalwa IFN (1.6×10^8 U/mg) we could also assess the IFN content in μg of protein. A comparison of results obtained by the biological and the non-biological methods showed that they were very similar (tab. I). Moreover, the specific activity of every one of the eight consecutive batches shown in the table was greater than 10^8 U/mg protein, i.e. the IFN was practically pure. Since then we have produced many more batches in the same way, all of similar quality.

Use of Monoclonal Antibodies in Purification

The purification of many substances, including the three types of human IFN (IFN-α, -β, -γ), has become easier with the advent of monoclonal antibodies (19). These offer many advantages over conventional polyclonal antibodies, mainly by virtue of their greater capacity and very high specificity. However, in the case of IFN-α, with its diversity of molecular species, one could hardly expect one monoclonal antibody to neutralise – or as an affinity column to retain-all the components. We have investigated three different monoclonal IFN-α antibodies (NK-2, Secher and Burke (27), HC-46 and HC-33, prepared by Dr. J. Iványi of our laboratories, Fantes et al. (9)), and have shown that not one of them retained all of the eight major components of our purified Namalwa IFN (3). Fig. 4 shows what happened when a preparation of pure Namalwa interferon was passed through monoclonal antibody columns HC-46. The three curves are absorption profiles (at 206 nm) after Sephadex G75 chromatography of pure input IFN (left); non-retained (middle) and retained and pH2 eluted portions (right), obtained from the column. Polyacrylamide gel electrophoresis of individual Sephadex G75 fractions showed that the NK-2 and HC-46 columns (which seem to be very similar or even identical) retained only six of the eight major components, the remaining two (identical from both columns and both contained in peak B) were found in the non-retained filtrate. Although column HC-33 again only retained six components, the distribution was different: one of the non-retained species was derived from peak A, the other one from peak B, but it was different from the flow-through components of NK-2 and HC-46.

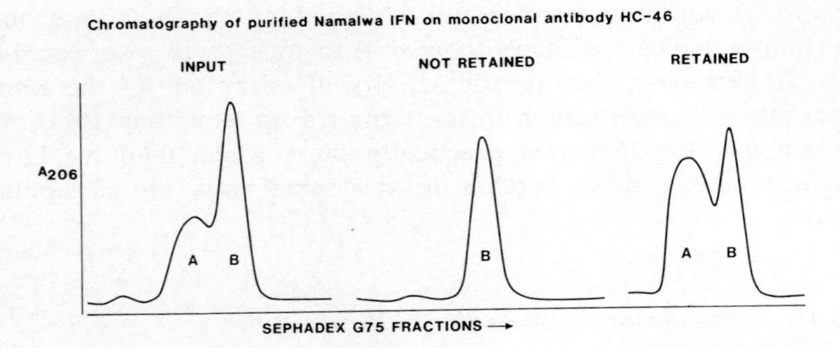

Fig. 4. Chromatography of purified Namalwa IFN on monoclonal antibody HC-46.
Purified Namalwa-IFN was passed through a monoclonal antibody (HC-46) column. An aliquot of the input material, and the non-retained and retained and eluted fractions were then analysed by chromatography on a Sephadex G75 column.

Multiple VS Single IFN-α Components

If the eight components all had identical biological properties, the loss of two of them during purification would be of little consequence. However, individual IFN-α species are by no means biologically equivalent: Yelverton et al. (35) found up to 200 fold differences in the antiviral activities of individual factors in a number of heterologous cell systems and Weck et al. (33) showed that even in the same type of cell different viruses exhibited different relative sensitivities to individual IFN-α components. An interesting difference between the multi-component Namalwa IFN and IFN-α_2, one of the genetically engineered components (31) was described by Taylor-Papadimitrou et al. (32). They found that based on an equal antiviral activity the former inhibited the growth of human breast tumour cells in the nude mouse more strongly than the latter. It is thus possible that a multicomponent preparation may have a wider application in the clinic, but this remains to be shown.

Safety of Clinical IFN

We have shown above (e.g. see tab. I) that IFN purified by our processes was substantially pure, but we nevertheless had to show that it was also free from potentially dangerous contaminants. The producing cells and hence the crude IFN could in theory harbour e.g. tumour-inducing viruses or their nucleic acids, hazardous cellular nucleic acids, bacteria or mycoplasmas. These, in traces, might find their way into the finally purified product. None of these agents could ever be demonstrated in the latter by conventional tests, or even in the crude IFN (with one exception – see below) but this is no absolute proof of their absence, as it is of course impossible to prove a negative. To strengthen our case we deliberately added to crude IFN high concentrations of a number of infections or radioactive marker substances, then subjected the mixtures to our purification process and finally tried to detect the additive in the purified product. In spite of the fact that the purification process at the same time effected a 100–1,000 fold concentration by volume, which should increase the sensitivity of the test, none of the added agents (see tab. II) could be found.

Traces of Namalwa cell nucleic acid were detected in only five of fifteen consecutive batches of crude IFN by "dot" hybridization on nitrocellulose membranes with highly ^{32}P-labelled (10^6–10^8 cpm/ /μg) nick-translated complementary DNA: concentrations ranged from 10 to 20 ng DNA per ml (M. Lockyer, personal communication). There was no obvious correlation between DNA content and IFN titre. In order to establish an upper limit for the presence of DNA in the purified IFN, we added nick-translated ^{32}P-labelled Namalwa DNA (6.95×10^7 cpm total) to 5 l of crude IFN and could show that it was reduced by a factor of 3.1×10^6 in the purified material (5 ml total, i. e. after 1,000 fold concentration by volume). From this one can calculate that one mega unit of purified IFN contained at most 1.0 to 2.0 pg Namalwa cell DNA.

As to EBV-DNA Dr. Raab-Traub, in professor J. Pagano's labor-

Table II. Agents deliberately added to crude interferon and eliminated by the purification process

Animal viruses	Concentration added	Bacteriophages & mycoplasmas	Concentration added	DNA Species	Concentration added
Polyoma	$10^{7.0}$ PFU	Phages	T3 $10^{9.5}$ PFU	Polyoma DNA	$10^{3.8}$ PFU
Mouse Cytomegalo	$10^{5.1}$ PFU		W31 $10^{8.6}$ PFU	EB-Virus DNA	$10^{5.6}$ CPM
Canine Hepatitis	$10^{7.7}$ PFU		BF23 $10^{11.3}$ PFU	Raji Cell DNA	$10^{6.9}$ CPM
Cat Herpes	$10^{4.8}$ PFU		QX174 $10^{11.5}$ PFU	Namalwa Cell DNA	$10^{7.8}$ CPM
Polio Type I	$10^{8.4}$ PFU		I-fil $10^{6.7}$ PFU		
Foot & Mouth Disease	$10^{8.3}$ PFU		MS2 $10^{10.2}$ PFU		
Swine Vesicular Disease	$10^{7.7}$ PFU				
ORF	$>10^{7.0}$ PFU	Mycoplasmas			
Newcastle Disease	$10^{11.3}$ EID_{50}	M. orale	$10^{8.8}$ CFU		
Sendai	$10^{7.8}$ EID_{50}	M. gallisepticum	$10^{8.6}$ CFU		
Avian Leukosis	$10^{6.8}$ FFU				
Scrapie Agent	$10^{7.0}$ Mouse LD_{50}				

atory at the university of North Carolina found, again by dot hybridization, that 1 mega unit of crude IFN contained less than 3 pg of the DNA. From these data one can calculate that one mega unit of purified IFN will contain less than 4×10^{-6} pg EBV-DNA, or fewer than 0.05 molecules EBV-DNA. Even if these minute amounts of viral DNA were present in a batch of purified IFN, they could not possibly constitute a hazard, as the transfection efficiency is very low.

To ensure that the cells themselves have not changed, by contamination with other cells, by mutation or selection, their karyology is being checked periodically, but no such changes have been observed in the last seven or eight years.

Biological, Non-Clinical Results

Namalwa IFN purified in the above way has been termed "Wellferon" and is a well characterized preparation consisting of a number of IFN-α components. As such it has been used by many investigators in a great variety of non-clinical biological studies. In general results obtained have resembled those from previous studies with much less pure leukocyte IFN, also a mixture of several components. Examples of a few such studies are outlined below.

Moore et al. (20) showed that Wellferon markedly enhanced the cytolytic activity of natural killer (NK) cells against five target cell lines, but that the latter themselves were protected when separately pretreated.

Apostolov and Barker (4) reported an effect of Wellferon on the lipid composition of cell membranes, manifested in a marked initial increase of the ratio of saturated to unsaturated fatty acid concentration, followed by a reversal some 24 hours later.

Balkwill et al. (5, 6) found that the IFN inhibited the growth of a number of human breast cancer and adenosarcoma cell lines (but not lung carcinoma oat cells), when grown as xenografts in nude mice. They also found that this action was synergistically enhanced by cyclophosphamide and by adriamycin (personal communication).

Flannery et al. (12) conjugated the IFN to a monoclonal antibody against human osteogenic sarcoma cells and showed that these complexes localized specifically in the human sarcoma xenografts in immunodeprived mice and that they increased NK-cell activity in the xenografts to a greater extent than free IFN.

Clinical Results

Many positive clinical results with various preparations of IFN-α against viral and neoplastic diseases have been reported in the literature. Mainly because of insufficient availability of suitable interferon, many of the earlier trials were uncontrolled and are therefore of dubious value. But confirmed results have been obtained against some diseases and these are listed in tab. III.

Table III. Viruses and tumours that have shown an objective response to interferons in patients

Hepatitis B	Osteosarcoma
Herpes simplex	Multiple myeloma
Herpes zoster	Malignant melanoma
Vaccinia	Breast cancer
Rhinoviruses	Non-Hodgkin's lymphoma
Warts	Metastatic renal cell carcinoma
Condyloma	AML, ALL, CML
Laryngeal papillomatosis	Adenocarcinoma of colon and prosta te
	Chondrosarcoma
	Malignant glioma
	Bladder carcinoma

IFN can be given topically or systemically and the size, number and frequency of doses depend on the nature of the disease and are still being established.

With Wellferon a number of phase I and phase II studies have been carried out or are in progress.

Phase I Studies

In phase I studies Priestman (22) found that daily intramuscular doses of $2.5–5.0 \times 10^6$ U/m^2 were well tolerated and appeared to be suitable for long-term administration. Rohatiner et al. (24) gave the

IFN by continuous i.v. infusion and showed that daily maximum doses of 100×10^6 U/m^2 could be given for up to seven days to fully hospitalised patients. Other investigators using yet different regimens proposed different maximum i.m. or i.v. doses, given at various intervals, and ranging from 10 to 100×10^6 U/m^2 (18, 14, 26, 28). All of these studies have shown side effects very similar to those associated with other IFNs and in particular include fever, fatigue, depression of blood counts and rise in hepatic enzymes. Such toxicities are normally readily reversible by cessation of therapy.

Phase II Studies

Oncological diseases that are currently being treated with Wellferon are shown in tab. IV, illnesses associated with virally-associated tumours in tab. V and viral diseases in tab. VI.

Table IV. Malignant disease types treated with Wellferon

Leukaemias	Carcinomas of	
AML	Kidney	Glioma
ALL	Breast	Malignant melanoma
CML	Ovary	Kaposi's sarcoma
CLL	Testis	
Hodgkin's disease	Stomach	
Non-Hodgkin's lymphoma	Colon	
Multiple myeloma	Lung	

Table V. Virally-associated diseases treated with Wellferon

Juvenile laryngeal papillomatosis
Nasopharyngeal carcinoma
Cervical dysplasia
Hepatoma

Table VI. Viral diseases treated with Wellferon

Condylomata acuminata
Skin warts
Chronic active hepatitis
Cytomegalovirus pneumonitis (treatment prophylaxis)
Rabies encephalitis

Results of Clinical Studies with Wellferon

Phase I studies have provided some anecdotal partial responses in anaplastic carcinoma and nodular poorly differentiated lymphoma (18), in diffuse hystiocytic lymphoma, Hodgkin's disease and renal cell carcinoma (26), in acute myeloid leukaemia (AML) (25), in chronic lymphocytic leukaemia (14) and in breast cancer (28).

Phase II studies in cancer have demonstrated minimal activity in cutaneous secondary deposits of malignant melanoma (23) and in AML (24), and no responses in small cell lung (17), gastric or colonic cancer. Partial responses (>50 % tumour regression) have been seen in breast cancer, ovarian cancer, glioma, malignant teratoma and renal cell cancer. Complete responses in patients with metastatic renal cell cancer have been observed.

In virally-associated diseases, studies in Australia, Eire, the United Kingdom and the U.S.A. have consistently shown marked responses in patients suffering from juvenile laryngeal papillomatosis. Regressions have been observed in cervical dysplastic lesions when the IFN has been injected directly around the lesions. Common warts, resistant to all other therapies, have been directly injected, often with striking clearances of these and even of some remote uninjected lesions. Venereal warts (Condylomata acuminata), (which may have a possible causative role in the aetiology of juvenile laryngeal papilloma), have regressed with intramuscular IFN, even in cases resistant to conventional treatments. Falls in the serum levels of alpha-foetoprotein have been recorded in cases of hepatoma.

Results of phase II trials in virus-induced diseases above have been encouraging. Intra-muscular treatment also in showing promising effects in patients with HBsAg positive chronic active hepatitis: Dane particle-associated DNA polymerase and HBV-DNA levels have fallen, indicating inhibition of viral replication (34). Wellferon had no effect in established cytomegalovirus pneumonitis, but a study is in progress to investigate whether there is a possible prophylactic effect in this disease.

Future Outlook

There is almost no doubt that IFNs will find a place in the clinic, either by themselves or in combination with other drugs. However, before their full potential can be realised, more will have to be learned about the most therapeutically efficacious dose and schedule to be employed in a particular disease. Account of actions IFN has on the immune response will possibly need to be remembered. By analogy with results from animal experiments one can also expect more favourable results, if treatment is started at the onset of a disease instead of treating only advanced or even terminal cases, as has been done in most instances so far.

References

1 Allen, G.; Fantes, K. H. (1980): Nature, London, 287, 408.
2 Allen, G. (1982): Biochem. J. 207, 397.
3 Allen, G.; Fantes, K. H.; Burke, D. C.; Morser, J. (1982): J. gen. Virol. 63, 207.
4 Apostolov, K.; Barker, W. (1981): FEBS Letters 126, 261.
5 Balkwill, F. R.; Moodie, E. M.; Freedman, V.; Fantes, K. H. (1982): Int. J. Cancer 30, 231.
6 Balkwill, F. R.; Moodie, E. M.; Mowshowitz, S.; Fantes, K. H. (1983): 2nd International TNO Meeting on the Biology of the Interferon System, Rotterdam.
7 Fantes, K. H.; Burman, C. J.; Ball, G. D.; Johnston, M. D.; Finter, N. B. (1980): In "Biochemical Characterization of Lymphokines", Acad. Press, 343.
8 Fantes, K. H.; Allen, G. (1981): J. Interferon Res. 1, 465.
9 Fantes, K. H.; Allen, G.; Iványi, J. (1982): 3rd Ann. Int. Congress of Interferon Res., Miami.
10 Finter, N. B.; Fantes, K. H. (1980): Interferon 2, 65.
11 Finter, N. B. (1981/1982): Texas Reports on Biology and Medicine 41, 175.
12 Flannery, G. R.; Pelham, J.; Gray, J. D.; Baldwin, R. W. (1983): Submitted to Eur. J. Cancer.

13 Goeddel, D. V., Leung, D. W,; Dull, T. J. et al. (1981): Nature, London, 290, 20.
14 Huang, A.; Laszlo, J.; Brenckman, W. (1982): Am. Assoc. Clin. Res. 27, 113.
15 Johnston, M. D. (1980): J. gen. Virol. 50, 191.
16 Johnston, M. D. (1981): J. gen. Virol. 56, 175.
17 Jones, D. H.; Bleehen, N. M.; Slater, A. J.; George, P. J. M.; Walker, J. R.; Dixon, A. K. (1983): J. Cancer, in press.
18 Knost, J. A.; Sherwin, S. A.; Abrams, P. G.; Ochs, J. J.; Foon, K. A.; Williams, R.; Tuttle, R.; Oldham, R. K. (1983): Clin Immunol. Immunopathol., in press.
19 Kohler, G.; Milstein, C. (1975): Nature, London, 256, 495.
20 Moore, M.; White, W. J.; Potter, M. R. (1980): Int. J. Cancer, 25, 565.
21 Nagata, S.; Mantei, N.; Weissmann, C. (1980): Nature, London, 287, 401.
22 Priestman, T. J. (1980): Lancet, 2, 113.
23 Retsas, S.; Priestman, T. J.; Newton, K. A.; Westbury, G. (1983); Cancer 51, 273.
24 Rohatiner, A. Z. S.; Balkwill, F. R.; Griffin, D. B.; Malpas, J. S.; Lister, T. A. (1982): Cancer Chemother. Pharmacol. 9, 97.
25 Rohatiner, A. Z. S.; Balkwill, F. R.; Lister, T. A. (1983): Cancer Chemother. Pharmacol., in press.
26 Sarna, G.; Figlin, R.; Bryson, Y.; Mauritzon, N.; Cline, M. (1982): Proc. Ann. Soc. Clin. Oncol. 18, 39.
27 Secher, D.; Burke, D. C. (1980): Nature, London, 285, 446.
28 Silver, H. K. B.; Connors, J.; Salinas, F. A. (1982): 3rd Ann. Int. Congr. Interferon Res., Miami.
29 Strander, H.; Cantell, K. (1966): Ann. Med. Exp. Biol. Fenn. 44, 265.
30 Strander, H.; Mogensen, K. E.; Cantell, K. (1975): J. Clin. Microbiol. 1, 116.
31 Streuli, M.; Nagata, S.; Weissmann, C. (1980): Science 209, 1343.
32 Taylor-Papadimitriou, J.; Shearer, M.; Balkwill, F. R.; Fantes, K. H. (1982): J. Interferon Res. 2, 479.
33 Weck, P. K.; Apperson, S.; May, L.; Stebbing, N. (1981): J. gen. Virol. 57, 233.
34 Weller, I. V. D.; Fowler, M. J. F.; Monjardino, J.; Carreno, V.; Thomas, H. C.; Sherlock, S. (1982): Phil. Trans. Roy. Soc. London, Series B. 299, 128.
35 Yelverton, E.; Leung, D.; Weck, P.; Gray, P. W.; Goeddel, D. V. (1980): Nucleic Acids Res. 9, 731.

K. H. Fantes, Wellcome Research Laboratories, Langley Court, Beckenham, Kent, BR3 3BS (U. K.)

II. Effect of Interferon on the Cell and the Consequences of Interferon Induction

Contr. Oncol., vol. 20, pp. 113—126 (Karger, Basel 1984)

Cellular Response to Varied Action of Interferon

T. Kuwata[a]; *Y. Tomita*[a]; *A. Fuse*[a]; *S. Sekiya*[b]

Departments of Microbiology[a], Obstetrics and Gynecology[b] School of Medicine, Chiba University, Chiba 280, Japan

Introduction

Since the discovery of interferon as an antiviral factor, its "non-antiviral functions" have been reported successively. Therefore, as

Table I. Pleiotypic actions of interferons

1	Suppression of virus multiplication (1957)
2	Priming action (1958)
3	Suppression of cell growth (1962)
4	Enhancement of phagocytosis (1971)
5	Enhancement of cytotoxicity of sensitized lymphocytes (1972)
6	Enhancement of toxicity of dsRNA (1972)
7	Enhancement of cell surface antigen expression (1973)
8	Suppression or enhancement of antibody formation (1974)
9	Enhancement of lectin binding to the cell membrane (1974)
10	Suppression of delayed type hypersensitivity (1975)
11	Enhancement of synthesis of prostaglandins (1977)
12	Enhancement of IgE-mediated histamine release by basophils (1977)
13	Suppression or enhancement of differentiation in leukemia cells (1977)
14	Enhancement of cytotoxicity of natural killer cells (1978, 1966)
15	Suppression or enhancement of plasminogen activator release (1978, 1981)
16	Suppression of cap formation in lymphocytes (1979)
17	Suppression of cell fusion by viruses (1979)
18	Suppression of cell motility (1980)
19	Alteration of microfilaments (1980)
20	Induction of ketosteroids production in adrenal cells (1980)

early as in 1966, Isaacs described it as follows: "How is interferon best defined? As knowledge of interferon has altered and it will surely alter again in the years ahead". In 1972 Lindahl, Leary and Gresser (37) reported that mouse interferon preparations enhanced the specific cytotoxicity of sensitized lymphocytes for allogenic target tumor cells. Their conclusion was that the factor responsible for the enhancement of cytotoxicity could not be dissociated from the antiviral activity of interferon. Thus, they suggested that the antiviral action of interferon may be only one expression of the effect of interferon cells. In 1973 De Clercq and Stewart (17) published the article entitled "The breadth of interferon action". Since then reports on varied actions of interferons increased up to the present time (25, 48) (tab. I). Thus, at present it may be appropriate to consider interferon as a kind of modulator of cell functions. Due to the space limitation, we will here describe chiefly the problems we have studied in our laboratory.

Inhibition of Cell Growth by Interferon

The problem of the cell growth inhibitory action of interferon has encountered "most persisted pessimism" (48). However, it was finally verified in 1979 by using electrophoretically pure mouse interferon, that such interferon not only inhibits viral multiplication, but also exerts a variety of important biological effects on cells (27). Several other group of investigators also reached the same conclusion.

Sensitivities of various cell strains to the cell growth inhibitory action of interferon are divergent. Adams and her associates examined 14 established lymphoid cell lines for their sensitivities to the cell growth inhibitory action of human IFN-α. Among these cell lines Daudi cells were found to be the most sensitive one. One IU/ml of IFN-α was sufficient to reduce the total number of living cells by 50 % after 21 days incubation period. One the other hand, Raji and Maku lines were resistant to 10,000 IU/ml of IFN-α (1).

We established human embryonic fibroblasts culture, which were transformed in vitro by Rous sarcoma virus and SV40, and isolated from them 2 clones RSa and RSb (34). These cells lines were incidentally found to be highly sensitive to cell growth inhibitory action of IFN-α and IFN-β (30). RSa and RSb were at the same time very sensitive to UV irradiation (49). However, UV-sensitive cells from xeroderma pigmentosum were not always sensitive to interferon (unpublished data). Generally interferon exhibits a cytostatic action and if it is removed from the culture medium, then cells propagate again in interferon-free medium. In the case of our RS cells, when cells were treated with 500 IU/ml of IFN-α or β for 3 days,

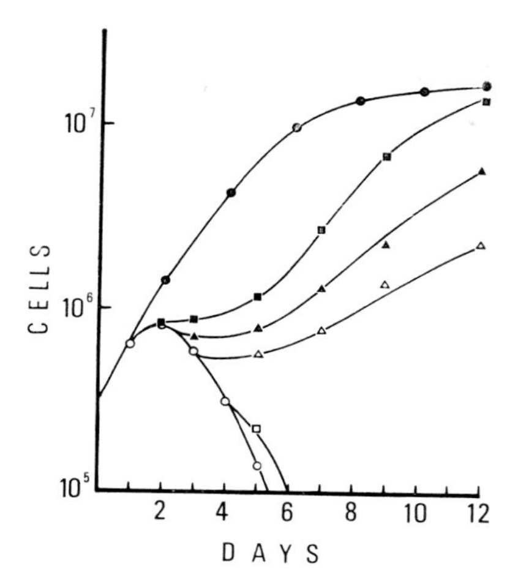

Fig. 1. Effect of duration of interferon treatment on the growth of RSa cells. Replicate cultures were made by using plastic dishes of 60 mm diameter. After interferon treatment for one to four days, cells were washed twice with MEN and them cell cultures were continued with growth medium. Cells not treated (●—●), treated with 500 IU/ml of IFN-α for one day (■—■), two days (▲—▲), three days (△—△), four days (□—□), and continuously treated (○—○).

then interferon acted as cytocidal (fig. 1). This action is not due to impurities which might be contained in the interferon preparation, since we could also observe such a cytocidal action by using recombinant type interferon, specific activity of which was 2×10^8 U/mg protein. Cytocidal effect of IFN-β on human carcinoma cells (RK-4) of urinary bladder was also reported by Ito and Buffett (29).

Cell growth inhibitory actions of IFN-α, β and γ are not the same. IFN-γ has been reported to exert strong cell growth inhibitory action relative to its antiviral action (14, 44). Direct cytolysis by partially purified preparations of mouse IFN-γ was also reported (57). However, sensitivity of cells to IFN-γ is variable as discussed in the later section of this paper.

Mechanisms of cell growth inhibitory action of interferon have not been fully clarified. In general, for the suppression of cell growth higher amounts of interferon are needed than for the suppression of virus growth and several cycles of cell division are necessary before interferon effect becomes evident. Synthesis of DNA and protein is markedly reduced in RSa cells in proportion to the concentration of interferon applied (20). Interferon treatment leads to accumulation of cells with increased volume which do not enter the division cycle. After reversal of thymidine block, cell growth resumes in a synchronous manner. When RSa cells were treated with interferon during late G1 phase, the peak of DNA synthesis in S phase disappeared and the subsequent mitosis did not occur, suggesting that interferon action takes place in late G1 phase or beginning of S phase (21) and results in critical events leading to the inhibition of DNA synthesis.

This high degree (90 %) inhibition of DNA synthesis was observed for concentrations of added interferon that produced more than twofold increase in the intracellular cAMP level. In the IF^r cell culture, which is relatively resistant to the anticell growth activity of interferon, considerably less inhibition of DNA synthesis and a slight increase in cAMP were observed (22). Tovey et al. (56) reported the increase of cAMP level in interferon-treated mouse cells, but they also found more rapid increase of cGMP level of L1210 cells in steady state chemostat culture. These results suggest that intracellular cyclic nucleotides may in part mediate the inhibitory effect of interferon on cell growth (46).

Prednisolone and indomethacin known as inhibitors of prostaglandin synthesis partially inhibited antiproliferative activity of interferon suggesting that alternatively interferon-inhibited cell growth was partly mediated by prostaglandins in some cell system (24). However, in another cell system, indomethacin and aspirin did not inhibit antiviral action of mouse interferon at concentrations which inhibited the formation of prostaglandins (55).

Interferon is known to induce several enzymes in cells (6). After interferon treatment, 2′—5′ oligoadenylate (2—5A) synthetase is induced in various cells and the 2—5A produced was reported to play a role in antiviral and also anti-cell growth actions of interferon although no clear genetic evidence for this role has yet been provided (35). IF^r cells were found to be equally well inducible as parent RSa cells for 2—5A synthetase and protein kinase. These results suggest that induction of these two enzymes, which may have important role in antiviral action, are not sufficient for the cell growth inhibitory activity of interferon (58). Thus, the roles of 2—5A or protein kinase in the antiviral and cell growth inhibiting actions of interferon is still not clearly understood. 2—5A introduced artificially into cytoplasma by microinjection induced antiviral state, but interferon did not induce antiviral state even if injected into cytoplasma (28).

Interferon-Resistant Cells

The problem of interferon resistance is important not only from the theoretical, but also from the practical standpoint. As there appear resistant variant cells against chemotherapeutic drugs, isolation of interferon resistant cells is to be expected after interferon therapy. Thus, Gresser et al. (26) selected with relatively high frequency sublines resistant to the growth inhibitory action of interferon from murine leukemia L-1210 cells. They concluded that interferon-resistant variants are originated by spontaneous random change rather than by induction by interferon. One of the interferon resistant cell lines, L-1210R cells are resistant not only to the cell growth inhibitory action, but also to the antiviral action of interferon. This fact was interpreted as the evidence that two actions of interferon preparation may be based on the action of the same molecule.

Affabris et al. (2) isolated interferon-resistant clones from Friend leukemia cells. Similarly to L-1210R cells, these clones were fully resistant to antiviral action of interferon against VSV and Mengo virus. On the other hand, we isolated 2 cell lines from RSa population IF^r and F-IF^r (30, 32), which were resistant to cell growth inhibitory action of IFN-α and IFN-β respectively. These cells showed sensitivity to the antiviral action of interferon despite their resistance to its anti-proliferative action. Cells of such type were also isolated from human tumor celli populations (36). Creasy et al. (15) described melanoma cell lines which showed resistance to antiproliferative effects of interferon, but were sensitive to antiviral action. Thus, it is apparent that antiviral and antiproliferative states are not always expressed in parallel in certain cells (tab. II). Fuchsberger, Borecký and Hajnická (19) treated L cells continuously with mouse interferon and isolated cell lines which were not inhibited in growth by interferon, but retained their sensitivity toward its antiviral action. They supported the view that interferon may induce an antiviral state in cells without suppressing cell multiplication.

Table II. Induction of different activities in interferon-treated cells[1]

Activity	RSa	IF^{r}	HEC-1C	YS-K	NUC	RD114-C1	Raji	A204
Antiviral activity	+	+	–	+	–/+[3]	–	–	+
Anticellular activity	+	–	–	–	–	–	–	+
2—5 synthetase activity	+	+	–	+	+	–	+	+
Phosphoprotein kinase activity	+	+	ND[2]	ND	ND	–	ND	+

[1] Modified the *Table II* of our previous paper (Vandenbussche et al., 1981).
[2] ND: Not done.
[3] Negative against VSV infection, and positive against EMCV.

On the other hand, we found a human uterine adenocarcinoma derived cell line (HEC-1), to be totally resistant to human IFN-α, β and γ (60, 61). These facts suggest that there may exist some interferon insensitive tumor cells, if we examine human tumor cells more widely. Insensitivity of HEC-1 cells to human IFN-γ was quite remarkable because L-1210R cells were reported to be as sensitive to mouse IFN-γ as L-1210S (3). However, HEC-1 cells are sensitive to cytotoxicity of natural killer cells (13). We also found among rhesus monkey cell lines one foreskin cell line, Rhfs-B7, to be insensitive to human IFN-α, β, γ and moreover resistant to rhesus monkey interferon (Kuwata et al., to be published).

Mechanisms Underlying Interferon Resistance

Interferon has been shown to exert its action by binding to receptor sites on the cell membrane. Then, as a cause of cellular changes from interferon sensitive cell to resistant cell, it is conceivable that in interferon resistant cells loss or decrease in the number of binding sites could occur. In fact in L-1210R cells loss of binding sites for interferon was demonstrated by Aguet (5). This cell line was also reported to be devoid of fatty acid cyclooxygenase activity (10). But deficiency of receptor may be the principal cause of interferon resistance. In two human cell lines, which were weakly sensitive to IFN-α or β, binding of labeled interferon was greatly reduced and Baglioni et al. (7) concluded that poor response of these cells was due to the presence of few interferon receptors. Human adenocarcinoma HEC-1 and its subclone HEC-1C were also verified to be defective in interferon binding sites (39, 62). However, loss of binding sites may not be all or none phenomenon. There should be cases of partial loss of binding sites. In such cases, both or one of antiviral and cell growth inhibiting actions of interferon may be expressed at reduced rate. WGA^r, F-IF^r and proably IF^r cells, which were isolated in our laboratory from RSa population, are supposed to be such cases. WGA^r cells are partially resistant to antiviral and cell growth inhibiting action of human interferon and by using internally radiolabeled IFN-α, their number of cell surface receptor sites was found to be fewer than that of parental RSa cells (Kuwata et al., to be published).

Binding of interferon to the plasma membrane is certainly an important step to express its biological effects on cells. However, binding of interferon per se does not mean expression of all biological effects of interferon. In Raji cells, when cells were treated with 50—200 IU/ml of IFN-α, enhancement of 2—5A synthetase was observed (50). This phenomenon may indicate binding of interferon to the cell surface, but the suppression of Raji cell growth was not observed. In a cell line selected from Daudi cells for their resistance to the cell growth inhibiting effect of interferon, 2—5A synthetase was induced to the same extent as in wild type Daudi cells (47). Thus again, interferon could apparently bind to these resistant variant cells, but, as observed in Raji cells and IF^r cells, this did not result in cell growth inhibition.

It was reported by Pfeffer et al. (43) that interferon treatment elicits increase of actin-containing microfilaments in human fibroblasts. We compared the effect of interferon on the actin organization in RSa and its interferon-resistant variant IF^r cells. When RSa cells were treated with IFN-α (250—500 IU/ml) or IF-γ (10—50 U/ml), for 3 days, appearance of microfilament bundles was noted. The same treatment to IF^r cells did not cause microfilament formation, suggesting that microfilament organization is associated with cell growth inhibitory effect of interferon (45).

Factors Which Modify Cellular Response to Interferon

Sensitivity of cells to interferon is genetically determined and its actions are species specific. Besides, various factors influence the effects of interferon. Here we consider the influence of human chorionic gonadotropin (HCG) on cell response to interferon. Besançon and Ankel (8) reported that preincubation of human interferon and HCG suppressed the antiviral action of interferon. We have examined 7 strains of choriocarcinoma derived cells which secrete variable amount of HCG, for their sensitivity to antiviral action of IFN-α. As shown in tab. III all these strains were resistant to antiviral action against VSV in comparison with HeLa cells.

The results also suggest that resistance of these choriocarcinoma cells to the cell growth inhibiting (data not shown) or antiviral action of interferon may be correlated with their level of HCG production.

Therefore, human IFN-α was preincubated with HCG for 1 hour at 37 °C, and mixtures of interferon and HCG were assayed on HeLa and YS-K, cells of yolk sac tumor origin. The results were negative since no reduction of interferon activity was observed. At present we have no clear explanation for this discrepancy between our results and these of Besançon and Ankel (8). Finally production of HCG in choriocarcinoma cells was not specifically suppressed by the interferon treatment (unpublished data).

Interferon resistance of various choriocarcinoma cells may be considered to be correlated with the degree of differentation of the cells. Burke et al. (9) first reported that mouse embryonal carcinoma (EC) cells fail to develop an antiviral state after interferon

Table III. Low sensitivity of choriocarcinoma cells to antiviral action of human α interferon

Interferon	HeLa <0.5[1]	Bewo 911.0	HCCM-5 2270.0	NUC-1 338.0	SCH 16.0	GCH-1 2.2	ENAMI-1 3.0	IMa 700.0
10 IU/ml	1.0[2]	0	0.2	0	0.7	0.8	0.5	0.5
100 IU/ml	2.5	0	0.2	0.3	2.1	1.5	1.0	0.5
1000 IU/ml	4.7	0.8	0.5	0.5	2.9	2.3	1.7	1.2

[1] Secretion of HCG (ng/10^6 cells/48 hrs).

[2] Yield reduction of VSV (log $TCID_{50}$/0.2 ml).

Cells were grown in 35 mm plastic dishes. Confluent cells were treated with interferon for about 20 hrs, once washed with MEM and infected with VSV for 1 hr at 37° C. Then, once washed with MEM, and 2 ml of RPMI medium with 5 % calf serum was added to each dish. Medium samples were taken 18 hrs after infection.

treatment. Differentiated lines derived from the EC cells are sensitive to mouse interferon. Thus, differentiation of EC cells in vitro is acompanied by development of sensitivity to interferon. In undifferentiated EC cells 2—5A synthetase was induced, but protein kinase was not induced (40). It is not clear whether this is the only reason of the low sensitivity of embryonal carcinoma cells to interferon treatment. In BeWo cells, however, the protein kinase is inducible and all the known elements of the 2—5A system are present (59).

Effect of Interferon on Cell Membrane

Various kinds of changes have been noticed in the cell membrane after interferon treatment. When mouse lymphocytes were pretreated with interferon, cap formation in cell membrane by antilymphocyte serum or concanavalin A was reduced (38). Interferon treatment increases the binding of concanavalin A and wheat germ agglutinin to the cell membrane (33). Because our RSb cells were found to be sensitive to the syncytium forming action of Mason-Pfizer monkey virus (41), we tested the effect of human interferon on the syncytium formation by the feline RD 114 virus and found an inhibition of cell fusion by interferon (51). Cell fusion by RD 114 virus was inhibitied by 50 IU/ml of either IFN-α or IFN-β to 1/3—1/9 of control. This fusion inhibition by interferon was neutralized by antiinterferon serum as for other interferon action. However, it was not due to the suppression of virus multiplication, because cell fusion caused by β-propriolacton-inactivated virus was also inhibited by interferon. Cell fusion inhibition by interferon was suppressed by cycloheximide as it is the case for the antiviral action of interferon. Syncytium formation by Sendai virus was also inhibited by pretreatment of cells with IFN-a, β (52—54) (fig. 2) and γ (to be published). In this case the inhibition of cell fusion was also suppressed by treating the cells with cycloheximide, which suggests the necessity of de novo protein synthesis for the establishment of

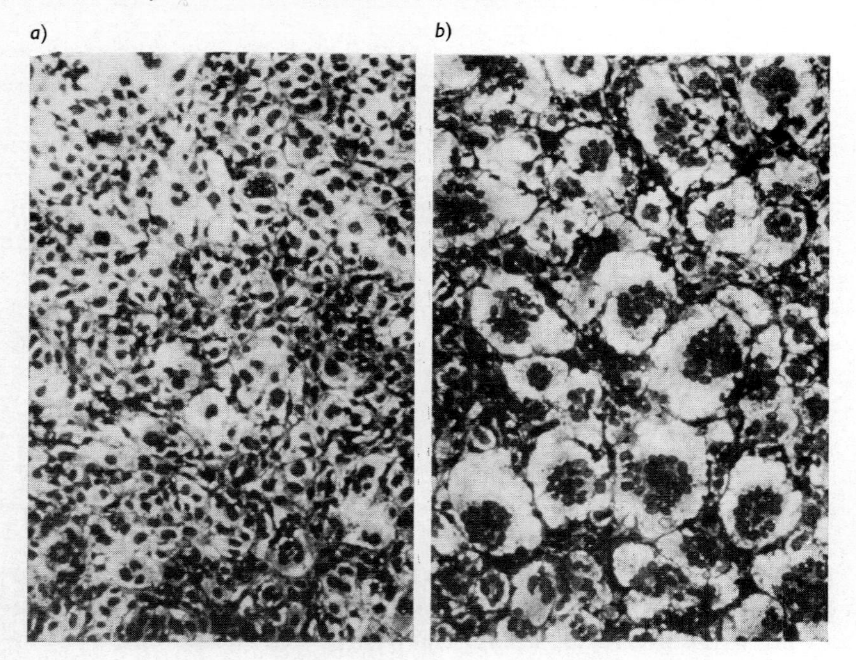

Fig. 2. Suppressive effects of interferon on cell fusion by Sendai virus. IF^r cells pretreated with or without IFN-α for 20 hrs were incubated on ice with UV-inactivated Sendai virus (1600 HAU/ml) for 20 min, then incubated for 2 hrs. These cells were stained with Giemsa solution.
a) Inhibition of syncytium formation by interferon treatment (250 IU/ml).
b) Syncytia in control cells after Sendai virus treatment.

cell fusion inhibition. Cell fusion by these viruses is elicited by interaction of viruses with cell membrane and fusion inhibition by interferon is apparently based on the modification of cell membrane. Chatterjee, Cheung and Hunter (12) observed a significant decrease in the membrane fluidity in interferon-treated cells. Concerning the change of cell membrane it was reported previously by Apostolov and Barker (4) and Chandrabose et al. (11) that interferon treatment elicits changes in phospholipid composition of plasma membrane. Treatment of HeLa-S3 cells with 640 IU/ml of IFN-β induced time-dependent structural changes in the plasma membrane lipid bi-layer and by 24 hrs after interferon treatment the rigidity of the membrane increased (42). Interferon treatment also increased a microfilament organization abnormally (43). Thus, such changes of cells may cause the decrease of fluidity of cell membrane and explain the subsequent inhibition of cell fusion by Sendai virus. Trapping of retroviruses at the cell surface after interferon treatment may be also related to this change of cell membrane (54).

Are Varied Actions of Interferon Based on the Same Mechanism?

Interferon exerts pleiotypic actions on cells (tab. I) but it is not clear whether such action are based on the same mechanism or not. We have studied the correlation between antiviral action and cell growth inhibiting action of interferon.

Inhibitors of protein synthesis such as cycloheximide suppress the antiviral action of interferon. However, treatment of cells with interferon and cycloheximide or puromycin at 0.5 to 5.0 μg/ml increased its cell growth inhibitory effects (31). Thus, interferon and these drugs acted rather additively. The same kind of results were obtained with cholera toxin and interferon on our RSa cells (23), though contradictory results were reported by using other cell systems (18) for unknown reason.

Anyway, our results may suggest that the mechanisms underlying antiviral and cell growth inhibiting action of interferon are clearly distinct.

On the other hand, Czarniecki et al. (16) tested sensitive to mouse interferon 2 clones of Swiss 3T3 cells infected with Moloney murine leukemia virus. They found that the production of infectious retrovirus was inhibited by interferon action in these 2 clones but the replication of VSV or EMCV was not inhibited in one of these clones. We also found RD 114 cells and subclone RD 114-C1 chronically infected with feline retrovirus to be resistant to antiviral action of human interferons against VSV and EMCV (54). In these cells both 2—5A synthetase and dsRNA-dependent protein kinase were not significantly induced. However, production of RD 114 virus was inhibited strongly since the treatment of cells with only 25 IU/ml of IFN-α inhibited the retrovirus production by over 80 %. Moreover, interferon treatment efficiently inhibited cell fusion induced by UV-inactivated Sendai virus. Thus the mechanisms underlying the anti-retrovirus and the anti-cell fusion activities of interferon may be closely related, and different from those of antiviral action against VSV and EMCV. Affabris et al. (2) reported that interferon-induced antiviral activity and its effect on erythroid differentiation are not coordinate features in Friend leukemia cells. They also postulated different mechanisms may be responsible for different effects of interferon.

References

1 Adams, A.; Strander, H.; Cantell, K. (1975): J. gen. Virol. 28, 207.
2 Affabris, E.; Jemma, C.; Rossi, G. B. (1982): Virology 120, 441.
3 Ankel, H.; Krishnamurti, C.; Besançon, F.; Stefanos, S.; Falcoff, E. (1980): Proc. Natl. Acad. Sci. USA 77, 2528.
4 Apostolov, K.; Barker, A. (1981): FEBS Letters 126, 261.
5 Aguet, M. (1980): Nature 284, 459.
6 Baglioni, C. (1979): Cell 17, 255.
7 Baglioni, C.; Branca, A. A.; D'Alessandro, S. B.; Hossenlopp, D.; Chadha, K. C. (1982): Virology 122, 202.
8 Besançon, F.; Ankel, H. (1976): C. R. Acad. Sci. D. 283, 1807.
9 Burke, D. C.; Graham, C. F.; Lehman, J. M. (1978): Cell 13, 243.
10 Chandrabose, A. K.; Cuatrecasas, P.; Pottathil, R.; Lang, D. J. (1981): Science 212, 329.
11 Chandrabose, K.; Quatrecasas, P.; Pottathil, R. (1981): Biochem. Biophys. Res. Comm. 98, 661.
12 Chatterjee, S.; Cheung, H. C.; Hunter, E. (1982): Proc. Natl. Acad. Sci. USA 79, 835.
13 Chen, H. Y.; Sato, T.; Fuse, A.; Kuwata, T.; Content, J. (1981): J. Natl. Cancer Inst. 61, 871.

14 Crane, J. L.; Glasgow, L. A.; Kern, E. R.; Younger, J. S. (1978): J. Natl. Cancer Inst. 61, 871.
15 Creasy, A. A.; Bartholomew, J. C.; Merigan, T. C. (1980): Proc. Natl. Acad. Sci. USA 77, 1471.
16 Czarniecki, C. W.; Sreevalsan, T.; Friedman, R. M.; Panet, A. (1981): J. Virol. 37, 827.
17 De Clercq, E.; Stewart, W. E. II (1973): Selective inhibitors of viral functions (W. Carter, ed.), p. 80 (CRC Press).
18 Degré, M. (1978): Proc. Soc. Exp. Biol. Med. 157, 253.
19 Fuchsberger, N.; Borecký, L.; Hajnická, V. (1974): Acta virol. 18, 85.
20 Fuse, A.; Kuwata, T. (1976): J. gen. Virol. 33, 17.
21 Fuse, A.; Kuwata, T. (1977): J. Natl. Cancer Inst. 58, 891.
22 Fuse, A.; Kuwata, T. (1978): J. Natl. Cancer Inst. 60, 1227.
23 Fuse, A.; Kuwata, T. (1979): Infect. Immun. 26, 236.
24 Fuse, A.; Mahmud, I.; Kuwata, T. (1982): Cancer Res. 42, 3209.
25 Gresser, I. (1977): Cell Immunol. 34, 406.
26 Gresser, I.; Bandu, M. T.; Brouty-Boyé, D. (1974): J. Natl. Cancer Inst. 52, 553.
27 Gresser, I.; De Maeyer-Guignard, J.; Tovey, M. G.; De Maeyer, E. (1979): 76, 5308.
28 Higashi, Y.; Sokawa, Y. (1982): J. Biochem. 91, 2021.
29 Ito, M.; Buffett, R. (1981): J. Natl. Cancer Inst. 66, 819.
30 Kuwata, T.; Fuse, A.; Morinaga, N. (1976): J. gen. Virol. 33, 7.
31 Kuwata, T.; Fuse, A.; Morigana, N. (1977): J. gen. Virol. 37, 195.
32 Kuwata, T.; Fuse, A.; Suzuki, N.; Morinaga, N. (1979): J. gen. Virol. 43, 435.
33 Kuwata, T.; Fuse, A.; Takayama, N.; Morinaga, N. (1980): Ann. N. Y. Acad. Sci. 350, 211.
34 Kuwata, T.; Oda, T.; Sekiya, S.; Morinaga, N. (1976): J. Natl. Cancer Inst. 56, 919.
35 Lebleu, B. J. (1982): Interferon 4, 447 (I. Gresser, ed.) Academic Press.
36 Lin, S. L.; Greene, J. J.; Tso, P. O. P.; Carter, W. A. (1982): Nature 297, 417.
37 Lindahl, P.; Leary, P.; Gresser, I. (1972): Proc. Natl. Acad. Sci. USA 69, 721.
38 Matsuyama, M. (1979): Exp. Cell. Res. 124, 253.
39 Morinaga, N. S.; Yonehara, S.; Tomita, Y.; Kuwata, T. (1983): Int. J. Cancer 31, 21.
40 Nilsen, T. W.; Wood, D. L.; Baglioni, C. (1980): Nature 286, 178.
41 Ogura, H.; Tanaka, T.; Ocho, M.; Kuwata, T.; Oda, T. (1978): Arch. Virol. 57, 195.
42 Pfeffer, L. M.; Landsberger, F. R.; Tamm, I. (1981): J. Interferon Res. 1, 613.
43 Pfeffer, L. M.; Wang, E.; Tamm, I. (1980): J. Cell. Biol. 85, 9.
44 Rubin, B. Y.; Gupta, C. (1980): Proc. Natl. Acad. Sci. USA, 77, 5928.
45 Sakiyama, S.; Kuwata, T. (1982): Proc. Japanese Cancer Ass. 41st Ann. Meeting, p. 186.

46 Schneck, J.; Rager-Zisman, B.; Rosen, O. M.; Bloom, B. R. (1982): Proc. Natl. Acad. Sci. USA 74, 1879.
47 Silverman, R. H.; Watling, D.; Balkwill, F. R.; Trowsdale, J.; Kerr, I. M. (1982): Eur. J. Biochem. 126, 333.
48 Stewart, W. E. II (1979): Interferon 1, 29 (I. Gresser, ed.), Academic Press.
49 Suzuki, N.; Nishimaki, J.; Kuwata, T. (1982): Mutation Res. 106, 357.
50 Tomita, Y.; Cantell, K.; Kuwata, T. (1982): Int. J. Cancer 30, 161.
51 Tomita, Y; Kuwata, T. (1979): J. gen. Virol. 43, 111.
52 Tomita, Y.; Kuwata T. (1980): Ann. N. Y. Acad. Sci. 350, 625.
53 Tomita, Y.; Kuwata, T. (1981): J. gen. Virol. 55, 289.
54 Tomita, Y.; Nishimaki, J.; Takahashi, F.; Kuwata, T. (1982): Virology 120, 258.
55 Tovey, M. G.; Gresser, I.; Rochette-Egly, C.; Begon-Lous-Guymarho, J.; Bandu, M.-T.; Maury, C. (1928): J. gen. Virol. 63, 505.
56 Tovey, M. G.; Rochette-Egly, C.; Castagna, M. (1979): Proc. Natl. Acad. Sci. USA, 76, 3890.
57 Tyring, S.; Klimpel, G. R.; Fleishmann, W. R.; Baron, S. (1982): Int. J. Cancer 30, 59.
58 Vandenbussche, P.; Divizia, M.; Verhaegen-Lewalle, M.; Fuse, A.; Kuwata, T.; De Clercq, E.; Content, J. (1981): Virology 111, 11.
59 Vandenbussche, P.; Kuwata, T.; Verhaegen-Lewalle, M.; Content, J. (1983): Virology, in press.
60 Verhaegen, M.; Divizia, M.; Vandenbussche, P.; Kuwata, T.; Content, J. (1980): Proc. Natl. Acad. Sci. USA 77, 4479.
61 Verhaegen-Lewalle, M.; Kuwata, T.; Zhang, Z.-X.; De Clercq, E.; Cantell, K.; Content, J. (1982): Virology 117, 425.
62 Yonehara, S.; Yonehara-Takahashi, M.; Ishii, A. (1983): J. Virol. 45, 1168.

T. Kuwata, M.D., Department of Microbiology, School of Medicine, Chiba University, Chiba 280 (Japan)

Contr. Oncol., vol. 20, pp. 127—133 (Karger, Basel 1984)

Primary Structure of Mouse Interferon-ß Deduced from cDNA structure; Considerations on Evolution of Interferon Genes

Y. Kawade[a]; *Y. Higashi*[a]; *Y. Sokawa*[a]; *T. Miyata*[b]

[a] Institute for Virus Research, Kyoto University, Kyoto, Japan; [b] Department of Biology, Faculty of Science, Kyushu University, Fukuoka, Japan

Detailed structural information on human IFN proteins and genes has recently been obtained through the use of recombinant DNA techniques. Comparable information on mouse IFNs will be valuable for its own sake, as well as for inquiries into the evolution of the IFN system and the structure-activity relationships of the molecules. Cloned human IFN DNAs were found to be cross-hybridizable to mouse DNA, and taking advantage of this, cloning of mouse DNAs coding for IFN-α (13) and γ (2) were recently reported. In collaboration with Dr. T. Taniguchi and his colleagues in Cancer Institute in Tokyo, we could clone cDNA for mouse IFN-β and deduce its complete amino acid sequence (5), as described below. Comparing the gene structures now available of human and mouse IFN-α and β, some inferences were made on the evolution of IFN genes.

cDNA for Mouse IFN-β

As the first step for the cloning of cDNA, IFN mRNA was prepared from virus-induced L cells, which contained messengers for both IFN-α and β, and reverse-transcribed into cDNA. A library of the cDNA was constructed using pBR322 in E. coli. The transformants were screened by colony hybridization using human IFN-β DNA as probe, and several positive clones were obtained. One of them had a 680 base pair cDNA insert, whose base sequence contained an open reading frame for 182 amino acids. The following findings on this clone indicated that it is an authentic cDNA for mouse IFN-β: a) the nucleotide sequence and the deduced amino acid sequence had high homologies (63 and 48 %, respectively) to the human IFN-β sequences (16), b) the first 24 amino acids after the putative signal sequence of 21 amino acids agreed completely with the N-terminal sequence reported previously for IFN-β from Ehrlich ascites tumor cell (15), and c) when the cDNA insert was ligated to an expression vector containing SV40 promotor and transfected to COS cells, it directed the synthesis of active mouse IFN-β molecules.

The complete amino acid sequence deduced from the nucleotide sequence is shown in fig. 1, together with those of mouse IFN-α and human IFN-α and β reported by others.

Using the cDNA as probe, we examined the number of IFN-β gene in the mouse genome by Southern blot analyses of mouse liver DNA digested with several restriction enzymes. The results indicated the presence of single IFN-β gene, and also the absence of introns in the gene, as in the case of human IFN-β gene. However, the possibility that the mouse genome has other IFN-β genes that do not cross-hybridize with our cDNA may not totally be excluded (14).

Amino Acid Sequence of Mouse IFN-β

The molecular weights of mouse IFN-β from L cell (36K) and other sources were known to be considerably greater than that of human IFN-β (22K) (7).

The number of amino acids in the mouse IFN-β polypeptide now determined from the cDNA structure is 161, 5 amino acids less than human IFN-β polypeptide. The larger molecular weight of natural mouse IFN-β is most likely due to a high carbohydrate content, since the sequence has three potential N-glycosylation sites (Asn-X-Ser or Thr) at Asn^{29}, Asn^{69} and Asn^{76}, in contrast to one in human IFN-β at Asn^{80}. IFN-β is sometimes heterogeneous in molecular size; the two electrophoretic components of IFN-β from Ehrlich ascites tumor cells (26K and 35K) were found to have identical N-terminal sequences of 24 amino acids (15), which in turn are

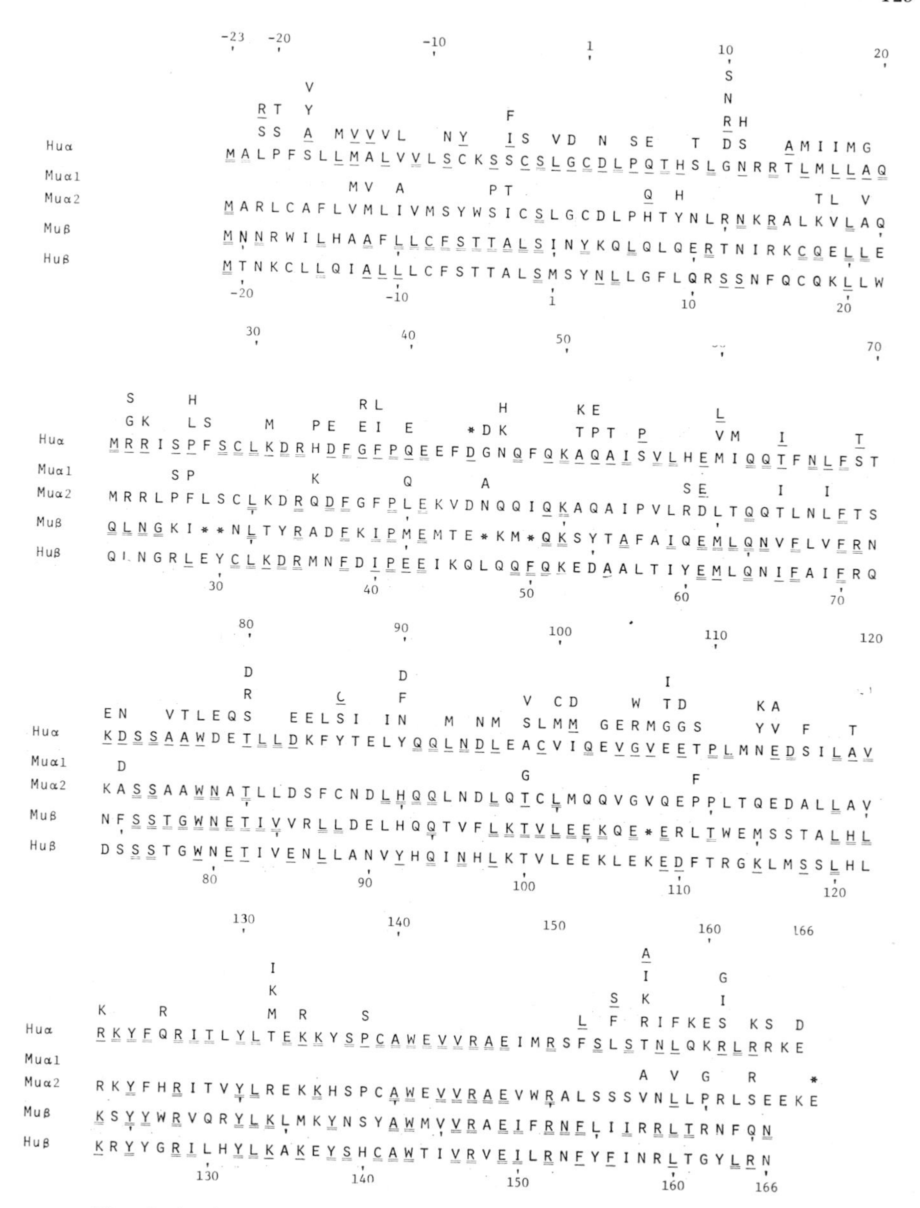

Fig. 1. Amino acid sequences of human and mouse α and β IFNs. One letter codes for amino acids are used. IFN-α sequences are from Saw et al. (1983) and IFN-β sequences, from Higashi et al. (1983) and Taniguchi et al. (1980). The human α sequence shows all amino acids which occur in each

identical to that of L cell IFN-β deduced here. This is consistent with the idea that these IFN-β proteins are the products of a single gene and the molecular size heterogeneity can arise from a heterogeneity in glycosylation.

There is only one Cys residue at position 17, indicating the absence of S-S bridges in the molecule, in contrast to three in human β (at 17, 31, 141).

The mouse IFN-β sequence is aligned in fig. 1 to the human IFN-β sequence, with 5 gaps introduced to maximize homology. They are identical in 78 positions out of 166 (47 %). The positions of identity appear to be scattered throughout the sequence, but relatively more frequent in the middle and C-terminal portions. When a hydrophilicity profile along the sequence was delineated by assigning a value of hydrophilicity index to each amino acid and calculating the sum of the values for six consecutive amino acids, it was found to be very similar to that of human IFN-β, suggesting that their three dimensional structure is conserved well (5).

The sequence homology between human and mouse IFN-β (47 %) is somewhat lower than that between their IFN-α (57 to 62 %) depending on the subtype; (13). Consistent with this, IFN-α crosses the human-mouse species barriers to some extent, but IFN-β does not appreciably do so (3). Also, weak but definite antigenic cross-reactions can be observed between human and mouse IFN-α, but not between their IFN-β (8).

The mouse IFN-β sequence is 24 % homologous to the two subtypes of mouse IFN-α so far sequenced. Similar degrees of homology are found between human IFN-α and β. Thus the α—β homology is much less than the human-mouse homology between the α's or between the β's, reflecting the earlier separation of the

---

position in 11 subtypes. For mouse IFN-α, the whole α2 sequence is indicated; for α1, only amino acids different from α2 are shown. Underlines indicate identities in the following way. Double underlines below human α indicate an identical amino acid to mouse α in all the subtypes of both; single underline, identical in one of them. Double underlines below mouse α, identical in both subtypes to mouse β; single underline, in one of the subtypes. Double underlines below mouse β, identical to human β. Double underlines below human β, identical to human α (all the subtypes); single underline, to one of the human α subtypes.

IFN-α and β genes than the human-mouse divergence. We then examined the gene structures of human and mouse IFNs more closely to gain information on their evolution.

Evolution of IFN Genes

The protein encoding regions of human and mouse IFN-β cDNAs appear to be markedly divergent, in comparison with other pairs of homologous proteins in the two species. Theoretically, two alternative interpretations may be possible for such marked divergence: (i) The two IFN-β genes of human and mouse are orthologous, i.e., the divergence of their DNA sequences reflects the time that passed since the separation of the two animal species, yet the amino acid sequences diverged at a rapid rate. (ii) They are paralogous, i.e., they separated long before the two species diverged. If the claim of the presence of a second (or more) human IFN-β gene, which shares little homology with IFN-β_1 gene (12, 18), is substantiated, the second case may become a real possibility. Close examination of the extent of sequence variation may provide a clue to decide which is the case. From the comparisons of DNA sequences between man and rodents for many genes, it has been shown that the degree of sequence difference per site (K_s) at the silent position (synonymous codon change) in the protein encoding region is large and, more importantly, is remarkably similar for different genes (9, 10, 11, 4), being on the average 0.50 ± 0.03. The latter implies that silent positions evolve at an even rate for different genes. This is in sharp contrast to the rate of protein evolution which varies greatly from protein to protein (1). Thus, when IFN-β sequences are compared between man and mouse, the silent position is expected to show a K_S value that is close to 0.50, if (i) is really the case. On the other hand, if (ii) is the case, the value will be larger than 0.50. Comparison of the IFN-β cDNA sequences of man and mouse supports the former view; the K_S value is 0.49, close to the average value for various gene pairs. That is, the present IFN-β genes of man and mouse are probably orthologous and the time of their separation just corresponds to the time of man/rodent divergence.

Assuming 75 million years (Myr) for the man/rodent divergence time, we determined the rate of evolution (v_A) at amino acid replacement positions (i.e., the rate of protein evolution) of IFN-β gene as 2.4×10^{-9}/site/year from the formula (Miyata & Hayashida, 1982): $v_A = -(3/4)\ln\,[1-(4/3)K_A]/2t$, where K_A is the nucleotide difference per site at replacement positions and t the divergence time of the species compared ($t = 75$ Myr for the present case). This figure is remarkably large, being comparable to that of Igε chain ($v_A = 2.6 \times 10^{-9}$), one of the most rapidly evolving protein molecules (Ishida et al., 1982), and about 3 times larger than the rate of α-globin. IFN-α was also shown to evolve at a rapid rate ($v_A = 1.7 \times 10^{-9}$), although it is slightly lower than the corresponding rate of IFN-β. Such rapid rates of protein evolution for both IFN-α and IFN-β may underlie the species specificity of the antiviral activity of IFN molecules.

The IFN-α and β genes must have descended from a common ancestral gene, but when were they separated? Taniguchi et al. (16) estimated the time to be 500—1,000 Myr ago, based on the assumption that interferon and globin genes evolved with the same rate. This estimate must be revised, since different proteins are now known to evolve with widely different rates. We compared six human IFN-α genes (A, C1, D, F, and J) and two mouse IFN-α genes (α1 and α2) with human and mouse IFN-β genes, and obtained an average value of 0.438 for K_A. This value leads to estimates of 163 Myr and 195 Myr for the time of separation of IFN-α and IFN-β, if v_A is assumed to be 2.0×10^{-9} (an average rate for IFN-α and IFN-β) and 1.7×10^{-9} (the rate for IFN-α), respectively. Thus the separation of interferon α and β is a much more recent event than has hitherto been thought, and possibly goes back only to about 200 Myr ago, a date around the divergence of mammals and birds-reptiles. It is of interest to know whether or not birds and reptiles have both IFN-α and -β genes.

References

1 Dayhoff, M. O. (1978): in "Atlas of Protein Sequence and Structure", ed. by M. O. Dayhoff, National Biomedical Research Foundation, Washington, DC, vol. 5, Suppl. 3, p. 1.
2 Derynck, R. et al. (1983): in "Biology of the Interferon System", ed. by H. Schellekens, Elsevier, Amsterdam (in press).
3 Fusijawa, J.; Kawade, Y. (1981): Virology 112, 480.

4 Hayashida, H.; Miyata, T. (1983): Proc. Nat. Acad. Sci. USA (in press).
5 Higashi, Y., Sokawa, S.; Watanabe, Y.; Kawade, Y.; Ohno, S.; Takaoka, C.; Taniguchi, T. (1983): J. Biol. Chem. (in press).
6 Ishida, N.; Ueda, S.; Hayashida, H.; Miyata, T.; Honjo, T. (1982): EMBO J. 1, 1117.
7 Kawade, Y. (1982): Texas Rep. Biol. Med. 41, 219.
8 Kawade, Y.; Watanabe, Y.; Yamamoto, Y.; Fujisawa, J.; Dalton, B. J.; Paucker, K. (1981): Antiviral Res. 1, 167.
9 Miyata, T.; Hayashida, H. (1982): Nature 295, 165.
10 Miyata, T.; Hayashida, H.; Kikuno, R.; Hasegawa, M.; Kobayashi, M.; Koike, K. (1982): J. Mol. Evol. 19, 28.
11 Miyata, T.; Yasunaga, T.; Nishida (1980): Proc. Nat. Acad. Sci. USA 77, 7328.
12 Sehgal, P. B.; Sagar, A. D. (1980): Nature 288, 95.
13 Shaw, G. D.; Boll, W.; Taira, H.; Mantei, N.; Lengyel, P.; Weissmann, C. (1983): Nucl. Acids. Res. 11, 555.
14 Skup, D.; Windass, J. D.; Sor, F.; George, H.; Williams, B. R. G.; Fukuhara, H.; Maeyer-Guignard, J. De; De Maeyer, E. (1982): Nucl. Acids Res. 10, 3069.
15 Taira, H.; Broeze, R. J.; Jayaram, B. M.; Lengyel, P.; Hunkapiller, M. W.; Hood, L. E. (1980): Science 207, 528.
16 Taniguchi, T.; Mantei, N.; Schwarzstein, M.; Nagata, S.; Muramatsu, M.; Weissmann, C. (1980): Nature 285, 547.
17 Taniguchi, T.; Ohno, S.; Fujii-Kuriyama, Y.; Muramatsu, M. (1980). Gene 10, 11.
18 Weissenbach, J.; Chernajovsky, Y.; Zeevi, M.; Shulman, L.; Soreo, H.; Nir, U.; Wallach, D.; Perricaudet, M.; Tiollais, P.; Revel, M. (1980): Proc. Nat. Acad. Sci. USA 77, 7152.

Y. Kawade, M.D., Institute for Virus Research, Kyoto University, Kyoto (Japan)

Contr. Oncol., vol. 20, pp. 134—141 (Karger, Basel 1984)

Formation of Human IFN-alpha Subtype Mixtures by Different Human Cells

G. Bodo; G. R. Adolf

Ernst Boehringer-Institut für Arzneimittelforschung, Vienna, Austria

Introduction

Interferon-alpha (IFN-α) made by human cells is heterogeneous. The heterogeneity is explained by the fact that multiple genes exist in human cells coding for IFN-α (10).

Methods to separate IFN-α subtypes include, for example, SDS-gel electrophoresis. Several groups have used this technique for the separation of human leukocyte-IFN (5, 9, 11). Human IFN-α from cultured human cells was also found to be heterogeneous. Thus, Allen (4) found up to 8 subtypes in Namalwa-IFN, using gel-filtration and SDS-gel electrophoresis. The elegant technique of high performance liquid chromatography was used by Rubinstein et al. (7) and 8 subtypes could be isolated from human leukocyte-IFN.

Since it becomes increasingly clear that different subtypes of human IFN-α have different properties with regard to antiviral and immunoregulating activity and may even potentiate each other's function, the analysis of subtype composition seems important. Therefore, we have analyzed a number of naturally formed human IFN-α mixtures using a new technique: Chromatofocusing on the carrier MONO-P (Pharmacia). The experiments presented in this paper demonstrate the usefulness of this technique for a fast and reliable subtype analysis of human IFN-α.

Materials and Methods

Cells

Namalwa clone 8 cells were kindly supplied by G. Klein, Stockholm. NC-37 cells were a gift from D. Neumann-Haefelin, Freiburg.

Wien-67 cells, a B-type lymphoblastoid human line, were obtained from P. Fischer, Vienna (1). All cells were grown in RPMI 1640 with 10 % fetal calf serum.

Interferons

Interferons were induced with Sendai virus, strain Cantell. In some experiments, cells were pretreated with 1 mM butyrate (8), or 10 μM dexamethasone (3) before infection. "Spontaneous-IFN" was obtained from uninfected Wien 67-cells pretreated with 50 μg/ml 5-bromo-2-deoxyuridine (BrdU) (1). Crude IFNs were kept at pH 2 over night, neutralized and precipitated with 65 % saturated ammonium sulfate. The pellets were taken up and dialyzed against pH 7,1-buffer (see below). Before use, IFN-solutions were clarified by ultracentrifugation.

Chromatofocusing

A column HR 5/20 of the carrier MONO-P (Pharmacia) was used in connection with the Pharmacia FPLC-System. The column was equilibrated with 0,025 M bis-tris buffered to pH 7,1 with imino-di-acetic-acid. IFN-samples were then applied to the column and the pH-gradient formed with Polybuffer 74: 10 ml Polybuffer 74 (Pharmacia) adjusted to pH 4,0 with 0,1 M imino-di-acetic-acid and diluted to 150 ml. Both bis-tris and Polybuffer 74 contained 25 % v/v 1,2-propanediol. The column eluate was monitored at 280 mm. Fractions of 1 ml were collected and pH and IFN-activity determined.

IFN-activity

Two test systems were used.

1. The IRMA test system for human IFN-α from Celltech (G. B.). This test employs a polyclonal anti-IFN-α and a radiolabeled monoclonal IFN-antibody (125 I-NK 2) (8).
2. A plaque-reduction assay using GL V3 Vervet monkey kidney cells and Vesicular stomatitis Virus as challenge.

Both assays were standardized with the IFN-α Research Standard B 69/19.

Results

Fig. 1 shows the analysis of Cantell-IFN. The sample was kindly provided by K. Cantell. 5 peaks of activity can be seen. From repeated analyses, their isoelectric points were at pH 5,8–5,6–5,2–5,0

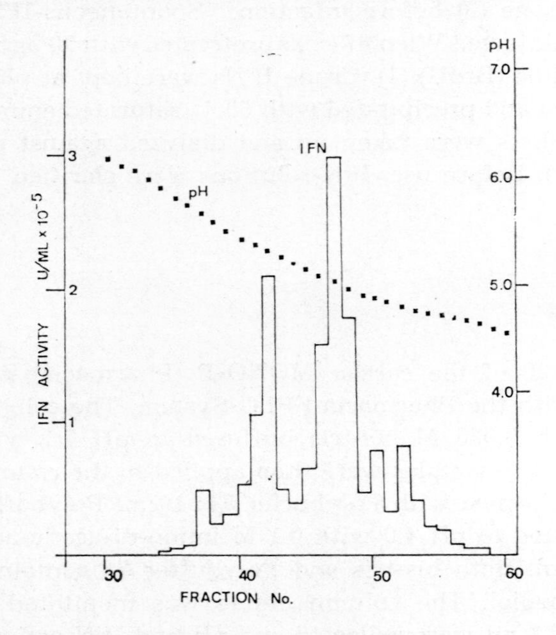

Fig. 1. Analysis of Cantell-IFN (PIF).
2.2×10^6 units (IRMA) were used for the analysis.
IFN-activity was determined by IRMA.

and 4,9. The minor peak focused at pH 5,8 could be identified as human IFN-α2, using a sample of authentic IFN-α2 cloned in E. coli (6). This IFN forms only one sharp peak at this pH (data not shown). Lacking other samples of homogeneous cloned material, the other peaks could not yet be classified. The major components in Cantell-IFN are focused at pH 5,6 and 5,2.

Fig. 2 shows the analysis of Namalwa-IFN from butyrate treated cells. The top panel gives IFN-activity as determined by IRMA, the lower panel as determined by plaque reduction. IRMA analysis

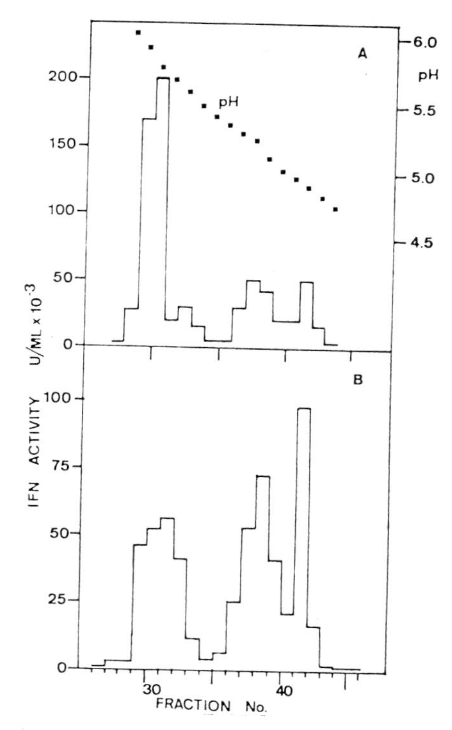

Fig. 2. Analysis of Namalwa-IFN from butyrate-treated cells. 920.000 units (IRMA) were used for analysis.
Panel A: IFN-activity by IRMA.
Panel B: IFN-activity by plaque-reduction.

shows the component focused at pH 5,8 as major one (60—70 % of total IRMA-IFN). This is IFN-α2. Although the two IFN-assays give peaks at the same isoelectric points, the relative intensities of the peaks differ. This reflects the different sensitivity of the 2 assays for different subtypes of IFN-α. Therefore, an absolute composition cannot be calculated from either test. However, the activity profile is quite reproducible from batch to batch, thus showing that subtype composition of different samples of IFN from butyrate treated Namalwa-cells is very similar. If compared to Cantell-IFN, however, there is a striking difference in composition: The IFN-α2 component focused at pH 5,8 is strongly enhanced, the component focused at pH 5,6, on the other hand, is the smallest in Namalwa-IFN and a major constitutent in Cantell-IFN.

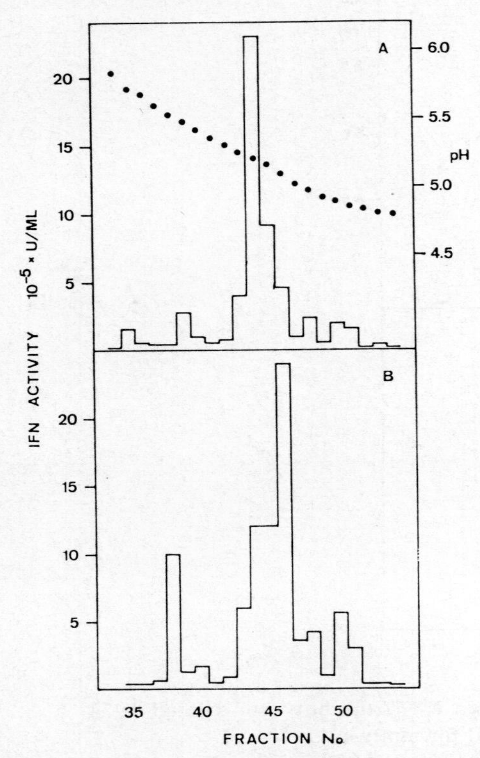

Fig. 3. Analysis of IFN from butyrate-treated NC-37 cells. 5.2×10^6 units (IRMA) were used for analysis.
Panel A: IFN-activity by IRMA
Panel B: IFN-activity by plaque-reduction.

Fig. 3 gives the result of an analysis of IFN from butyrate-treated NC-37 cells. Again, 5 peaks at the isoelectric points pH 5,8—5,6—5,2—5,0 and 4,9 are seen. This indicates that the same subtypes are present also in this IFN-α. However, their relative amounts differ again from the foregoing preparation shown: The pH 5,2-component is by far the major one, by IRMA about 70—75 % of total IFN. In this figure, there is an indication that the peak focused at pH 5,2 is still heterogeneous: IRMA- assay and plaque reduction assay do not show the same maximum. Thus, more than 5 subtypes of IFN-α may be present in this IFN.

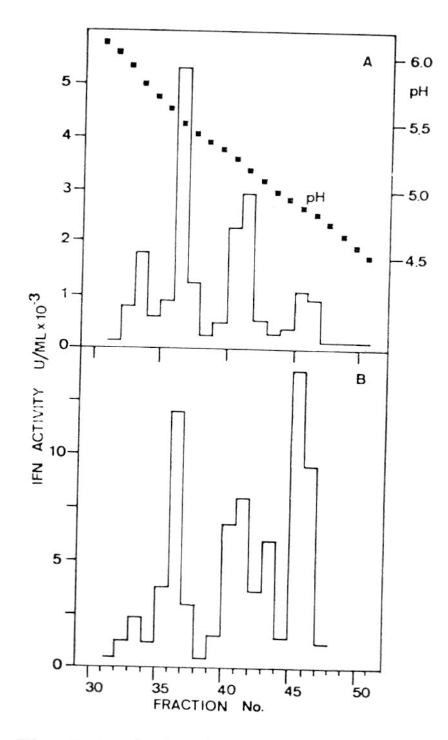

Fig. 4. Analysis of IFN from Sendai virus-infected Wien-67 cells. 25,000 units (IRMA) were used for analysis.
Panel A: IFN-activity by IRMA
Panel B: IFN-activity by plaque-reduction.

Fig. 4 shows the result of an analysis of Wien 67-IFN induced with Sendai virus. Again, 4 peaks at pH 5,8—5,6—5,2/5,0 and 4,9 are present. The analysis by plaque-reduction shows splitting of the peak at pH 5,2/5,0 into two. Thus, the same qualitative pattern of subtype-composition is seen again. The relative amounts are again different from the examples shown so far: A dominant peak at pH 5,6 (IRMA), making this IFN more similar to Cantell-IFN than Namalwa-or NC 37-IFN.

Fig. 5 shows the analysis of IFN from uninfected Wien-67 cells, a socalled "spontaneous IFN-α", the formation of which was enhanced by treatment of cells with BrdU. By both IRMA and plaque-

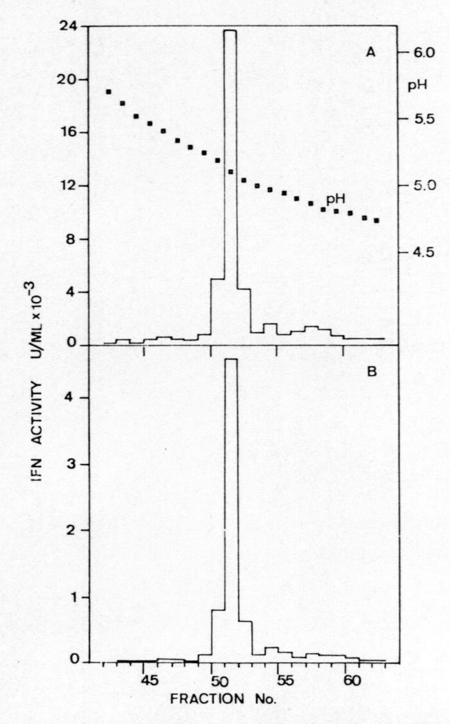

Fig. 5. Analysis of "spontaneous-IFN" from BrdU-treated Wien-67 cells. 60,000 units (IRMA) were used for analysis.
Panel A: IFN-activity by IRMA.
Panel B: IFN-activity by plaque-reduction.

reduction, the only major component is focused at pH 5,1, representing about 75 % of total IFN-activity. Thus, this IFN-α is more homogeneous than the others investigated so far.

Summary

Chromatofocusing of human IFN-α on MONO-P columns in the pH-range 7—4 separates the subtypes according to their isoelectric points. Up to 5 peaks can be found in this pH-range. Separation is, however, probably not complete and more than 5 subtypes may be present.

Subtype composition is characteristic for the producer cell-line, the IFNs of different producer cell-lines giving clearly different subtype patterns. The subtype composition of Cantell-IFN and of IFN from butyrate-treated Namalwa-cells is rather different, a fact that may be of significance for their clinical use.

A distinct advantage of the chromatofocusing technique over other methods for the separation of IFN-α subtypes is, that reproducible separations can be obtained in a very short time (2—3 hours), and that fractions obtained may immediately be used for further characterizations. This offers a possibility to analyze the subtype mixtures of naturally occuring IFN-α, for example the IFN produced by various human cells in culture, as well as the IFN found in sera of patients with viral infections or immunological disorders.

References

1 Adolf, G. R. et al. (1982): Arch. of Virology 72, 169—178.
2 Adolf, G. R.; Swetly, P. (1979): Virology 99, 158—166.
3 Adolf, G. R.; Swetly, P. (1982): J. Interferon Res. 2, 261—269.
4 Allen, G. (1982): Biochem. J. 207, 397—408.
5 Berg, K.; Heron, I. (1980): Scand. J. Immunol. 11, 489—502.
6 Dworkin-Restl, E.; Dworkin, M. B.; Swetly, P. (1982): J. Interferon Res. 2, 575—585.
7 Rubinstein, M. et al. (1981): Arch. Biochem. Biophys. 210, 307—318.
8 Secher, D. S.; Burke, D. C. (1980): Nature 285, 446—450.
9 Stewart II, W. E.; Desmyter, J. (1975). Virology 67, 68—73.
10 Weissmann, C. et al. (1982): UCLA Symposia on Molecular and Cellular Biology XXV, 295—326 (Academic Press 1982).
11 Zoon, K. C. et al. (1982): J. Interferon Res. 2, 253—260.

G. Bodo, M.D., Ernst Boehringer-Institut für Arzneimittelforschung, 1120 Vienna, Dr. Boehringer-Gasse 5—11 (Austria)

Contr. Oncol., vol. 20, pp. 142—159 (Karger, Basel 1984)

Effect of Interferon on Monoclonal Antibody Producing and Proliferating Capacity of Cell Hybridomas

M. Novák; L. Borecký; K. Poláková[a]*; G. Russ; Z. Sekeyová; M. Lipková; N. Fuchsberger; F. Čiampor; P. Kontsek*

Institute of Virology, Slovak Academy of Sciences, Bratislava
[a] Institute of Experimental Oncology, Bratislava, Czechoslovakia

Introduction

It is widely accepted that, in addition to their antiviral activities, the interferons (IFN) have antiproliferative effects in vitro on both diploid and transformed cells and cell lines, including those from osteo- and soft sarcomas (16), lymphomas (1), myelomas (3), lymphoid leukemias (6), ovarian carcinoma (7), or, methylcholanthrene transformed embryonal cells (2). Since the antiproliferative activity has been observed using highly purified interferons (14) as well as interferons produced by recombinant DNA techniques (10), the growth inhibition is considered a property of interferon per se (2). However, the effect of IFN on the proliferative and antibody secreting capacity of the monoclonal antibodies (MnAb) producing hybridomas has not been reported. For this reason, it seemed interesting to test the effect of various interferons on the proliferating and MnAb producing capacity of hybridomas secreting IgM, IgG1, IgG2a, IgG2b and IgG3 monoclonal antibodies.

Materials and Methods

Viruses

The BaI-A strain of avian myeloblastosis virus (AMV) was purified from the plasma of leukemic chicken by differential centrifugation, followed by equilibrium centrifugation in a 10 to 45 per cent sucrose gradient.

The Newcastle disease virus (NDV), Hertfordshire strain, was used for IFN induction. The virus was maintained as described elsewhere in this volume (9). The Indiana strain of Vesicular stomatitis virus (VSV) was used as challenge in antiviral IFN-assays.

Interferons

Mouse interferons β[1] and γ were prepared as described in our previous paper in this volume (9). Human IFN-α was produced in leukocytes obtained from healthy donors (4).

Preparation of hybridoma cell lines

Hybridoma cell lines were prepared by fusion of a mouse myeloma cell line Sp 2/0 — Ag 14 with spleen cells of AMV immunized Balb/c mice in serum free medium with polyethylene glycol (1540 m.w.). The immunization protocol, detection of monoclonal antibodies, characterization of the monoclonality of hybridoma antibodies and their specificity were in detail described elsewhere (13).

Cultivation of Hybridomas

The hybridomas were cultivated in a standard medium RPMI 1640 supplemented with L-glutamine (2 mM), sodium puryvate (1 mM), 2-mercaptoethanol (0,028 mM), 10 mM HEPES (N-2-hydroxyethyl-piperazine-N-2/ethane-sulphonic acid), streptomycin (50 μg/ml), penicillin (100 units/ml) and 10 % heat-inactivated (56 °C

[1] The mouse interferon β used in this study is a mixture of α and β interferons (see Kawade et al. in this volume).

for 30 min.) normal horse serum. All hybridomas were cultivated at 37 °C in a humidified atmosphere containing 7,5 per cent CO_2.

Chromosome Analysis

The chromosomes of hybridoma cells were analyzed by the method of Rothfels and Siminovitch (14).

Inhibitory Effect of Interferons on the Monoclonal Antibody Production

Five 3-times recloned hybridomas producing antibodies of IgM, IgG1, IgG2a, IgG2b and IgG3 classes and/or subclasses were used in tests. Six two-fold dilutions of from 10,000 up to 312 units of each IFN (α and β) were tested in triplicate. Six replicate wells with hybridoma cells without IFN were used as controls in each experiment. The hybridoma cells in RPMI 1640 medium with 10 per cent normal horse serum were seeded in amounts of 4×10^4 cells in 100 μl per well in U-bottomed microcultivation plates (Flow Laboratories) on day 0. Subsequently, tested dilutions of IFN-preparations in 100 μl volume per well in RPMI 1640 medium with 10 % normal horse serum were added to each cell suspensions. The production of monoclonal antibodies (MnAb) was determined after 24, 48 and 72 hrs in the supernates from hybridoma cells by solid phase radioimmunoassay (RIA) as described (13). Briefly, 25 μl of purified AMV (4 μg per ml) were added to each well and the plates were kept at 37° C overnight. The dried wells were saturated with bovine serum albumin. The wells were thoroughly washed, incubated with hybridoma culture fluids for 1.5 hrs, again washed, and subsequently treated with ^{125}I-labelled rabbit anti-mouse $F(ab)_2$. The incorporation was determined in a gamma-counter (Packard, PRIAS). The maximal MnAb producing potency (MPP) of hybridomas was estimated from the control cell-cultures after 24, 48 and 72 hrs of cultivation. The results are presented as averages of counts per minute (CPM) from 6 parallels. The inhibition of MPP by IFNs is presented in per cent of controls. As negative controls, myeloma cells Ag 8,653 and Sp 2/0—Ag14, respectively, were included in tests.

Antiproliferative Effect of Interferons α, β and γ on Hybridomas

The ^{3}H-TdR incorporation assay was used in the tests aimed at measuring the antiproliferative (AP) effect of IFNs on hybridomas (9). Each tested IFN preparation (IFN-α and -β had a starting potency 800 U/100 μl; IFN γ 500 U/100 μl) was diluted in two-fold steps in RPMI 1640 medium supplemented with 10 per cent normal horse serum in U-type microplates. To these dilutions, 40,000 hybridoma cells suspended in RPMI 1640 with 10 per cent normal horse serum (NHS) were added per well in 100 μl volumes. The microplates were incubated for 48 hrs at 37° C in a humidified atmosphere with 7.5 CO_2. ^{3}H-TdR in amount of 1 μCi per well was added to suspension of cells for the last 12 hrs of incubation. Subsequently, the cells were harvested from wells with a 12-channel semiautomatic harvestor and the incorporated radioactivity was determined in a beta-counter (Pacard, model 3390).

The incorporation of ^{3}H-TdR in control hybridoma cells has been considered as 100 per cent. Each tested IFN dilution was evaluated in 3 parallel cultures. In control, 6 parallel hybridoma cultures without IFN were tested and the results were averaged.

Results

The antiproliferative effect of human IFN-α and mouse IFNs-β and γ on the proliferating capacity of hybridomas producing different classes and/or subclasses of immunoglobulins is shown on figures 1 to 5 and summarised in tab. 1. The results indicate that a) the most pronounced inhibitory effect on cell proliferation was exerted by IFN-γ. b) The hybridomas IgG2a, IgG3 and IgG2b after IFN treatment had a 70 per cent lower proliferating capacity than the untreated control hybridomas. c) The hybridomas producing IgM antibodies were less sensitive to IFN-γ than the other tested hybridomas. The decrease of incorporation was 44.7 per cent in comparison with control hybridomas. d) The hybridomas producing IgG1 antibodies showed a relative low sensitivity to IFN-γ (19.8 per cent inhibition), a very high sensitivity to the antiproliferative effect of IFN-β (86.4 per cent inhibition), and an insensitivity to

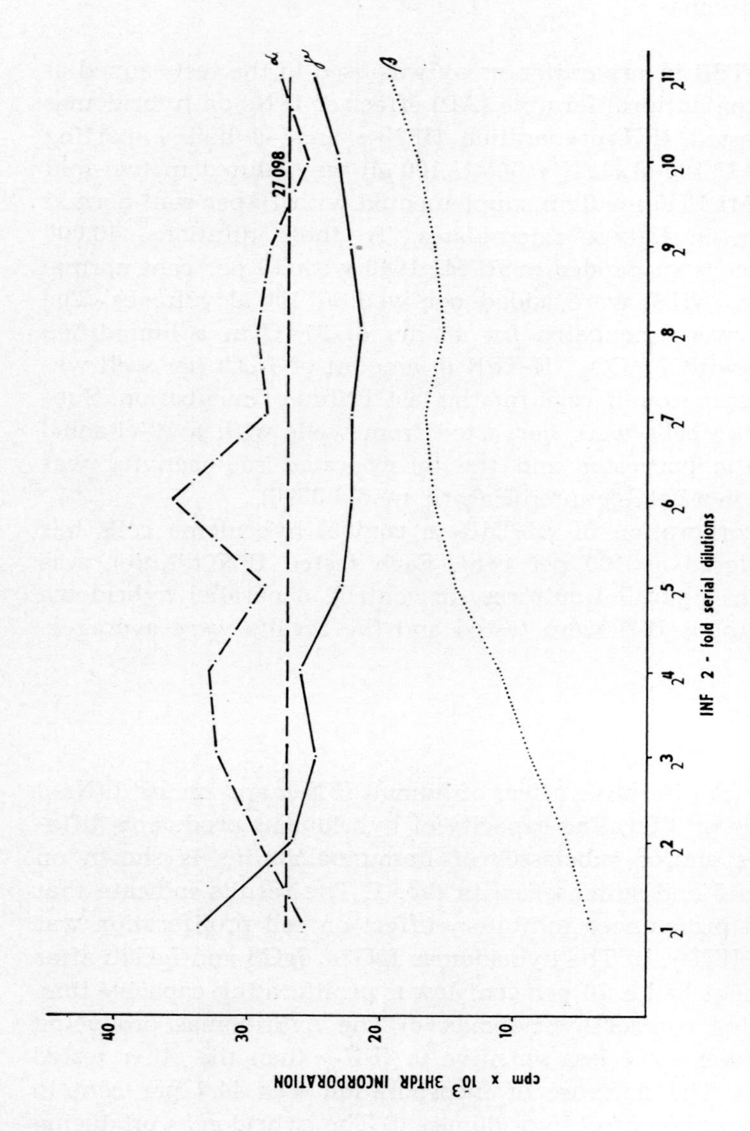

Fig. 1. Antiproliferative activity of interferons α, β and γ on hybridomas secreting IgG1 monoclonal antibodies.
Abscissa: Two-fold dilutions of interferon tested.
Ordinate: Incorporation of ^{3}H-TdR in CPM$\times 10^3$.
— — —: Incorporation into control (untreated) cells in CPM$\times 10^3$.

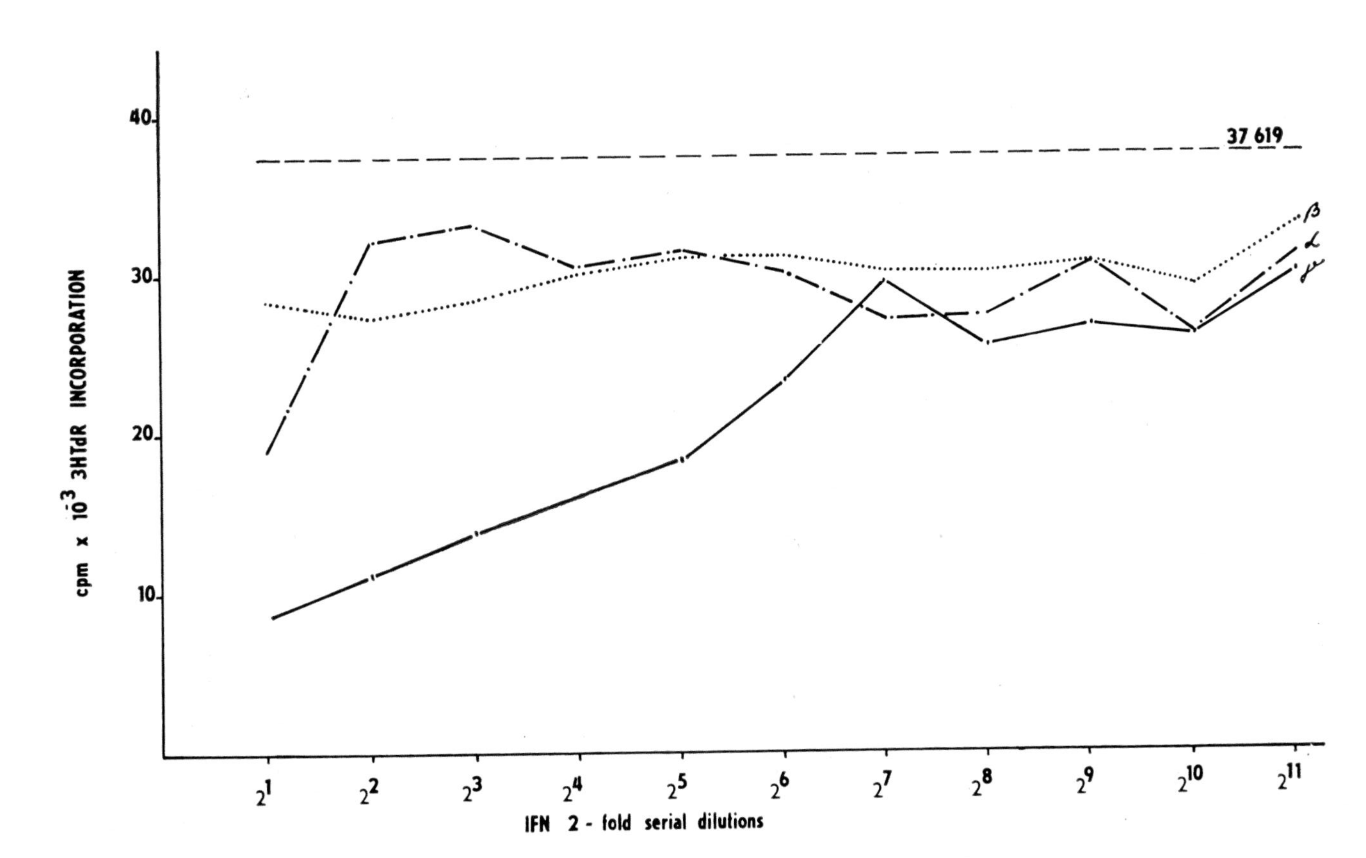

Fig. 2. Antiproliferative activity of interferon α, β and γ on hybridomas secreting IgG2a monoclonal antibodies. For explanation see Fig. 1.

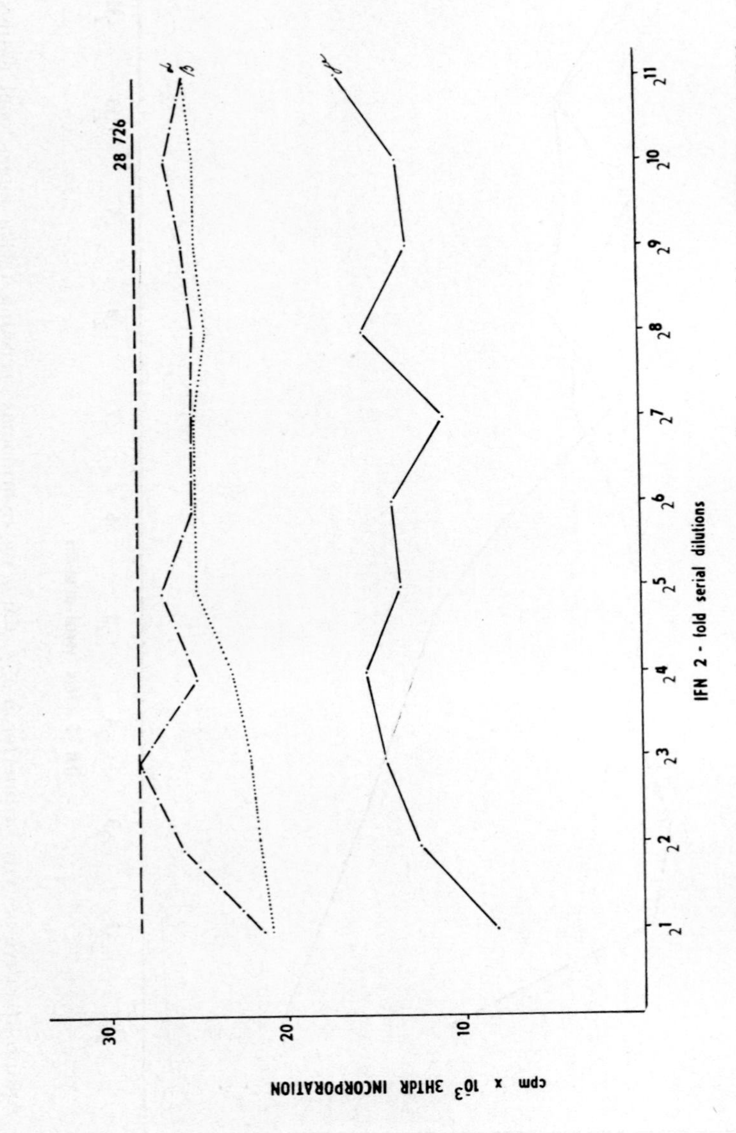

Fig. 3. Antiproliferative activity of interferons α, β and γ on hybridomas secreting IgG2b monoclonal antibodies.
For explanation see Fig. 1.

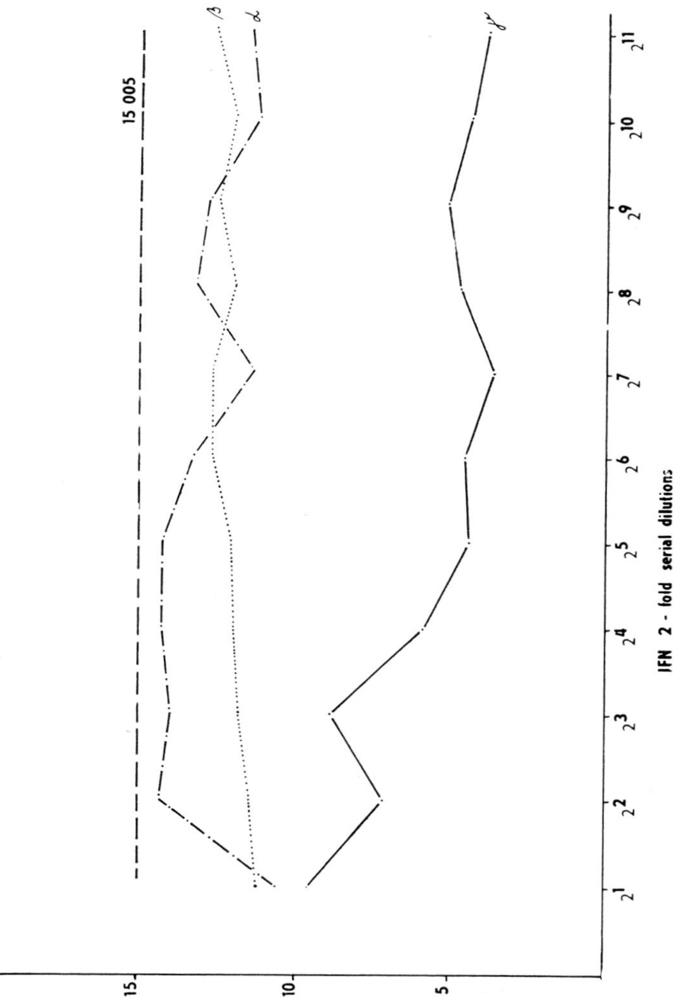

Fig. 4. Antiproliferative activity of interferons α, β and γ on hybridomas secreting IgG3 monoclonal antibodies.
For explanation see Fig. 1.

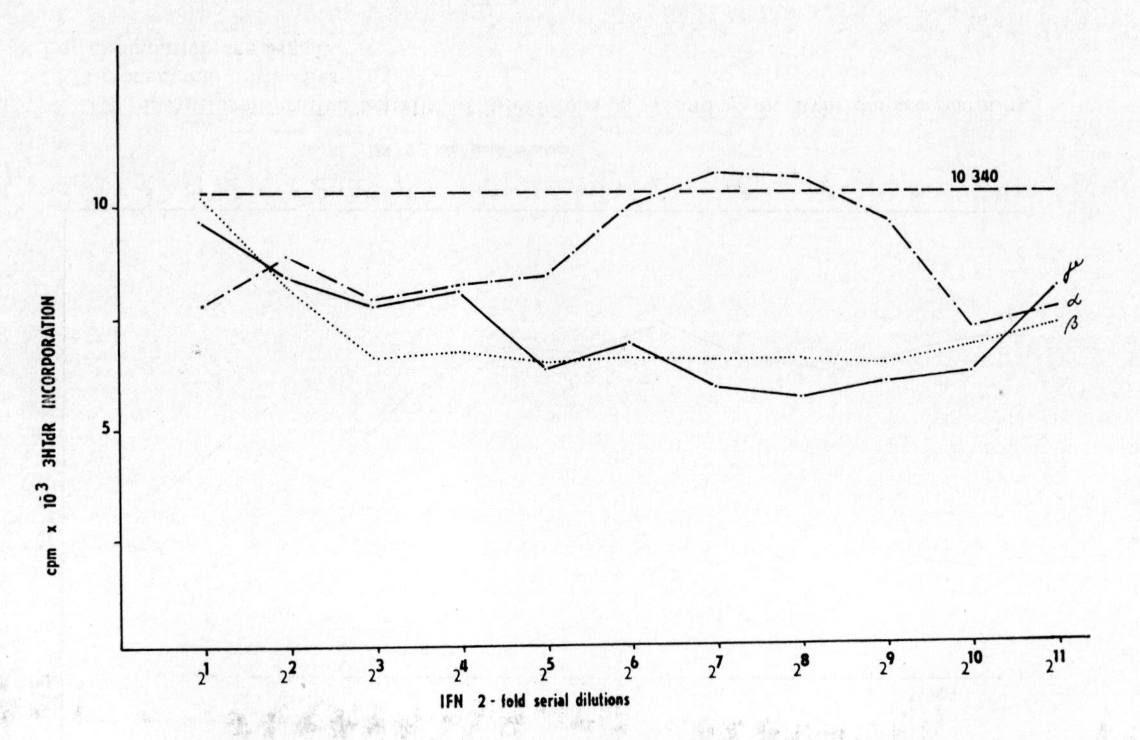

Fig. 5. Antiproliferative activity of interferons α, β and γ on hybridomas secreting IgM monoclonal antibodies.
For explanation see Fig. 1.

Table I. The antiproliferative (AP) activity of interferon upon hybridomas

Hybridomas producing antibodies	Maxim. AP activity in %	Type of IFN and concentr. (See Fig. 1—5)
IgM	30.3	α 2^{10}
IgM	38.2	β 2^{9}
IgM	44.7	γ 2^{8}
IgGI	5.6	α 2^{1}
IgGI	86.4	β 2^{1}
IgGI	19.8	γ 2^{8}
IgG2a	48.5	α 2^{1}
IgG2a	27.4	β 2^{3}
IgG2a	77.0	γ 2^{1}
IgG2b	25.5	α 2^{1}
IgG2b	26.1	β 2^{1}
IgG2b	72.6	γ 2^{1}
IgG3	29.4	α 2^{1}
IgG3	26.0	β 2^{1}
IgG3	76.7	γ 2^{7}

IFN-α (5.6 per cent decrease of ^{3}H-TdR incorporation). e) The hybridomas producing IgM antibodies can be considered "universally sensitive" since they showed an approximately equal decrease of proliferating capacity after treatment with the IFNs tested. The degree of ^{3}H-TdR incorporation in IgM producing hybridoma cells after treatment with IFN-α was 30.3 per cent, after IFN-β 38.2 per cent, and after IFN-γ 44.7 per cent. f) With the exception of the hybridoma producing IgM, the other tested hybridomas were usually more sensitive to one type of IFNs tested, and, showed higher sensitivity to the antiproliferative effect of homologous IFN than to heterologous types.

The fact that the antibody producing cell can not survive more than 20 to 30 divisions in vitro, makes the assay of antibody production in such cells after IFN treatment difficult. This difficulty has been overcome by estimating the antibody production in the supernates of proliferating hybridomas harvested 24, 48 and 72 hrs after starting the cultivation. As negative controls, non-antibody synthesizing myeloma cells of 2 lines were used. In these cells the binding of ^{125}I-labelled anti-mouse F(ab)$_2$ in the RIA test never exceeded 87 to 111 cpm. In contradistinction, the maximal binding of ^{125}I-labelled rabbit anti-mouse $F(ab)_2$ in supernates of IgG1 producing hybridoma line showed 6,100 cpm values.

The figures 6 to 9 show the inhibitory effect of 6 various dilutions of interferons α and β, respectively, on the monoclonal antibody producing activity of hybridomas. During the first 24 hrs the production of antibodies in cells treated with IFNs was only partially depressed. The incorporation of ^{125}I was about 13.7 per cent lower than in controls. The inhibitory effect was more evident after 48 hrs (the incorporation of ^{125}I was depressed by 23.4 per cent), and it reached its maximum at 72 hrs, when the hybridomas

Table II. Karyotypes of hybridomas

Hybridoma	IFN	N	Range	x	±s
IgM	0	20	73—84	79.1	3.1
IgM	α	20	72—86	79.8	5.0
IgM	β	20	72—84	78.3	4.1
IgM	γ	20	74—90	80.5	5.2
IgGI	0	20	52—62	57.7	3.6
IgGI	α	not determined			
IgGI	β	20	54—64	60.9	2.7
IgGI	γ	20	56—66	60.6	3.0
IgG2a	0	20	60—79	69.3	5.3
IgG2a	α	20	64—77	69.2	3.5
IgG2a	β	20	64—76	70.4	4.0
IgG2a	γ	20	64—77	69.9	4.3
IgG2b	0	20	86—100	92.6	4.0
IgG2b	α	20	84—96	88.8	3.7
IgG2b	β	20	83—100	90.7	4.5
IgG2b	γ	20	83—93	88.1	3.3
IgG3	0	20	73—94	82.3	6.1
IgG3	α	20	73—89	80.7	6.1
IgG3	β	20	74—90	80.9	5.0
IgG3	γ	20	73—88	80.9	5.0

Explanation: 0 = hybridoma cells without IFN, N = Number of mitoses counted, x = Average number of chromosomes, s = Standard deviation, α = Hu IFN-α(Le) 800 U/ml (72 hrs), β = Mu IFN-β /F/800 U/ml /72 hrs/, γ = Mu IFN-γ/Le/ 500 U/ml (72 hrs).

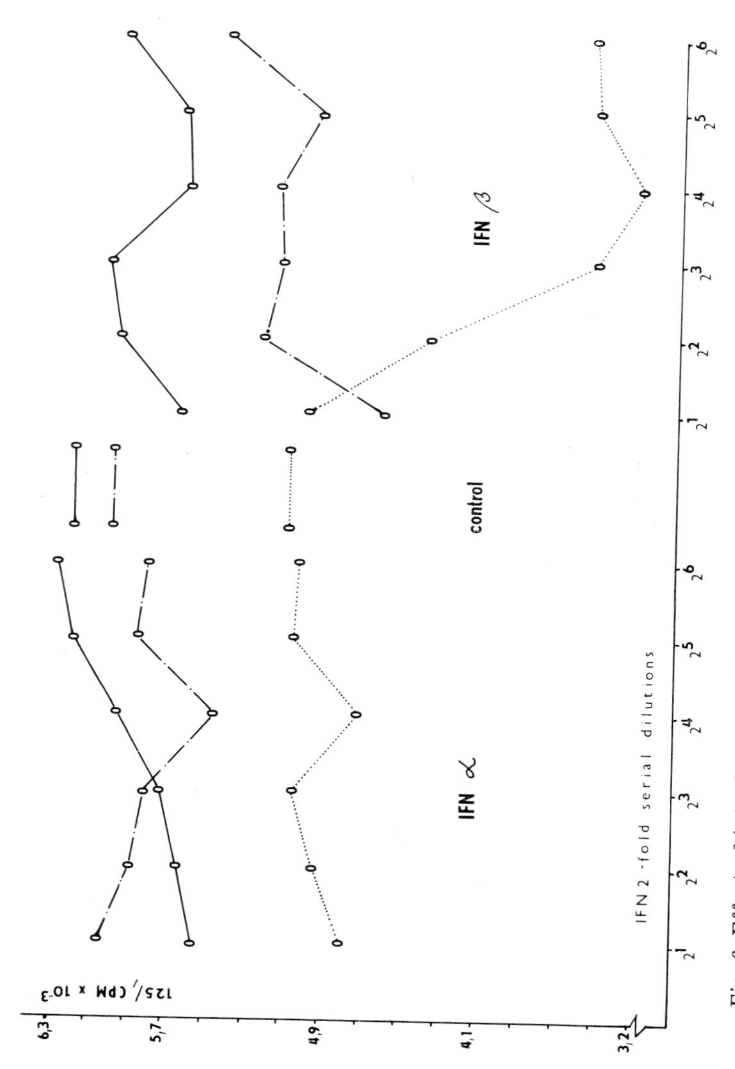

Fig. 6. Effect of interferons on the monoclonal IgG1 production.
Abscissa: Two-fold dilutions of interferons α and β.
Ordinate: Binding of ^{125}I labelled anti-mouse $F(ab)_2$ in CPM$\times 10^3$ after 24 (O—O), 48 (O—.—.—O) and 72 hrs (O......O) of cultivation.

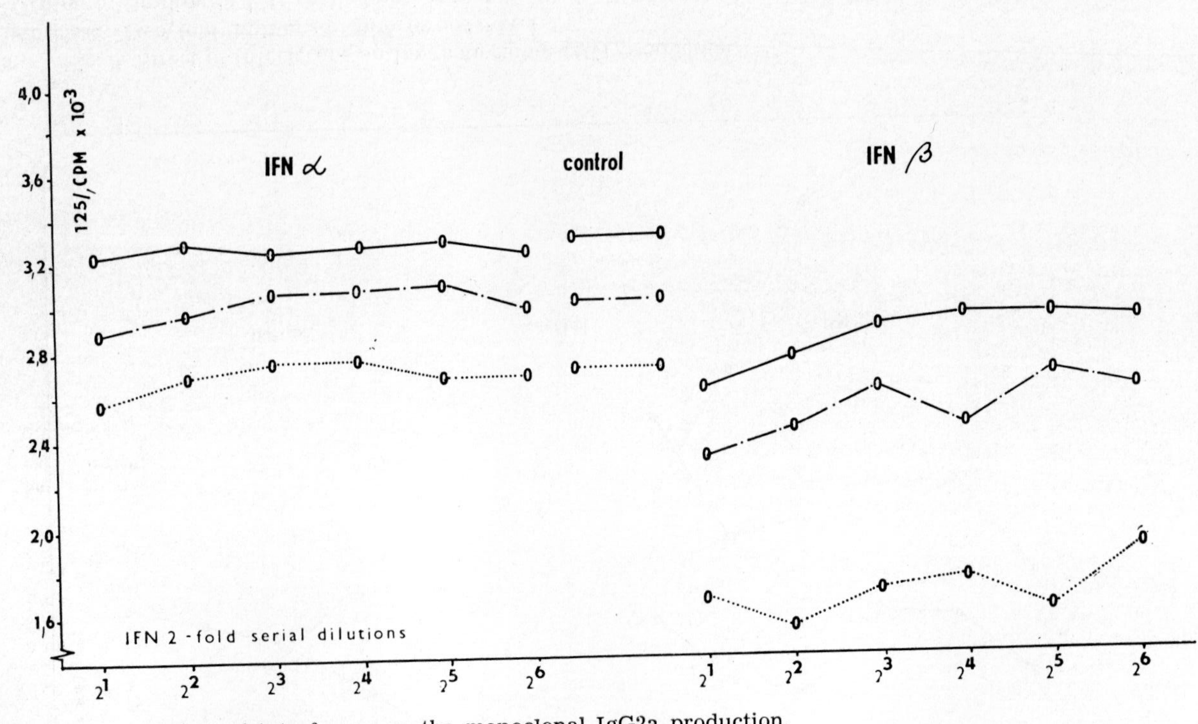

Fig. 7. Effect of interferons on the monoclonal IgG2a production.
For explanation see Fig. 6.

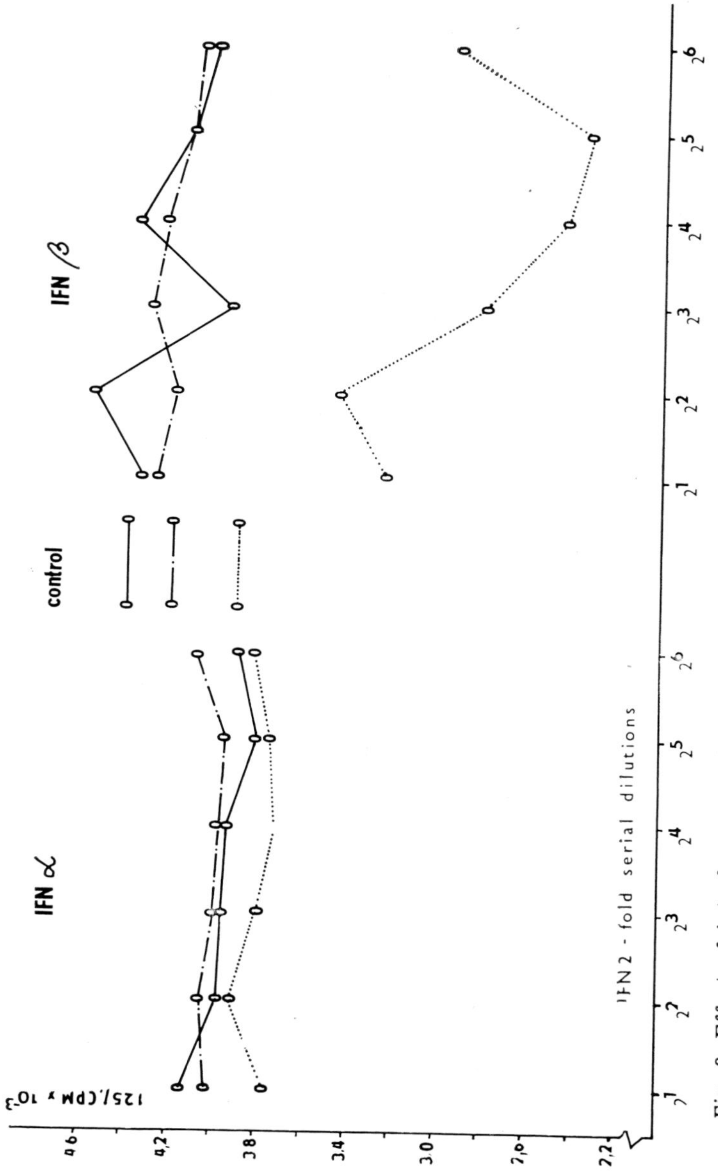

Fig. 8. Effect of interferons on the monoclonal IgG3 production. For explanation see Fig. 6.

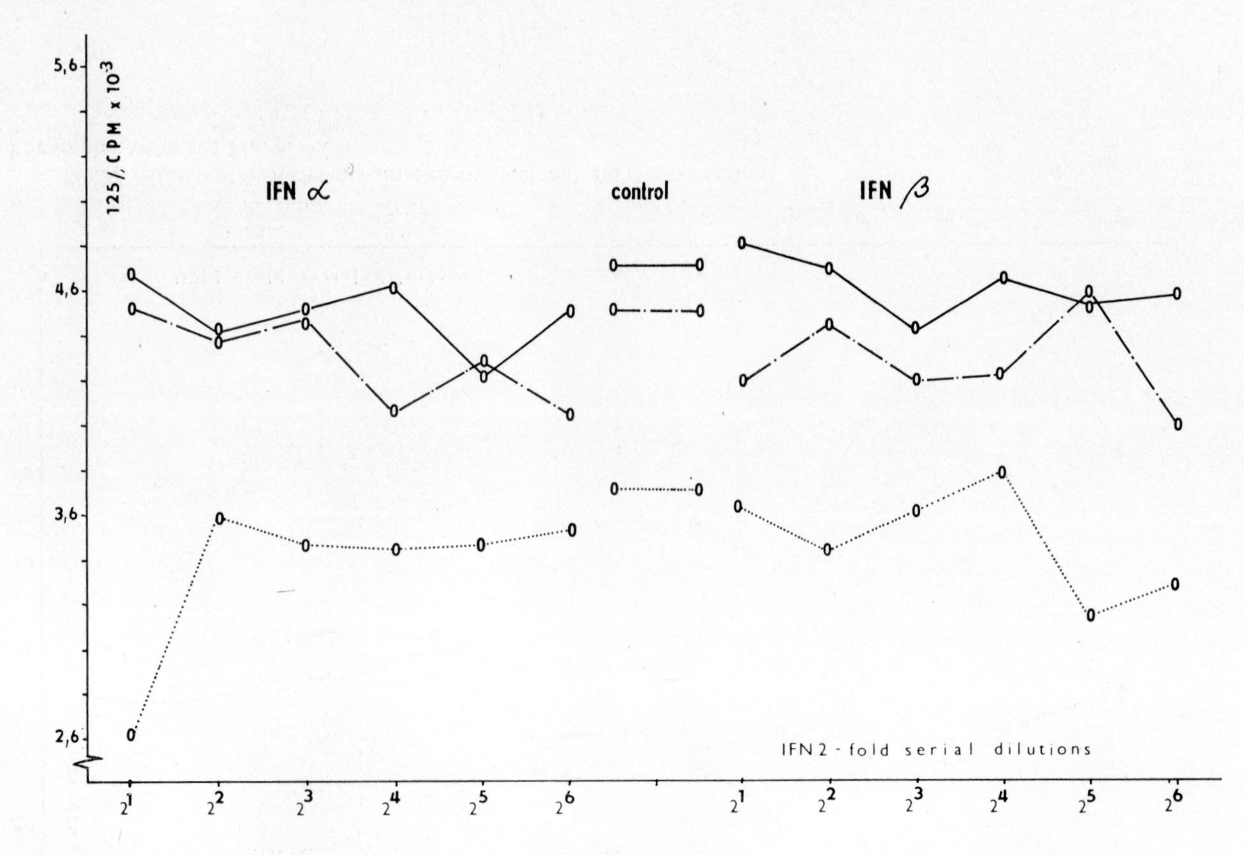

Fig. 9. Effect of interferons on the monoclonal IgM production.
For explanation see Fig. 6.

treated with IFN-β showed a 40 per cent depression in incorporation of ^{125}I. Again, the inhibitory effect of IFNs tested was higher in homologous (32.1 per cent) than in heterologous cell-IFN systems (11.7 per cent).

Since the loss of chromosomes in hybridoma cells was described by several authors (14, 8), in addition to the effect of IFNs on the antibody production, also the chromosomal stability of IFN treated hybridomas was studied. No deleterious effect of IFNs on the number or morphology of chromosomes could be observed in these tests (tab. II).

Discussion

The chromosome analysis and the continuous registration of antibody producing activity of hybridomas producing monoclonal antibodies during a 72hrs period suggests that IFN exerts its inhibitory effect on antibody synthesis without damaging the chromosomes carrying gens involved in antibody production. Of interest

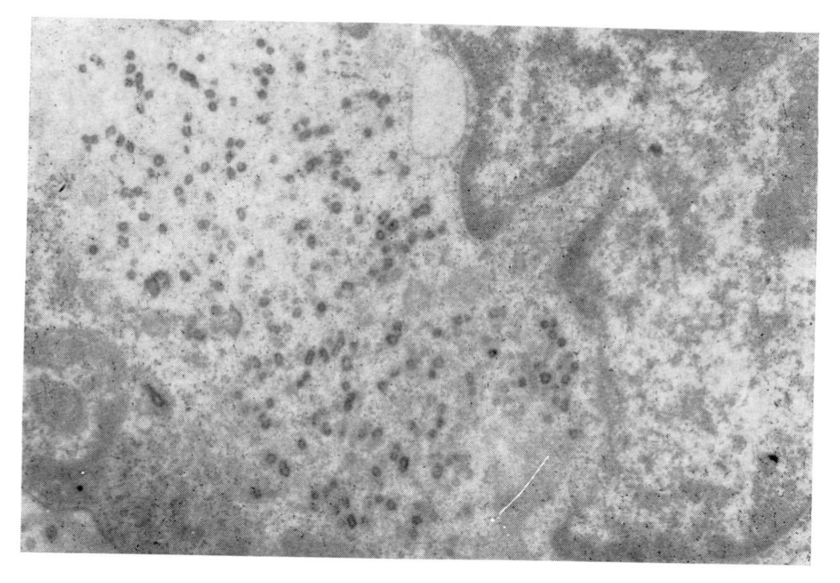

Fig. 10. Electron micrograph of NS-I myeloma cell.
Retrovirus A-type particles are synthesized in the cytoplasma of the cell. UA and Pb-citrate staining. Magn. 20 000×.

is also the observation that the antiproliferative effect of IFN preceded by about 24 hrs the clear inhibitory effect on the antibody secretion. This observation supports the view that the synthetic activity of the cell continues for some time after the proliferating capacity of the cell was stopped (11). Also, it is known that the "producer-component" of the hybridomas i.e. the lymphocyte B differentiates during antibody production to the plasma-cell which continues to secrete antibodies about 2 hrs without proliferation (15). However it seems useful to look for explanation of the antibody inhibiting effect of IFNs also in the "non-producer component" i.e. myeloma cell which usually contains high numbers of virus particles with unknown activity in this respect (fig. 10).

The species-specific effect of IFNs, as registered in this study suggests the possibility that also histocompatibility antigens may participate in effects of IFNs.

Finally, we realize that partially purified IFNs were used in this study and the effect of "impurities" could not be excluded (12). However, the recent studies indicate that the ratio between the antiviral and the antiproliferative effect is independent of purity (1, 5).

Summary

We have studied the effect of 3 types of IFNs (α, β and γ) on proliferation, and, of 2 types of IFNs (α and β) on antibody producing capacity of hybridoma cells secreting monoclonal antibodies of 2 classes and 4 subclasses (IgG1, IgG2a, IgG2b, IgG3 and IgM). The inhibitory effect of IFNs on the proliferation of hybridomas ranged between 5.6 per cent and 86.4 per cent. The inhibitory effect of IFNs on antibody production by hybridomas was IFN-type dependent and reached its maximum after 72 hrs (40 per cent). It was not accompanied by chromosome damage.

References

1 Adams, A.; Strander, H.; Cantell, K. (1975): J. gen. Virol. 28, 207.
2 Borecký, L.; Hajnická, V.; Fuchsberger, N.; Kontsek, P.; Lackovič, V.; Russ, G.; Čapková, J. (1980): Ann. N. Y. Acad. Sci. 350, 188.

3 Bradley, E. C.; Ruscecetti, F. W. (1981): Cancer Res. 41, 244.
4 Cantell, K. (1970): In International Symp. on Interferon and Interferon Induces. Symp. Ser. Immunobiol. Standards 14, 6. Karger, Basel, Switzerland.
5 Cassingena, R.; Chany, C.; Vignal, M.; Suarez, H.; Estrada, S.; Lazar, P. (1971): Proc. Nat. Acad. Sci. USA 68, 580.
6 Einhorn, S.; Strander, H. (1977): J. Gen. Virol. 35, 573.
7 Epstein, L. B.; Shen, J. T.; Abele, J. S.; Reese, C. C. (1980): Ann. N. Y. Acad. Sci. 350, 228.
8 Kennett, R. G.; Mc Kearn, T. J.; Bechtold, K. B. (1980): Monoclonal Antibodies, Plenum Press, New York and London.
9 Lipková, M.; Borecký, L.; Novák, M.; Sekeyová, Z.; Fuchsberger, N. (1983): in this volume.
10 Melchers, F.; Potter, M.; Varner, N. L. (1978): Curr. Top. Microbiol. Immunol. 81, 1.
11 Novák, M.; Grešíková, M.; Sekeyová, M.; Russ, G.; Zikán, J.; Pospíšil, M.; Čiampor, F. (1983): Acta virol. 27, 34.
12 Paucker, K.; Cantell, K.; Henle, W. (1962): Virology 17, 324.
13 Poláková, K.; Russ, G. (1983): Arch. Virol. (in press).
14 Rothfels, K. H.; Siminovitsch, L. (1958): Stain. Technology 33, 73.
15 Šterzl, J. (1976): Progr. Immunol. 113.
16 Strander, H.; Einhorn, S. (1977): Int. J. Cancer 19, 468.

M. Novák, M.D., Department of Immunology, Institute of Virology, Slovak Academy of Sciences, 01703 Bratislava 9 (Czechoslovakia)

Contr. Oncol., vol. 20, pp. 160—168 (Karger, Basel 1984)

A ß-Interferon-Inducing Lymphokine?

J. van Damme; A. Billiau; M. De Ley & P. De Somer

Department of Human Biology, The Rega Institute, University of Leuven, Belgium

One of the lymphokines produced by antigen or mitogen-stimulated lymphocytes is an interferon (for review see ref. 5, 2), which has been called HuIFN-γ (2). Several of its physicochemical and biological properties allow it to be distinguished from other molecular types of interferon. However, its classification as γ, as opposed to the α- and β-types of interferon, is essentially based on its reactivity in neutralization tests with corresponding specific antisera (3, 6, 11). Cells induced to synthesize interferon often produce a mixture of different types. For instance, blood leukocytes or cultured lymphoblastoid cells stimulated with virus produce predominantly HuIFN-α but in addition also produce HuIFN-β (7). Thus, we were not surprised to find that peripheral blood leukocytes induced with Concanavalin A (ConA), released, in addition to HuIFN-γ, an antiviral factor of molecular weight 22,000 (22K factor), serologically characterizable as HuLFN-β-like (4, 14). In the present study we found that 22K meets all the available criteria for qualification as an interferon, and resembles HuIFN-β by its molecular weight and resistance to acid and SDS, but differs by its chromatographical behaviour on ConA-Sepharose and Zn-chelate columns. Unexpectedly, 22K failed to bind to immobilized antibody against HuIFN-β although its biological activity was completely neutralized when such antibody was included in the bioassay medium. Moreover, a specific antiserum raised against 22K did not neutralize classical HuIFN-β. These two observations led us to formulate the hypothesis that 22K is not an interferon but an interferon-inducer.

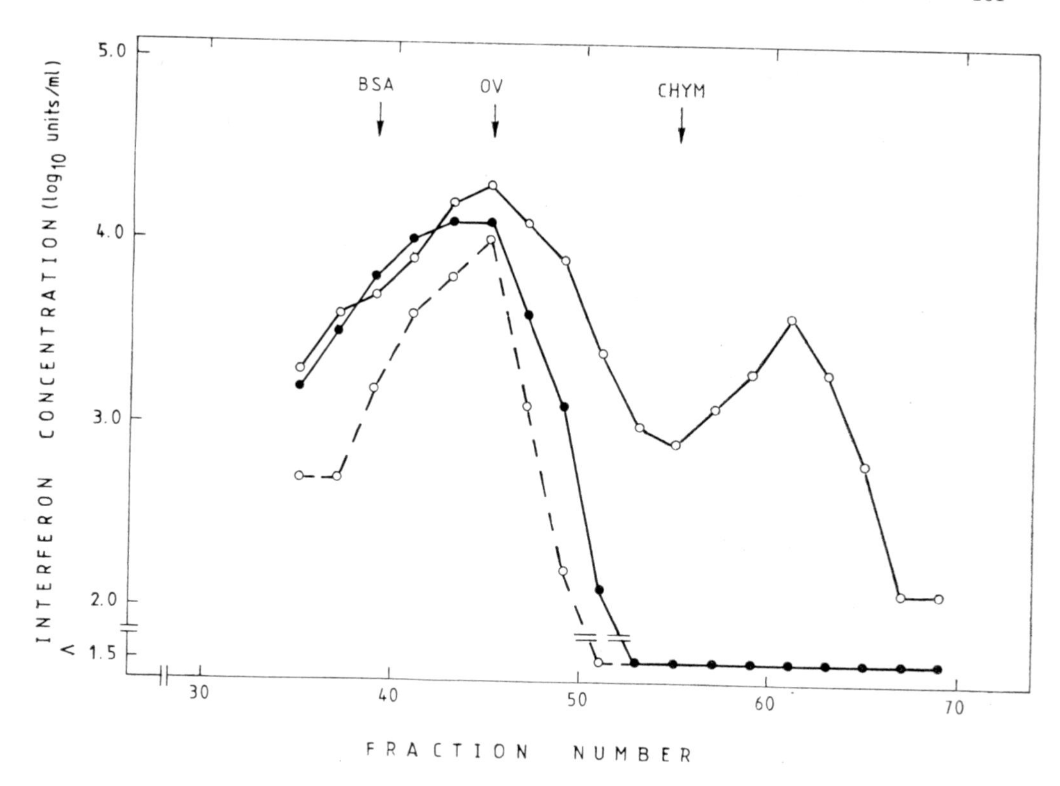

Fig. 1. Crude interferon was prepared by stimulation of blood leukocytes with ConA.

Leukocytes were obtained by ammonium chloride treatment of pooled buffy coats from blood donations. Cells were suspended (5×10^6 ml) in Eagle's minimum essential medium with spinner salts containing 1 % fetal calf serum (FCS). The cell suspension was stirred in spinner culture vessels and induced with ConA (10 μg/ml). After 48 h crude interferon was harvested, clarified (2000 rpm, 10 min) and further processed for partial purification and concentration by adsorption to silicic acid. Interferon, eluted from silicic acid by a 50 % ethylene glycol solution in 1.4 M NaCl, was further concentrated by dialysis against 15 % polyethylene glycol (type 20,000). The concentrate was fractionated by gel filtration on a Sephacryl S-200 column (2.6×100 cm, 33 ml/h). The antiviral activity was eluted in 18 % (v/v) ethylene glycol, 1.55 M NaCl, 8 mM phosphate, pH 7.2 (5 ml fractions). Molecular weight markers: bovine serum albumin (BSA), ovalbumin (OV), chymotrypsinogen (CHYM). Fractions were assayed on HEp-2 cells (●—●) and on diploid fibroblasts in the presence (○——○) or absence (○—○) of antibody against HuIFN-β.

Suspension cultures of peripheral blood leukocytes obtained from pooled buffy coats were stimulated with ConA (for specifics see ref. 4 and 14, and legend to fig. 1). The crude interferon obtained was concentrated and partially purified by adsorption to silicic acid and desorption in 50% ethylene glycol. It was then fractionated by gel filtration. As shown in fig. 1, the antiviral activity (as titrated on diploid human cells) eluted in 2 peaks with apparent molecular weights of ~ 45,000 daltons (45K component) and ~ 22,000 daltons (22K component).

The 45K component had the essential characteristics of HuIFN-γ: no serological relationship to HuIFN-α and -β, neutralization by a homologous specific antiserum, sensitivity to acid (pH 2) and no activity on heterologous cells (14). The 22K component, on the other hand, was neutralizable by a specific anti-HuIFN-β antiserum as shown in fig. 1 by the absence of the second peak when the column fractions were titrated in the presence of such antibody. Pooled fractions of the 22K peak were used for all further characterization studies.

As documented by tab. I, the biological activity of the 22K component was associated with material of proteinaceous nature. The molecular weight (22,000) also resembled that of known human interferon molecules. The antiviral effect was directed against viruses of different taxonomic groups (non-virus specificity criterion for interferon) but was restricted by species of origin of the cells (species specificity criterion for interferon). Establishment of the antiviral state induced by 22K required intact cellular mRNA synthesis, another critical condition for a protein to be considered as an interferon. Finally, in a single cycle RNA virus infection system, 22K treatment inhibited viral RNA synthesis, concordant with the mechanism of action of known interferons on RNA viruses.

Whereas these characteristics allowed us to identify 22K, with high probability, as an interferon, some other characteristics allowed to differentiate it from known human interferons. Specifically, the biological activity of 22K was not neutralizable by antibodies against HuIFN-α or HuIFN-γ (see tab. II). Neutralization of the antiviral activity by a specific anti-HuIFN-β serum strongly suggested identity with, or relatedness to HuIFN-β. Although the physicochemical characteristics summarized in table 1 are compatible with such relatedness to HuIFN-β, they eliminate complete identity:

Table I. Physico-chemical characteristics of the 22K component of ConA-induced interferon

Characteristic	Test system	Results
1 Proteinaceous nature	Digestion by trypsin (400 μg/ml, 1 h at 37° C)	99 % inactivation of biological activity
2 Virus specificity	Parallel titration with several unrelated viruses: VSV, SFV, EMC and mengo-virus	Dose-dependent inhibition of all viruses
3 Species specificity	Titration on cells of non-human origin	Strict species specificity (see *table III*)
4 Dependance of antiviral activity on cellular metabolism	Pretreatment of test cells with actinomycin D prior IFN assay (3 μg/ml, 1 h at 37° C)	99 % inhibition of antiviral activity
5 Inhibition of intracellular step of viral replication	Infection of cells with high m.o.i. (> 10) and labeling with 3H-uridine in the presence of actinomycin D	IFN dose dependent inhibition of virus specific RNA synthesis
6 Stability against acid	Dialysis against 0.3 M glycine-HCl, ph 2.0 (48 h at 4° C)	80 % residual activity
7 Stability against SDS	Incubation with 0.1 % SDS (1 h at 37° C)	~20 % residual activity (similar to HuIFN-β)
8 Affinity for Zn-chelate	Chromatography on Zn-chelate column (13)	100 % non-adsorption as opposed to complete adsorption in case of classical HuIFN-β
9 Affinity for Con-Sepharose	Chromatography on Con-Sepharose column	100 % non-adsorption as opposed to complete adsorption in case of classical HuIFN-β

Table II. Serologic characterization of the 22K component of ConA-induced interferon

Antiserum added[a]		Residual antiviral activity ($\log_{10}$ U/ml) o			
Specificity	$\log_{10}$ dilution	HuIFN-α	HuIFN-β	HuIFN-γ	22K
Control (no antiserum)		3.6	4.8	3.0	3.3
Anti HuIFN-α[b]	—4	2.3	4.6	2.8	3.2
Anti HuIFN-β[c]					
serum no 1	—2	3.7	2.0	2.6	<1.0
	—3	4.0	2.8	2.6	<1.0
	—4	4.1	4.1	3.0	3.2
serum no 2	—3	3.7	<2.0	2.6	<1.0
	—4	3.7	2.7	2.9	1.4
serum no 3	—1	3.4	<2.0	2.3	<1.0
serum no 4	—2	3.7	2.4	2.5	<1.0
serum no 5	—2	—	2.3	2.7	1.3
Anti HuIFN-γ[d]	—2	3.5	4.8	<1.5	3.4
Anti 22K[e]	—2	3.6	4.9	2.9	<1.0

[a] Constant concentration of antibody added to each dilution of interferon assay (2 h/37° C).

[b] Serum provided by Dr. B. Dalton (Wistar Institute, Philadelphia). Origin: sheep immunized with purified HuIFN-α.

[c] Origin of anti-HuIFN-β sera: 1. goat immunized with fibroblast interferon of specific activity (s.a.) = 10^6 units/mg; 2. idem, s.a. = 10^6 units/mg; 3. rabbit immunized with fibroblast interferon, serum obtained from Dr. J. Vilcek, New York University, New York; 4. sheep immunized with fibroblast interferon of s.a. = 10^6 units/mg, serum obtained from Dr. B. Dalton; 5. rabbit immunized with HuIFN-β from E. coli carrying plasmid with inserted HuIFN-β gene.

[d] Serum prepared by M. De Ley from a rabbit immunized with 45K component of human immune interferon (3).

[e] Serum prepared by J. van Damme from a rabbit immunized with 22K component of human immune interferon.

Table III. Antiviral activity of 22K on homologous and heterologous cells incomparison with known IFN-types

Cell type	Antiviral activity[a]				IFN-producing capacity[b]
	IFN-α	IFN-β	IFN-γ	22K-IFN	
Human diploid fibroblasts, high passage level	100	100	100	100	100
Human diploid fibroblasts, low passage level	320	45	25	<1	—
Human trisomic-21 cells LiR	1070	2075	860	<8	64
Human trisomic-21 cells RoL	1065	1040	329	229	100
Human amnion cells (WISH)	63	49	60	5	8
Human epidermal carcinoma cells (HEp-2)	71	26	28	<2	5
Human osteosarcoma cells (MG-63)	29	21	68	61	286
Monkey kidney cells (VERO)	40	40	2.5	0.3	<1
Porcine kidney cells (PK)	400	10	<1.0	<0.5	—
Primary calf kidney cell culture	320	40	<0.5	0.4	—
Primary rabbit kidney cell culture	—	0.3	<0.6	<0.1	—

[a] Figures are averages of 3 separate determinations, expressed as percentages of the IFN titers on human diploid fibroblasts of low passage level; all IFN preparations used have a specific activity of $\geq 10^8$ U/mg.

[b] Cells were tested for interferon production using the superinduction method with Poly rI : rC (50 μg/ml), cycloheximide (10 μg/ml) and actinomycin D (1 μg/ml) as described previously(10).

like HuIFN-β, 22K was found stable in acid or SDS-containing buffers, but unlike HuIFN-β it had no affinity for zinc-chelate or ConA-sepharose columns.

Analysis of the degree of species and cell specificity of 22K revealed another difference with HuIFN-β (tab. III). Whereas classical HuIFN-β is active on cells of various non-human mammalian species, 22K was found to be strictly species specific, thereby resembling HuIFN-γ. Moreover, when tested in different human cell types, 22K showed a remarkable cell specificity. It was found inactive in HEp-2 (see also fig. 1) and WISH cells which are otherwise sensitive to classical HuIFN-β, as well as to HuIFN-α and -γ. It also failed to act in certain disomic and trisomic cells which were otherwise highly sensitive to HuIFN-β. This pattern of responsiveness of human cells to 22K was reminiscent of their inducibility by double-stranded RNA to produce HuIFN-β. Thus, low passage diploid cells (9, 13) and MG-63 cells (1, 13) are known as the best producers of HuIFN-β and are also very sensitive to 22K (tab. III). Also, the 22K-sensitive trisomic RoL-cells were good producers while the 22K-insensitive WISH and HEp-2 cells were low producers (tab. III). The only exception to the parallelism were the trisomic LiR cells which were relatively good producers of HuIFN-β but were poorly sensitive to 22K. The relatively good parallelism between inducibility and sensitivity to 22K led us to formulate the hypothesis that 22K is not an interferon but rather an interferon-inducing lymphokine. None of the characteristics of 22K studied up to that point could contradict this hypothesis. In particular, the neutralization of the biological activity of 22K by anti-HuIFN-β serum might be explained if the interferon induced by 22K were HuIFN-β. To simulate this system we treated diploid cells with dilutions of the double-stranded RNA, polyriboinosinic-ribocytidylic acid, and then challenged them with Vesicular stomatitis virus. Addition of anti-HuIFN-β to the titration neutralized the antiviral activity of the inducer as if it were a HuIFN-β ($< 1\,\%$ residual activity).

The hypothesis of 22K being a HuIFN-β-inducing lymphokine might easily be refuted if 22K would not only be neutralizable by anti-HuIFN-β antibody, but would also bind to the insolubilized antibody. Experiments done to investigate this showed that 22K failed to absorb to a column of anti-HuIFN-β antibody fixed on

Sepharose beads, although this column did adsorb near 100 % of classical HuIFN-β and released this interferon on elution with acid. This suggested that, indeed, 22K might be serologically quite different from classical HuIFN-β. To further support this notion an antiserum against 22K was prepared by immunizing a rabbit with the corresponding peak fractions of gel filtration runs. Tab. II shows that the antibody so-obtained was quite active in neutralizing the biological activity of 22K, but was inactive against classical HuIFN-β.

Attempts to obtain direct evidence for the inducer hypothesis by demonstrating the production of HuIFN-β in diploid cultures after "induction" with 22K were always unsuccessful: supernatants and cell lysates of such cultures were always devoid of detectable amounts of HuIFN-β. Thus, if our hypothesis is correct the amounts of interferon induced by 22K should be small, but still sufficient to make the cells virus-resistent.

In summary, the data obtained in the present study identify 22K as a biologically active protein produced together with HuIFN-γ by mitogen-stimulated leukocyte cultures. The protein has some apparent serological relationships with classical HuIFN-β, but differs from HuIFN-β by several characteristics. Two alternative explanations are available for the findings: 1. 22K is an interferon whose molecular structure resembles that of the known HuIFN-β but is not identical to it. For instance, the primary structure of 22K might be identical to that of HuIFN-β, but glycosylation may be different. 2. 22K is not an interferon but a lymphokine that can induce the production of HuIFN-β by certain lines of fibroblastoid cells, and thereby can mimic HuIFN-β in certain but not all assay systems depending on their inducibility.

Experiments are in progress to distinguish between these two possibilities.

Acknowledgements

This study was supported by the Cancer Research Foundation of the Belgian A.S.L.K. (General Savings and Retirement Fund) and by the "Geconcerteerde Onderzoeksacties". Marc de Ley is Bevoegdverklaard Navorser at the Belgian NFWO (National Fund for Scientific Research). The authors thank their colleagues Dr. C. Vermylen, M. Peetermans, H. Claeys and

L. Muyle from the Blood Transfusion Service of Leuven and Antwerp, for providing buffy coats. The excellent technical assistance of R. Conings, J. P. Lenaerts, W. Put and I. Ronsse, as well as editorial help of C. Callebaut are gratefully acknowledged.

References

1 Billiau, A.; Edy, V. G.; Heremans, H.; van Damme, J.; Desmyter, J.; Georgiades, J.; De Somer, P. (1977): Antimicrob. Agents Chemother. 12, 11—15.
2 Committee on Interferon Nomenclature. (1980): Nature 286, 110.
3 De Ley, M.; van Damme, J.; Biliau, A.; De Somer, P. (1981): J. Virol. Methods 3, 149—153.
4 De Ley, M.; van Damme, J.; Claeys, H.; Weening, H.; Heine, J. W.; Billiau, A.; Vermylen, C.; De Somer, P. (1980): Eur. J. Immunol. 10, 877—883.
5 Epstein, L. B. in Biology of the Lymphokines (eds. Cohen, S., Pick, E. & Oppenheim, J. J.) 443—514 (Academic Press, New York, 1979).
6 Havell, E. A.; Berman, B.; Obgurn, C. A.; Berg, K.; Paucker, K.; Vilcek, J. (1975): Proc. natl. Acad. Sci. USA 72, 2185—2187.
7 Havell, E. A.; Yip, Y. K.; Vilcek, J. (1977): J. gen. Virol. 38, 51—59.
8 Heine, J. W.; van Damme, J.; De Ley, M.; Billiau, A.; De Sommer, P. (1981): J. gen. Virol. 54, 47—56.
9 Horoszewicz, J. S.; Leong, S. S.; Ito, M.; Di Berardino, L.; Carter, W. A. (1978): Infect. Immun. 19, 720—726.
10 Mikulski, A. J.; Heine, J. W.; Le, H. V.; Sulkowski, E.: Preparative Biochemistry 10, 103—119.
11 Osborne, L. C.; Georgiades, J. A.; Johnson, H. M. (1980): Cell. Immunol. 53, 65—70.
12 Stewart, W. E. (1981): The Interferon System (Springer-Verlag, Wien, New York.
13 van Damme, J.; Billiau, A. (1981): Methods in Enzymology 78, 101—119.
14 van Damme, J.; De Ley, M.; Claeys, H.; Billiau, A.; Vermylen, C.; De Somer, P. (1981): Eur. J. Immunol. 11, 937—942.

J. van Damme, M.D., Department of Human Biology, The Rega Institute, University of Leuven, Minderbroedersstraat 10, Leuven (Belgium)

Contr. Oncol., vol. 20, pp. 169—175 (Karger, Basel 1984)

Regulation of 2-5A-Dependent RNase Levels during Interferon-Treatment, Growth Inhibition and Cell Differentiation

R. H. Silverman[a]; *D. Krause*[a]; *H. Jacobsen*[b]; *S. A. Leisy*[a]; *D. P. Barlow*[c]; *R. M. Friedman*[a]

[a] Department of Pathology, Uniformed Services University of the Health Sciences, Bethesda, USA,
[b] Institut für Virusforschung, Deutsches Krebsforschungszentrum, Heidelberg, FRG, and
[c] Imperial Cancer Research Fund Mill Hill Laboratories, London, England.

Introduction

Investigations into biochemical mechanisms of interferon action led Kerr and collaborators to discover the unusual oligonucleotide series, ppp(A2′p)$_n$A, n = 2 to ≧ 4, 2—5A (6). The 2-5A-system, as it is currently recognized, contains three enzymes: 1. 2-5A-synthetase, a double-stranded (ds)RNA-dependent enzyme that polymerizes ATP into 2—5A and whose cellular levels increase after interferon-treatment by between three- to several thousand-fold depending on the type of cell; 2. 2′—5′-phosphodiesterase that degrades 2—5A to ATP and AMP (18) and whose cellular levels after interferon-treatment were reported to either remain constant (18, 17), or increase by between 2- to 4-fold (13, 7), and 3. 2—5A-dependent RNase (3), an endoribonuclease of 77,000—85,000 daltons that, when activated by 2—5A, cleaves RNA on the 3′-side of UpN sequences. Cellular levels of the nuclease after interferon-treatment were reported either to remain constant (11) or to increase by about two-fold (14). Here, however, we report that the regulation of 2—5A-dependent RNase is more complex than previously thought. In various murine cell lines, substantial fluctuations in levels of the nuclease are demonstrated as a function of interferon-treatment, growth rate and cell differentiation.

Results and Discussion

Interferon-Induced Synthesis of 2–5A-Dependent RNase in Murine JLS-V9R Cells.

2–5A-dependent RNase levels were measured throughout this study by a specific radiobinding assay (8) that employs ppp(A2′p)$_3$-A(^{32}P)pCp (14) as a probe. Probe binding activity was reported to copurify with 2–5A-dependent RNase activity (20) and we assume that these are different activities of the same protein.

In the course of surveying various murine cell lines, JLS-V9R cells were found to contain very low levels of the nuclease (5–10 % of that found in control Ehrlich ascites tumor cells). However, following interferon-treatment (2,000 units ml^{-1}) of JLS-V9R cells, there was a time-dependent increase (10- to 20-fold by 72 hrs) in levels of the nuclease. In contrast there was no increase in the nuclease in the control cells. The induction began between 3 and 6 hrs after interferon-treatment and was completely prevented by simultaneously treating the cells with both actinomycin D (2.0 $\mu g\ ml^{-1}$) and interferon (2,000 units ml^{-1}). Therefore, the induction was probably occurring de novo. The level of the nuclease obtained in JLS-V9R cells using interferon at 500 units ml^{-1} was about half that found with 2,000 units ml^{-1} (4). The radiobinding activity was almost completely displaced with 10 nM of 2–5A thus demonstrating the specificity of the binding of probe to the nuclease (4).

Further evidence for the induction by interferon of the nuclease in JLS-V9R cells was obtained by two additional types of assays (4). The first involves forming a covalent linkage between oxidized p_n(A2′p)$_3$A(^{32}P)pC, n = 1 to 3, and the nuclease in cell extracts under reducing conditions (20). The complex is detected as a protein of about 85,000 daltons on autoradiograms of sodium dodecyl sulfate (SDS)/10% (w/v) polyacrylamide gels. An interferon-induced synthesis of the nuclease from JLS-V9R cells was observed by affinity labeling using either postmitochondrial supernatant fractions or whole cells lysates (4). Therefore, the increase in the nuclease could not be due to an intracellular redistribution of the nuclease (for instance from the nuclei to the cytoplasm). The second additional assay for the nuclease, the rRNA cleavage assay (20), was performed to demonstrate an induction by interferon of functional 2—5A-dependent RNase. Addition of 2—5A to cell-free systems from untreated JLS-V9R cells did not result in cleavage of rRNA. However, specific rRNA fragments were produced in response to added 2—5A in cell-free systems from interferon-treated JLS-V9R cells (4). The cleavage products were found after incubation of the cells with 500 units ml^{-1} and were increased with 2,000 units ml^{-1} of interferon (4).

The results indicate an induction of functional 2—5A-dependent RNase by interferon in JLS-V9R cells. The extent of the induction (10- to 20-fold) was several times that reported for human Daudi cells (14). The reason for this may be related to the very low basal level of the nuclease in JLS-V9R cells.

Induction of 2—5A-Dependent RNase in JLS-V9R Cells during Growth Inhibition

Levels of 2—5A-synthetase in various lines of cells were reported by Stark et al. (16) to increase during confluency. In that report, 2—5A-synthetase levels were also found to increase in chick oviducts after withdrawal from stimulation by estrogen. Conditions other than interferon-treatment, therefore, regulate cellular levels of the synthetase. Because interferon controls the synthesis of 2—5A-dependent RNase in JLS-V9R cells, it was of interest to determine if this enzyme is also regulated as a function of the rate of growth. JLS-V9R cells were seeded at subconfluent densities and then grown to a confluent, stationary phase. The cells reached confluency at about 50 hours after subculturing; this was followed by increases in both the 2—5A-binding activity (6- to 8-fold) and 2—5A-synthetase (46-fold). The binding of the probe was specific for 2—5A because the addition of unlabeled 2—5A (250 nM) in the assays abolished the binding activity. The cells were inhibited in their rate of growth after reaching confluency (50 hours), and cease to divide after 100 hours in culture. The major part of the increases in nuclease and synthetase occurred between 50 and 100 hours. In addition, the induction of the enzymes and the inhibition of cell growth coincided with a marked decrease in the incorporation of ^{3}H-thymidine by the cells into

DNA (5). Induction of the nuclease during confluency in JLS-V9R cells was confirmed by both affinity labeling and rRNA cleavage assays. In a separate experiment the cells were seeded at a lower density and the nuclease remained at basal levels in the control cells at 72 hrs; presumably because those cell layers were still subconfluent.

Because the nuclease is induced by interferon in subconfluent JLS-V9R cells, it was possible that the increase in the enzyme in the untreated confluent cells was due to the spontaneous production of interferon. However, addition of sheep anti-murine $(\alpha + \beta)$ interferon globulin (a gift of I. Gresser), sufficient to neutralize > 3,000 units ml^{-1} of interferon, to the culture medium did not prevent the increase in the confluent cells (5). Furthermore, no antiviral activity (using murine L-cells and Encephalomyocarditis virus) was found in the culture supernatants from late-confluent JLS-V9R cells (5). The assay employed will detect < 5 units ml^{-1} whereas addition of > 500 units ml^{-1} of interferon were required to induce the nuclease to the levels found in late-confluent JLS-V9R cells (4). Therefore, spontaneous production of interferon was probably not responsible for the induction of the nuclease during confluency.

2—5A-Binding Activity Correlates with Cell Differentiation and Interferon-Sensitivity

Several lines of murine teratocarcinoma cells were analyzed to determine if 2—5A-dependent RNase varied with cell differentiation. Furthermore, because undifferentiated or embryonal carcinoma (EC) and differentiated teratocarcinoma cells are interferon-sensitive and -resistant, respectively (2, 19, 1), they offer an opportunity to evaluate the significance of the 2—5A and protein kinase (12,9) systems. There was little or no effect of interferon on the growth of the undifferentiated cells, F9, Nulli 2A or PC13. In contrast, the differentiated cell lines, F9-clone 9, PYS and PSA-5E were substantially inhibited (50 to 75 %) in their growth following interferon-treatment. In this regard, the PSA-5E cells were slightly more sensitive to interferon than the F9-clone 9 and PYS cells.

The sensitivities of the teratocarcinoma cells to the antiviral activity of interferon was determined. The undifferentiated cells were resistant to protection by interferon against Encephalomyocarditis virus (EMCV), Vesicular stomatitis virus (VSV) and Sindbis virus (SBV). As expected, the differentiated cells were all relatively sensitive to interferon action as compared to the undifferentiated cells. In two conflicting studies, the EC cell line, PCC4, were reported to be either sensitive (10) or resistant (1) to interferon action against EMCV. Our results with EMCV are consistent with the latter report.

The interferon and dsRNA-stimulated protein kinase (12, 9) was measured in extracts from control and interferon-treated cells to determine if there was a correlation with interferon-sensitivity. Wood and Hovanessian (19) reported that interferon failed to induce the kinase in two EC cell lines but interferon did induce the enzyme in their differentiated derivative cell line. In agreement with their results, there was no detectable dsRNA-dependent protein kinase in the undifferentiated cell lines, F9, Nulli 2A or PC13. However, of the three differentiated cell lines, only the PSA-5E cells developed detectable levels of the protein kinase following interferon-treatment. There was, therefore, no consistent correlation between levels of the protein kinase and interferon-sensitivity.

Levels of the 2—5A-synthetase and 2—5A-binding proteins in extracts of the teratocarcinoma cells were measured to determine whether the 2—5A-system correlated with interferon sensitivity. In agreement with previous reports (19, 1, 10), the 2—5A-synthetase was greatly increased (50- to 200-fold) in both undifferentiated and differentiated cell types following interferon treatment. The EC cell lines are not, therefore, negative for the cell surface receptors for interferons. In addition, Aguet et al. (1) reported that PCC4 EC cells had as many interferon receptors as differentiated cells and that the receptors had a similar affinity for interferon.

The presence of induced levels of 2—5A-synthetase only indicates a capacity of the cells to synthesize 2—5A. To be active, however, the 2—5A must bind to the 2—5A-dependent RNase. The undifferentiated cells had only very low basal levels of the nuclease as estimated by the radiobinding assay, about 27- to 40-fold less than untreated Ehrlich ascites tumor (EAT) cells. In contrast, the differentiated cells contained several-fold higher basal levels of 2—5A-binding proteins compared to the undifferentiated cell lines;

furthermore, 2—5A-binding activity increased following interferon-treatment in each of the differentiated, but not in the undifferentiated cell lines. As a result there was 10- to 25-fold more 2—5A-binding activity in interferon-treated, differentiated cells than in similarly treated undifferentiated cells.

These results were confirmed by estimating levels of the nuclease by the affinity labeling method. There was, therefore, a correlation between the levels of the nuclease and interferon-sensitivity. The differentiated cell lines were relatively sensitive to interferon action and contained much greater levels of the nuclease than did the undifferentiated, interferon-resistant cell lines. The results implicate the 2—5A-system, in particular the 2—5A-dependent RNase, in the mechanism of action of interferon.

We have shown that, in certain systems, the nuclease is regulated during interferon-treatment, cell growth and cell differentiation. Control of the nuclease levels by interferon was prevented by actinomycin D and may, therefore, involve transcriptional regulation. It is not known, however, if the increases in the nuclease as a function of growth rate or cell differentiation involve a similar mechanism to that following interferon-treatment. The possibility of a common intracellular pathway for the regulation of the 2—5A-dependent RNase gene, however, is intriguing.

Acknowledgements

Part of this work was completed by Robert M. Friedman while he was a Royal Society visiting fellow at the Biochemistry Department of the University of Warwick, Coventry, U. K.

References

1 Aguet, M.; Gresser, I.; Hovanessian, A. G.; Bandu, M.-T.; Blanchard, B.; Blangy, D. (1981): Virology 114, 585—588.
2 Burke, D. C.; Graham, C. F.; Lehman, J. M. (1978): Cell 13, 243—248.
3 Clemens, M. J.; Williams, B. R. G. (1978): Cell 13, 565—572.
4 Jacobsen, H.; Czarniecki, C. W.; Krause, D.; Friedman, R. M.; Silverman, R. H. (1983a): Virology, in press.
5 Jacobsen, H.; Krause, D.; Friedman, R. M.; Silverman, R. H. (1983b): Manuscript in preparation.

6 Kerr, I. M.; Brown, R. E. (1978): Proc. natn. Acad. Sci. USA 75, 256—260.
7 Kimchi, A.; Shulman, L.; Schmidt, A.; Chernajovsky, Y.; Fradin, A.; Revel, M. (1979): Proc. natn. Acad. Sci. USA 76, 3208—3212.
8 Knight, M.; Cayley, P. J.; Silverman, R. H.; Wreshchner, D. H.; Gilbert, C. S.; Brown, R. E.; Kerr, I. M. (1980): Nature (Lond.) 288, 189—192.
9 Lebleu, B.; Sen, G. C.; Shalia, S.; Cabrer, B.; Lengyel, P. (1976): Proc. natn. Acad. Sci. USA 73, 3107—3111.
10 Nilsen, T. W.; Wood, D. L.; Baglioni, C. (1980): Nature 286, 178—180.
11 Nilsen, T. W.; Wood, D. L.; Baglioni, C. (1981): J. Biol. Chem. 256, 10751—10754.
12 Roberts, W. K.; Hovanessian, S.; Brown, R. E.; Clemens, M. J.; Kerr, I. M. (1976): Nature (Lond.) 264, 477—480.
13 Schmidt, A.; Chernajovsky, Y.; Shulman, L.; Federman, P.; Berissi, H.; Revel, M. (1979): Proc. natn. Acad. Sci. USA 76, 4788—4792.
14 Silverman, R. H.; Cayley, P. J.; Knight, M.; Gilbert, C. S.; Kerr, I. M. (1982): Eur. J. Biochem. 124, 131—138.
15 Silverman, R. H.; Wreschner, D. H.; Gilbert, C. S.; Kerr, I. M. (1981): Eur. J. Biochem. 115, 79—85.
16 Stark, G. R.; Dower, W. J.; Schimke, R. T.; Brown, R. E.; Kerr, I. M. (1979): Nature (Lond.) 278, 471—473.
17 Verhaegen-Lewalle, M.; Content, J. (1982): Eur. J. Biochem. 126, 639—643.
18 Williams, B. R. G.; Kerr, I. M.; Gilbert, C. S.; White, C. N.; Ball, L. A. (1978): Eur. J. Biochem. 92, 455—462.
19 Wood. J. N.; Hovanessian, A. G. (1979): Nature (Lond.) 282, 74—76.
20 Wreschner, D. H.; James, T. C.; Silverman, R. H.; Kerr. I. M. (1981): Nucleic Acid Res. 9, 1571—1581.
21 Wreschner, D. H.; Silverman, R. H.; James, T. C.; Gilbert, C. S.; Kerr, I. M. (1982): Eur. J. Biochem. 124, 261—268.

R. H. Silverman, M. D., Department of Pathology, Uniformed Services University of the Health Sciences, 4301 Jones Bridge Road, Bethesda, Maryland 208 14 (USA)

Contr. Oncol., vol. 20, pp. 176—184 (Karger, Basel 1984)

Antiviral and Physico-chemical Properties of Macromolecular Complex Isolated from the Virus—Infected Cell Extract Treated with Ribonuclease

V. I. Votyakov[a]; *M. E. Khmara*[a]; *A. A. Akhrem*[b]; *S. G. Moroz*[b]

[a] Byelorussian Research Institute of Epidemiology and Microbiology
[b] Institute of Bioorganic Chemistry of the BSSR Academy of Sciences, Minsk, USSR

A cell can trigger specific mechanisms of defense under the influence of interferon shortly after virus penetration. One of the mechanisms is associated with the protein kinase activation and phosphorylation of the initiation factor e-IF-2. This leads to inhibition of the synthesis of viral proteins. Another mechanism involves activation of 2′—5′A-polymerase. Subsequent 2′—5′A-oligonucleotide synthesis is followed by the local activation of endonuclease and selective degradation of viral mRNA. These two enzymes (2′—5′A-polymerase and protein kinase) are activated by viral double-stranded RNA. The mechanism of discrimination and degradation of viral mRNAs can be envisioned as follows: 2′—5′A-polymerase forms a complex with double-stranded RNA and becomes capable of polymerizing 2′—5′A-oligonucleotide (2′—5′A) from adenosine. Based on high affinity for each other, 2′—5′A forms a complex with endonuclease and activates it. Excess of 2′—5′A is degraded by 2-phosphodiesterase (1, 2, 3, 7, 9, 10, 12, 13, 15, 16, 20). The mechanism of selective degradation of viral mRNAs seems to be based on certain structural and functional control in the activation of different enzymes and other products, or their association into certain structural and functional complexes. Owing to the supposition that such complexes are present in infected cells we made an attempt of their isolation and studying their antiviral and physico-chemical properties (19). Since macromolecules are the basis of such complexes, they will be designated in this communication as antiviral macromolecular complexes (AMC).

Materials and Methods

Isolation of Crude AMC

Primarily trypsinized chick embryo fibroblast cells (CEF) were grown in the medium containing 5 % hemohydrolyzate and 10 % bovine serum (the products of the Byelorussian Research Institute of Epidemiology and Microbiology). After the growth medium had been replaced by serum-free maintenance medium 199, the monolayer was infected with Venezuelan equine encephalomyelitis (VEE) virus at 0.1 $TCID_{50}$ per cell. After the cell destruction (virus titer 7.5—8.5 log $TCID_{50}$/0.1 ml), the entire contents of the flask were frozen six times to − 20° C and thawed at 37° C. Then the supernatant liquid was separated by centrifugation at 3,000 g for 10 min and treated with 0.15—0.17 % pancreatic ribonuclease (RNA-ase). The percentage of RNA-ase was calculated for the amount of lysate. The treatment was made at 37° C, then at 56° C, and finally 95° C. Duration of the treatment with RNA-ase at these temperatures differed in each experiment, but the following procedure was used. The treatment was started at 37° C and during 20—25 min the temperature reached 56° C, then eventually 95° C. This temperature was maintained for 30—35 min. The supernatant, together with the precipitate, was kept at 4° C for 24 h (pH 6.8—7.0). Flaky precipitates formed during this time were removed by centrifugation at 3,000 g during 10 min. The supernatant (AMC) was used as the final crude product. Control products were also obtained by this technique from uninfected cells treated with RNA-ase and from infected ones but without treatment with RNA-ase. Crude AMC, control products and RNA-ase were used at a dose of 300—350 μg of protein per 1 ml of maintenance culture medium when studying their antiviral action. The dose ten times larger was still innocuous for cells (RNA-ase excluded). Freshly prepared crude AMC was used as a rule; but lyophilized and subsequently dissolved preparations were also tested. When studying "pure" AMC, the same dose was used.

Purification of AMC

With the purpose of its concentration, crude AMC was preliminarily lyophilized and then dissolved in a less volume in 10 mM tris-HCl buffer, pH 7.0. The material was applied to a column of Sephadex G-50 (1 cm × 60 cm), the elution rate was 30 ml/h, fractionation was made at 20° C in the same buffer and at the same pH value. A high-molecular-weight fraction was used for further studying its properties since it revealed antiviral action.

A Study of the Composition of the High-Molecular-Weight AMC

The material preliminarily concentrated as described above (the protein concentration was 6 mg/ml) was placed on a column of Sephadex G-100 (1.7 cm × 70 cm) both in 10 mM tris-HCL buffer (pH 7.0) and in 1 M NaCl solution.

Gel electrophoresis of the high-molecular-weight fraction of AMC was carried out by the method of Summers et al. (1965). Scanograms of bands (densitometer ERI-65) are shown in fig. 1c. Standard proteins used for determining mol. wts. were: catalase (232,000); aldolase (160,000); hemoglobin (68,000); peroxidase (40,000); trypsin (23,000); and cytochrome C (130,000).

A Study of Other Properties of AMC

RNA-ase activity of AMC was revealed by the method of Davidson et al. (1957). The activity was expressed in μg of hydrolyzed RNA (during 20 min at 37° C) per mg of protein. The protein concentration was determined by the method of Lowry. Sensitivity of AMC to trypsin was determined using 0.25% trypsin at 37° C during 30 min. The enzyme and AMC were mixed in the ratio of 1 : 100.

The rate of RNA synthesis in infected cells was determined by ^{3}H-uridine incorporation (8).

Virus and Cells

A stock of VEE virus (attenuated strain 230), titering 7.5—8.5 lg $TCID_{50}$/0.1 ml was used in these studies. 48 hour cultures of primarily trypsinized CEF were prepared according to standard techniques.

Reagents

The source of our reagents was as follows: Pancreatic RNA-ase (Reanal); RNA (Sigma); serum albumin (Reanal); 2-mercaptoethanol; sodium dodecylsulfate (SDS) and Coomassie G-250 (Serva); trypsin (Worthington); Sephadexes G-50 and G-100 (Pharmacia); ^{3}H-uridine (Isotope). Reanal apparatus and a standard set of reagents were used for gel electrophoresis.

Results and Discussion

A Study of Structural and Functional Properties of AMC

The supernatant obtained after the treatment with pancreatic RNA-ase of VEE virus infected cells (crude AMC) was preliminarily purified on a column of Sephadex G-50. Subsequently this high-molecular-weight fraction ("pure" AMC) was used as it possessed antiviral properties (see below). "Pure" AMC was eluted as one mol. wt. 97,000 peak on a column of Sephadex G-100 in 10 mM tris-HCl buffer (pH 7.5) (fig. 1a). Gel filtration on the column of Sephadex G-100 in the presence of 1 M NaCl led to dissociation of AMC into three polypeptides of mol. wts. 68,000, 15,800, and 13,800 respectively (fig. 1b). These data suggest to us that AMC consists of at least 3 subunits whose bonds are broken at high ionic strength (1 M NaCl). The high-molecular-weight fraction of the supernatant obtained upon RNA-ase treatment of uninfected cells contained two polypeptide chains — 68,000 and 13,800, which at low ionic strength (10 mM tris-HCl buffer) did not form a stable complex. The peak of mol. wt. 15,800 was not detected in the preparation from uninfected cells. Since the protein of mol. wt. 68,000 was present in the preparation both from infected and uninfected cells which had not been treated with RNA-ase it might be supposed that peak 13,800 contained RNA-ase used for obtaining the preparations.

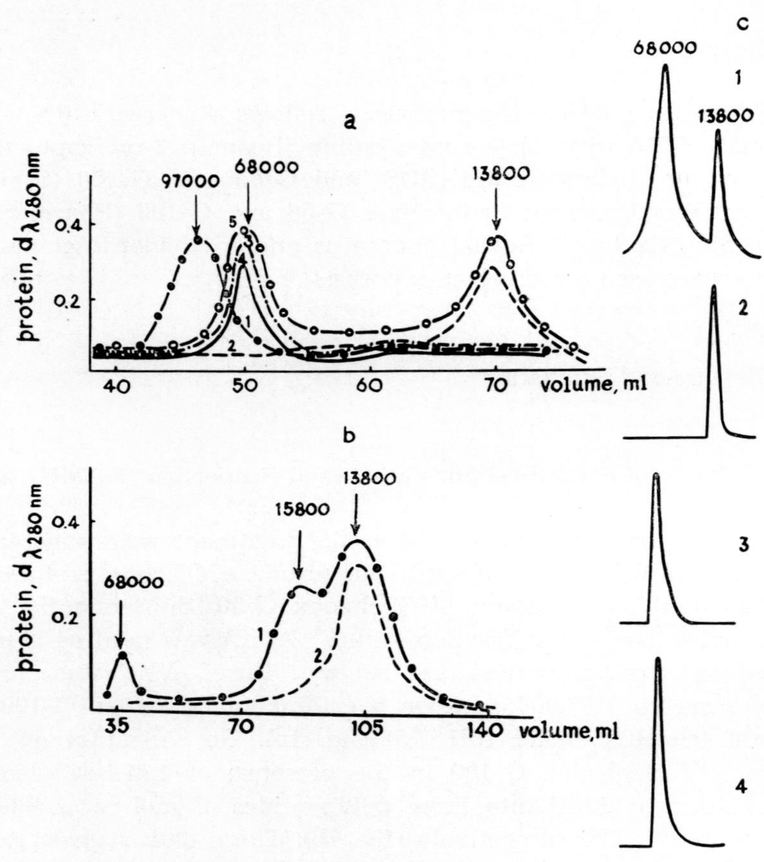

Fig. 1. Gel filtration and gel electrophoresis of AMC.
a — fractionating on a column of Sephadex G-100, tris HCl buffer: 1-AMC; 2-RNA-ase; 3, 4-supernatants of normal and infected cells exposed to heating without RNA-ase; 5-control preparation (RNA-ase + uninfected cells) exposed to heating; b — fractionating on a column of Sephadex G-100, 1 M NaCl; 1-AMC; 2-RNA-ase; c — gel electrophoresis in PAAG: designations are the same as in a/1.

In gel electrophoresis under denaturing conditions (in the presence of SDS and mercaptoethanol) AMC was only found in the form of two bands – 68,000 and 13,800 (fig. 1c), but the protein peak of mol. wt. 15,800 was not observed. In this connection, it is possible to suppose that this peak consists of two subunits – 13,800 and some unidentified material of mol. wt. approximately 2,000 which is dissociated with SDS treatment. AMC appears to consist of protein 68,000, two molecules of molecular weight typical for the RNA-ase (13,800) and a low-molecular-weight subunit (2,000) of an oligonucleotide or an oligopeptide nature.

Antiviral and RNA-ase Properties of AMC

From tab. I is seen that both crude AMC and its high-molecular-weight fraction ("pure" AMC) exert an antiviral action both in protection of the cellular monolayer, and in inhibiting the virus reproduction. In the presence of actinomycin D inhibition of virus

Table I. Antiviral effect and RNA-ase activity of AMC

Groups	Titer of VEE virus ($lgTCID_{50}$/0.1 ml)	RNA-ase activity (%)
AMC (crude)	1.50 ± 0.50[1]	99
AMC (crude) + actinomycin D	2.00 ± 0.25	—
AMC (high-molecular-weight fraction)	1.00 ± 0.35[1]	92
AMC (crude) + trypsin	8.00 ± 0.43	37
AMC after gel-filtration in 1 M NaCl and dialysis:		
68,000	7.50 ± 0.30	3
15, 800	8.00 ± 0.43	42.2
13, 900	8.00 ± 0.35	34
RNA-ase	8.50 ± 0.35	100
RNA-ase + uninfected cells	8.50 ± 0.35	—
Infected cells without RNA-ase	8.50 ± 0.35	—
RNA-ase + trypsin without virus	—	75
Medium 199 (virus control)	8.50 ± 0.35	—

[1] Inhibition of virus reproduction is accompained by monolayer protection.

of actinomycin D. Therefore, the antiviral action of AMC is characterized by: inhibition of virus-induced RNA syntheses; preservation of total RNA syntheses of infected cells; decrease of virus reproduction in the presence or absence of actinomycin D, and prevention of the virus cytolytic action.

Thus, these results show that the heat-stable macromolecular complex with expressed antiviral activity was isolated from CEF cells infected with VEE virus. Three fractions were found in its composition: 68,000, 15,800 and 13,800. Fraction 15,800 appears to consist of a protein molecule 13,800 and some other unidentified material of mol. wt. approximately 2,000. It is noteworthy that subunits 68,000 and 13,800 are also present in normal cells treated with RNA-ase, but do not form the complex between each other. Therefore, one could suppose that formation of the complex between AMC components is accomplished by virus-induced material. For example, polyamines or some proteins (histones) which are known to form de novo in virus-infected cells may play a role, since they have molecular weights from 2,000 to 12,000, are sensitive to trypsin, and are able to interact well with RNA-ase (5, 11, 14). High sensitivity of AMC to trypsin; isolation of proteins of mol. wts. of 13,800 and 15,800 in 1 M NaCl gel filtration and, more significantly, increasing of one of the peaks by 2,000 daltons favour this hypothesis (17). However, the role of nucleotides can not be excluded as they are also able to form complexes with the enzyme (4).

The absence of antiviral action in some AMC fractions (68,000, 15,800 and 13,800) after gel filtration in 1 M NaCl and subsequent dialysis (RNA-ase properties being present) as well as the loss of antiviral activity of AMC treated with trypsin allow us to suggest that antiviral effect cannot be achieved by RNA-ase. It is quite likely that under these conditions the loss or destruction of the active part of the complex occurs.

It could be that the hydrolytic properties of RNA-ase ensures release of the antiviral factor (factors) from virus infected cells. Based on the available data (1, 3) one could suggest even more complicated mechanisms of RNA-ase action. On the one hand, RNA-ase "cleans" virus-specific ds-RNA replicative complexes from single-stranded RNAs thus promoting adsorption and activation of 2′—5′A-polymerase and subsequent synthesis of 2′—5′A-oligonucleotides. On the other hand, the affinity of RNA-ase bind-

reproduction is also observed. The control preparations including RNA-ase did not block the virus cytolytic action and its reproduction. No antiviral action was found in any control preparations and AMC subunits obtained after gel filtration in 1 M NaCl and subsequent dialysis. Incubation of AMC together with trypsin resulted in complete loss of its antiviral properties which were not reversible in spite of RNA-ase addition after trypsin heat denaturation. Crude AMC possessed ribonuclease activity similar to that of pancreatic RNA-ase used for its preparation. RNA-ase activity of AMC treated with trypsin decreased by 63 %. At the same time the activity of RNA-ase (by itself) heated to 95° C only decreased by 25 %. Subunits obtained after gel filtration of AMC in the presence of 1 M NaCl also possessed RNA-ase activity. RNA-ase activity of the 68,000 mol. wt. peak contained 3 % of the initial activity of RNA-ase used for preparing crude AMC. The two other peaks of mol. wts. 15,800 and 13,800 contained 42.2 % and 34.0 % of the RNA-ase activity, respectively. RNA-ase activity of AMC appears to be largely due to pancreatic RNA-ase contained in the complex since cellular protein possessed low nuclease activity.

Table II. Virus-induced and cell RNA synthesized in the presence of AMC or pancreatic RNA-ase

Groups	^{3}H-uridine incorporation imp/min		
	medium 199	RNA-ase	AMC
1. Infected cells	4,100 ± 900	2,800 ± 600	24,000 ± 3,600
2. Uninfected cells	20,000 ± 1,350	21,080 ± 1,200	28,000 ± 2,320
3. Infected cells + + actinomycin D	6,000 ± 970	250 ± 70	1,000 ± 200

Tab. II presents the data on the influence of AMC and RNA-ase used for its production on total cellular and virus-induced syntheses of VEE virus RNA. As can be seen, AMC prevents the decrease of total cellular synthesis and inhibits virus-induced RNA synthesis. RNA-ase does not prevent inhibition of total cellular syntheses, though it inhibits virus-induced synthesis. As mentioned above, AMC in addition to inhibiting virus-induced syntheses also inhibited the virus reproduction by 6.5 $\log_{10}$ $TCID_{50}$ in the presence ing to 2′–5′A-oligonucleotides promotes their retention in AMC.

It is also noteworthy that complex-forming factors appear to condition the structure organization necessary for realization of the antiviral action. We are continuing to study the antiviral properties of AMC.

References

1 Baglioni, C. (1979): Cell 17, 255.
2 Baglioni, C.; Minks, M. A.; Marvney, P. A. (1978): Nature 273, 684.
3 Baglioni, C.; Nilsen, T. W. (1981): Amer. Sci. 69, 392.
4 Benz, F.; Robert, G. (1973): Phys. Chem. Prop. Nucl. Acids. 3, 77.
5 Bush, G. (1967): in "Histone and other nuclear proteins", Moscow.
6 Davidson, J.; Watson, D.; Roseman, S. (1957): Nature 179, 965.
7 Ehrenfeld, E.; Hunt, T. (1971): Proc. natn. Acad. Sci. USA 68, 1075.
8 Ershov, F. I.; Berezina, O. N.; Sokolova, T. M.; Tazulakhova, E. B. (1973): Vop. virusol. SSSR 6, 743.
9 Farell, J.; Balkov, K.; Hunt, T.; Jachson, R. (1977): Cell 11, 187.
10 Hovonessian, A. G.; Brown, R. E.; Kerr, I. M. (1977): Nature 268, 537.
11 Johnson, M.; Morkham, R. (1962): Virology 17, 276.
12 Kerr, I. M.; Brown, R. E (1978): Proc. natn. Acad. Sci. USA 75, 256.
13 Lebleu, B.; Sen, G. C.; Shaila, S.; Cobrer, B.; Lengyel, P. (1976): Proc. natn. Acad. Sci. USA 73, 3107.
14 Levy, C.; Hieter, P.; Le Gendre, S. (1974): J. Biol. Chem. 21, 6762.
15 Nilsen, T. W.; Maroney, P. A.; Baglioni, C. (1983): J. Biol. Chem. (in press).
16 Roberts, W. K.; Hovanessian, A.; Brown, R. E.; Clemens, M. J.; Kerr, I. M. (1976): Nature 264, 477.
17 Samsonov, G. V. (1979): Vysokomolek. soedineniya SSSR 4, 723.
18 Summers, D.; Maizel, J.; Darnell, J. (1965): Proc. natn. Acad. Sci. USA 54, 505.
19 Votyakov, V. I.; Khmara, M. E.; Tkach, V. M. (1980): Author's certificate No. 909814, SSSR.
20 Williams, B. R. G.; Golgher, R. R.; Brown, R. E.; Gilbert, C. S.; Kerr, I. M. (1979): Nature 282, 582.

V. I. Votyakov, Institute of Microbiology and Epidemiology, Nogina, 3, 220050 Minsk (USSR)

Contr. Oncol., vol. 20, pp. 185—195 (Karger, Basel 1984)

Biochemical and Biological Characterization of Tumor-Associated Lymphokine-Like Factors

R. S. Fritsch; B. Fahlbusch; I. Schumann; D. Zippel; P. Zöpel

Central Institute of Microbiology and Experimental Therapy, Jena, Research Centre of Molecular Biology and Medicine, Academy of Sciences of the GDR, GDR

Introduction

Previously, Zschiesche et al. (19) described soluble factors in cell-free ascites from rat Zajdela hepatoma which were capable of altering functions of macrophages (MØ) in vitro. Similar lymphokine-like factors have been found in supernatants of tumor cell cultures (1), in blood and body fluids, e.g. ascites of tumor-bearing animals (10) as well as in body fluids of tumor patients (3). That such factors may play a role also in vivo has been shown by Pike and Snyderman (11).

The activities which have been characterized in Zajdela hepatoma ascites fluid comprise

- a MØ migration inhibitory factor (MIF);
- a microbial growth inhibitory factor (MGIF) mediating bactericidal activity of MØ against Corynebacterium murium kutscheri;
- a factor which enhances the nonspecific cytotoxicity of MØ against certain tumor target cells (MCF).

Since tumor ascites fluid is available in relatively high amounts, it provides a more favourable source for the isolation of those factors than the minute amounts of fluid from lymphocyte supernatants.

The experiments were performed using chromatographically purified preparations showing activities of MGIF or MCF, both in mouse assay systems. The aim of the experiments was to characterize some aspects of the mode of action of MØ stimulated by either factor. In the case of MGIF we applied methods of quantitative stereology (morphometry) on transmission electron micrographic (TEM) pictures, in experiments with MCF we used conventional TEM and scanning electron microscopy (SEM).

Purification Procedures and Chemical Characterization of Lymphokine-Like Activities

Crude cell-free ascites fluid may contain not only lymphokine-like factors but also endotoxins, prostaglandins, cyclic nucleotides, enzymes, and other biologically active substances which are able to modify MØ functions. We applied, therefore, conventional purification methods in order to isolate and characterize the 3 factors acting on MØ functions.

The cell-free ascites fluid was purified in three consecutive steps including ion exchange chromatography, gel filtration and isoelectric focusing (fig. 1). Since ion exchange chromatography allows the separation of relatively high amounts of proteins this procedure was applied first. Elution was acomplished with a linear gradient of NaCl. Significant MIF activity was found in the fractions eluting at 0,04–0,10 M (fraction I) and 0,17–0,25 M NaCl (fraction II), respectively (6). MGIF activity eluted in fraction I up to 0,05 M NaCl and in fraction II from 0,11 to 0,18 M NaCl (5). Fractions I and II were purified further by gel filtration on Sephadex G-100 to determine the molecular weights (MW) and to separate lymphokine-like activities from contaminating proteins. By means of this procedure, MIF activity could be resolved into 3 subfractions with MW of about 10,000–15,000, 20,000 and 45,000, respectively, whereas MGIF activity was found in one fraction with MW of 19,000–25,000. In this fraction MCF activity was present, too. For further characterization of lymphokine-like factors the isoelectric points (pI) of biological activities were determined. As demonstrated in figure 1, MIF was found in 3 fractions while MGIF activity was present only in 2 fractions.

As shown by physicochemical and biochemical data, MIF and MGIF derived from rat Zajdela hepatoma ascites appear heterogeneous with respect to net charge and MW. Similar heterogeneity has been described also for "classic" lymphokines from supernatants of stimulated lymphocytes (18). According to Salvin et al.

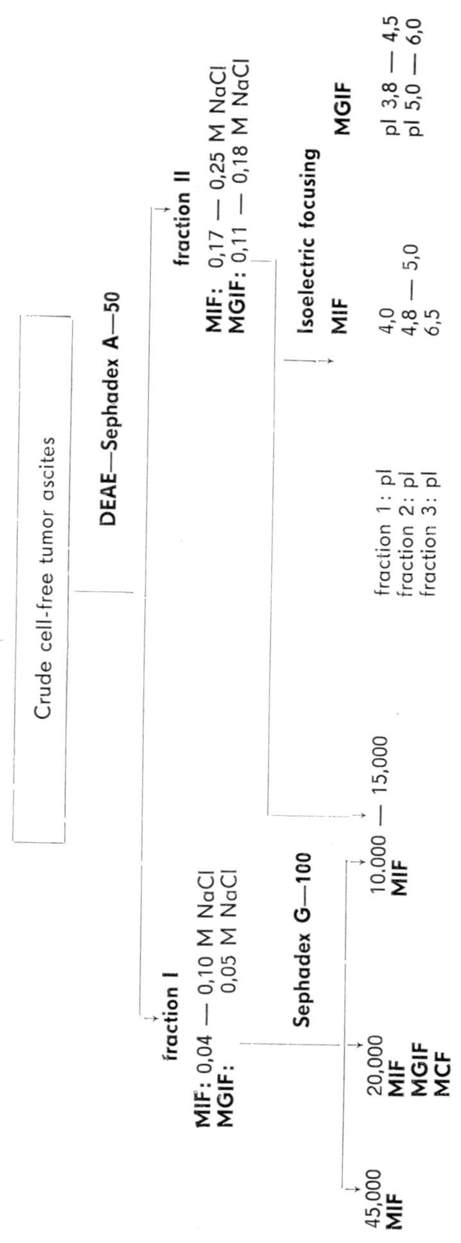

Fig. 1. Scheme of fractionation of cell-free ascities from Zajdela hepatoma.

(13) an unequivocal separation of MIF and MGIF has not been achieved by physico-chemical techniques. Contrary to this, we suggest that MIF and MGIF are different, individual factors. Consistent with this interpretation is the fact that MGIF from rat Zajdela hepatoma ascites is less heterogeneous than MIF. Furthermore, observations on the different kinetics of generation of MIF and MGIF strongly suggest the existence of separate factors (15, 19). Clearly, the arguments cited above indicate caution because the identity of the factors under question could conclusively be verified only by appropriate monoclonal antibodies.

Studies on the Mode of Action of MGIF

A preparation exhibiting a bactericidal activity of 52 per cent after 4 h was used. The test system for evaluation of the bactericidal activity has been described previously (14). Briefly, 4 d afer intraperitoneal (ip) injection of mice with 2 ml of a 3 per cent thioglycollate solution peritoneal exudate cells were harvested and MØ were separated by allowing them to adhere on a plastic surface for 3 h at 37° C. After removal of nonadherent cells, MØ were cultivated for 24 h, then infected for 30 min with corynebacteria (4×10^5 bacteria per 10^6 MØ) and washed thoroughly to remove free bacteria. Cell samples were cultivated with culture medium only (controls) or with medium containing MGIF for 4 h at 37° C. After counting of MØ cells were lysed. After plating on agar and cultivation for further 36 h bacterial colonies were counted and bacterial activity was calculated (4).

Murine adherent MØ prepared as described above were studied by TEM 1, 2 and 4 hrs after infection with corynebacterial followed by cultivation in medium without (controls) or containing MGIF. Morphometric estimations were performed on TEM graphs of MØ (16) and altogether 28 stereological parameters were calculated for each MØ, its compartements and organelles: phagosomes, lysosomes, rough-surfaced endoplasmic reticulum (rER), Golgi apparatus, ingested bacteria (7).

The following changes could be related to the action of MGIF after 4 h in comparison to control MØ (fig. 2). Phagosomes in MGIF-treated MØ were slightly reduced in number with reduction of their volume fraction, mean size and surface density but an increase of specific surface area. The rER showed a slightly reduced absolute volume and volume fraction after 4 h cultivation with MGIF in comparison to control MØ.

	1 h	2 h	4 h
Phagosomes			
Number	↑	↓	↓
Volume Fraction	-	-	↓
Mean Size	-	-	↓
Spec. Surface Area	-	-	↑
rER			
Content	-	-	↓
Volume Fraction	-	-	↓

Fig. 2. Changes of stereological parameters of MØ in comparison to controls (out of 28 parameters).

	1 h	2 h	4 h
Mean Cell Size	-	-	↑
Lysosomes			
Number	-	-	↓
Golgi			
Volume Fraction	-	-	↓
Mean Size	-	-	↓
Surface Density	-	-	↓

Fig. 3. Changes in MØ both after MGIF and in controls (in comparison to 0 h).

Both MGIF-treated and control MØ showed an insignificant increase of mean cell size and a decrease of the number of lysosomes as well as a decrease of volume fraction, mean area and surface density of Golgi apparatus (fig. 3).

No significant influence of MGIF treatment could be observed on the degree of ruffling of MØ plasma membrane, the nuclear volume fraction, the volume fraction and surface density of rER and the volume fraction of ingested bacteria (cf. 7).

In conclusion, our results on MGIF-treated MØ after a 4 h period of MGIF action, when the bactericidal effect is distinctly marked in the bacterial assay, only indicated some changes which could be interpreted as an increased digestive activity in phagosomes with some emptying of rER but without any morphologically detectable damage of ingested bacteria. No indications for the induction of synthetic ativities by MGIF could be observed which would be directed towards enzyme production and formation of lysosomes, and there were no signs detectable for an increased fusion of lysosomes with phagosomes in MGIF-treated MØ. Studies of Römer (12) with MGIF showed an early stimulation of cAMP metabolism after 2 to 3 h treatment of MØ resulting in an elevated level of cAMP and increased activities of adenylate cyclase as well as of cAMP phosphodiesterase. MIF-containing fractions, on the contrary, showed no influence on the cAMP system. These early influences of MGIF on the cyclic nucleotide metabolism support the suggestion that the induction of membrane changes including those of membrane receptors and changes of the activity of membrane-associated enzyme systems may be involved in the early events of bactericidal cytokine effects rather than an induction of synthetic processes, as they have been observed after nonspecific stimulation of MØ (7).

Studies on the Cytotoxic Action of BCG-Activated MØ

MCF was chromatographically isolated from rat Zajdela hepatoma ascites fluid (cf. fig. 1), but the cytotoxicity enhancing activity was biochemically not separable from MIF in this MW fraction. In short, the cytotoxic effect was investigated as follows. Murine peritoneal MØ activated by 2 ip injections of BCG with an interval of 4 weeks were harvested 4 d after the last BCG injection and were permitted to adhere to glass coverslips, nonadherent cells were removed by washing. RAB I ascites tumor cells (8) labelled with ^{51}Cr were added as target cells to the activated MØ at a ratio of 1:5. Cells were incubated in RPMI 1640 medium with 5 per cent FCS (control) or the same medium completed by MCF. The cytotoxic activity was evaluated by means of ^{51}Cr release assay.

MØ-target cell interactions were examined by means of TEM

and SEM after 16 h incubation, at which time also the ^{51}Cr release was measured. In SEM studies a mordanting method was used for cell preparation which nearly totally avoids shrinkage of cells which otherwise show an extreme shrinkage with loss of fine structure of cell surface even after applying critical point drying (17).

Tumor cells were surrounded by a number of MØ thus producing a rosette-like appearance. The plasma membrane of tumor cells exhibited structural disintegration with areas showing pits and

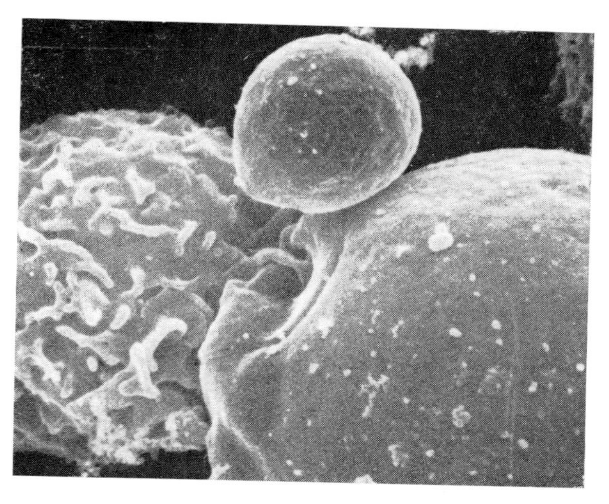

Fig. 4. Close contact of MØ to tumor cell with a finger-like process of MØ extending into a corresponding hole of the target cell. SEM 9/0010. 4,800×.

holes (fig. 4) up to gross defects of the plasma membrane with discharge of the cell content (fig. 5). In the intercellular space between MØ and tumor cells numerous blebs and vesicles were visible, in TEM as a rule with homogeneous or electron-opaque content without discernible cytoplasmic organelles (fig. 6, cf. fig. 5). Greater distances between tumor cells and MØ were occasionally bridged over by long filopodia originating from MØ and touching tumor cells (fig. 7). From the base up to the distal and of filopodia bead-like bulges were visible resembling vesicles in the contact zone of MØ and tumor cells described above. In TEM corresponding

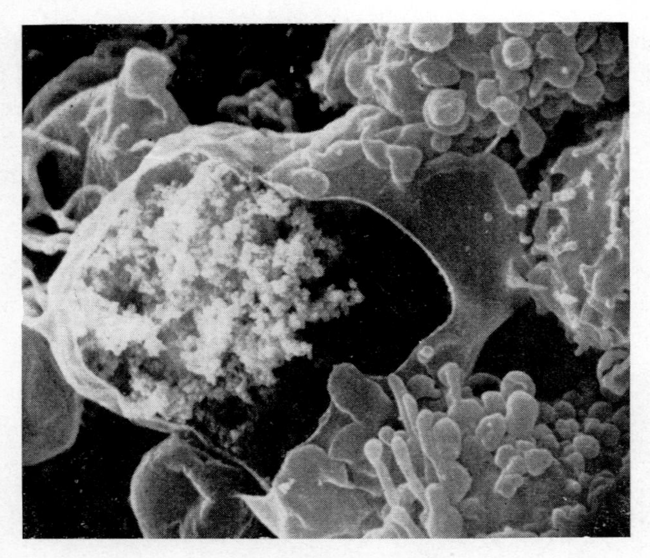

Fig. 5. Rosette-like arrangement of MØ around a tumor cell with a gross plasma membrane defect and discharge of cell content.
Note the finger-like processes and vesicular structures on the surface of MØ. SEM 64/0041. 4,100×.

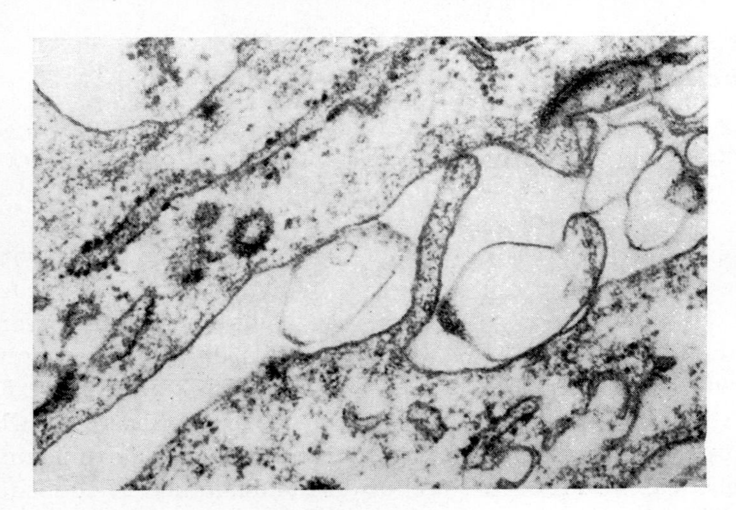

Fig. 6. Vesicular structures in the intercellular space between MØ and tumor cell. TEM 30/073. 36,000×.

vesicular structures could be shown from the base (fig. 8a) up to the distal end of filopodia (fig. 8b).

It should be emphasized that the MØ-target cell interactions described above could be found both in MCF-containing cultures and controls. Thus MCF increases only quantitatively the cytotoxic effect of activated MØ without qualitative changes of the cytotoxic

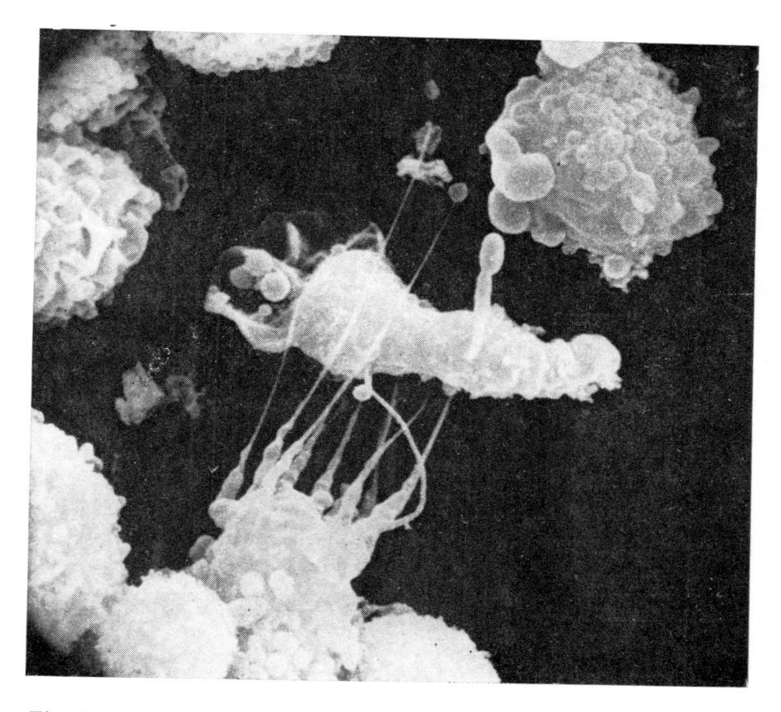

Fig. 7. Long filopodia originating from the MØ at the bottom directed to distant tumor cells.
Bead-like bulges on the base of filopodia and single vesicles of similar size up to the distal end of filopodia. SEM 58/0035. 2,450×.

action. Acording to other authors (9), our findings support the view of a two-step mechanism of cytotoxic action: a first step of specific contact between MØ and tumor target cell and a second step with liberation of cytotoxic substances in the immediate vicinity of tumor cells which substances, in this way, are accumulated in high concentration at the tumor cell surface. We could not observe an injection of lysosomes into the tumor cell cytoplasm as has been described by Bucana et al. (2). Furthermore, we were not able to demonstrate acid phosphatase activity in the vesicular structures of

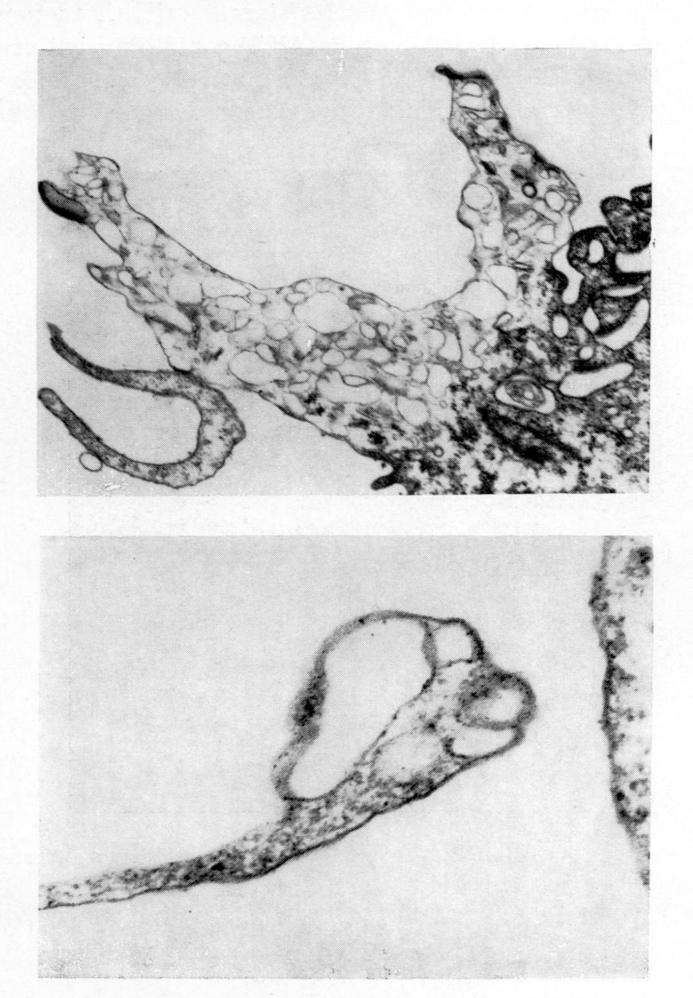

Fig. 8a, b. Numerous vesicles at the base of filopodia of an activated MØ (a), similar vesicles at the distal end of a filopodium (b). TEM 50/236 (a), 69/1375 (b). 16,000× (a), 63,000× (b).

MØ (unpublished observations) which findings suggests that we were not dealing with "customary" lysosomes. Further studies are in progress which shall elucidate the nature of cytolytic substances.

References

1 Bigazzi, P. E.: In: Biology of the lymphokines. Eds. S. Cohen, E. Pick, J. J. Oppenheim. Academic Press, New York, 1979, pp. 243—257.
2 Bucana, C.; Hoyer, L. C.; Hobbs, B.; Breesman, S.; McDaniel, M.; Hanna, M. G. (1976): Cancer Res. 36, 4444—4458.
3 Fahlbusch, B.; Dornberger, G.: Immunobiology, submitted for publication 1983.
4 Fahlbusch, B.; Schumann, I. (1979): Acta biol. med. germ. 38, 813—819.
5 Fahlbusch, B.; Schumann, I.; Zschiesche, W.; Borecký, L.; Lackovic, V. (1982): Acta virol. 26, 382—389.
6 Fahlbusch, B.; Zschiesche, W. (1981): Acta biol. med. germ. 40, 793—800.
7 Fritsch, R. S.; Augsten, K.: Munich Symposia on Microbiology: Biological Products for Viral Diseases. Edit. P. A. Bachmann. London: Taylor and Francis Ltd., 1981, pp. 229—249.
8 Fritsch, R. S.; Heinecke, H.; Jungstand, W. (1969): Neoplasma 16, 195—204.
9 Hibbs, J. B.: In: The Macrophage in Neoplasia. Edit. M. A. Fink. Academic Press, New York, 1976, pp. 83—112.
10 Neta, R.; Salvin, S. B. (1981): Lymphokines 2, 295—308.
11 Pike, M. C.; Snyderman, R. (1976): Immunol. 117, 1243—1249.
12 Römer, W. (1981): Diss. B. Acad. Sci. GDR.
13 Salvin, S. B.; Nishio, I.; Shormard, J. T. (1974): Infect. Immunity 9, 631—635.
14 Schumann, I.; Fahlbusch, B.; Vasiljeva, I. G. (1979): Acta biol. med. germ. 38, 807—811.
15 Simon, H. B.; Sheagren, J. H. (1972): Infect. Immunity 6, 101—103.
16 Weibel, E. R.; Kistler, G. S.; Scherle, W. F. (1966): J. Cell. Biol. 30 23—38.
17 Wollweber, L.; Stracke, R.; Gothe, U. (1981): J. Microsc. 121, 185—189.
18 Yoshida, T.: In: Biology of the lymphokines. Edit. S. Cohen, E. Pick, J. J. Oppenheim. Academic Press, New York, 1979, pp. 259—290.
19 Zschiesche, W.; Fahlbusch, B.; Schumann, I.; Lackovic, V.; Borecký, L. (1980): Acta virol. 21, 37—44.

R. S. Fritsch, M.D., Central Institute of Microbiology and Experimental Therapy, Jena (GDR)

III. Consequences of Interferon Administration to the Animal Organism

Contr. Oncol., vol. 20, pp. 196—204 (Karger, Basel 1984)

Interferon Induced Disease

M. G. Tovey; I. Gresser

Laboratory of Viral Oncology, Institut de Recherches Scientifiques sur le Cancer, Villejuif, France

Interferons constitute a group of closely related cellular proteins which exhibit potent antiviral activity. There is abundant experimental data indicating that the production of interferons in the course of viral infections is an integral part of the host response, and interferon treatment of animals and man is associated with prophylactic and therapeutic antiviral effects. Although interferon was originally considered to inhibit viral multiplication within the cell without affecting host cell metabolism or function, it is now widely accepted that interferons can also affect cell division and function both in cell culture and in the animal. If this is so, we might expect to find instances in which too much interferon, instead of being beneficial, might even prove inimical to the host. This is the subejct of our presentation.

When newborn mice were inoculated daily from birth subcutaneously or intraperitonealy with potent mouse interferon preparations[1] there was a progressive inhibition of growth in the first week of life; the mean weight of interferon treated mice was 60 to 80 % of the weight of control inoculated litter mates. When interferon treatment was continued for the next few days, all of the mice died (tab. I) (11). At autopsy, the most striking abnormality was a pale grey liver. Light microscopic examination revealed a marked steatosis, large areas of cell degeneration and necrosis without any inflammatory reaction. The only other abnormalities on light microscopic examination were some diminution of the cortical region of

[1] (0.05 ml of interferon having a titer of 8×10^{-5} reference units).

Table I. Effects of treatment of suckling mice with mouse interferon

	Mouse strain	No. experiments[1]	Injected with: Mouse interferon[2]	Control preparation	Inactivated mouse interferon	Human interferon	Uninoculated
Lethality of mouse interferon preparations	Swiss	5	107/107[3]	0/52	0/15	0/23	0/142
(Daily interferon treatment continued)	C_3H	1	15/15	0/30	0/15	NT	0/18
Development of glomerulonephritis[4]	Swiss	9	60/60[5]	0/50	0/18	1/18	2/65
(Daily interferon treatment stopped at day 6 to 8)	C_3H	1	13/17	—	—	—	0/13

[1] For experimental details see references (1, 2).

[2] Mice injected daily subcutaneously with 0.05 ml of a C-243 cell interferon preparation (titer 3.2×10^{-6} reference units).

[3] Number of mice dead/total number of mice injected. Mean day of death = 11 days.

[4] Results pooled — mice sacrified between 20 and 274 days of life.

[5] No. of mice with histologic confirmation of glomerulonephritis per total number of mice killed.

the thymus in some mice, and poorly differentiated germinal follicles in the spleens of most mice. No renal lesions were present. Examination of liver sections under the electron microscope showed a marked accumulation of fat in hepatocytes. No viral particles were observed. Kinetic studies have shown that the first abnormality seen in the liver is a marked steatosis which is followed by extensive necrosis of the liver parenchyma. This necrosis is at first subcapsular but then extends to involve the whole liver. The extent and diffuse nature of this necrosis is suggestive of an "ischaemic necrosis". However, the blood vessels appear to be uninvolved. Jill Moss and her colleagues have described the appearance of characteristic tubular aggregates associated with endoplasmic reticulum of liver cells from interferon treated mice (13). Morphologically there is a considerable similarity between these tubular aggregates and the inclusions observed in glomerular endothelial cells of patients with systemic lupus erythematosus.

In the experiments described above interferon treatment proved lethal only when begun on the day of birth and continued daily for the first week of life. When treatment was initiated on day 6 and continued daily for 14 days, mice grew normally and liver lesions were not observed. The effect of interferon on newborn mice depended, therefore, not only on the amount of interferon administered but also on the immaturity of the host.

What happens to these mice when interferon treatment, begun at birth, was stopped on the 7th day of life? (a time when steatosis and discrete foci of liver cell necrosis were present). The majority of these mice appeared to recover and they gained weight. However, in the ensuing months a number of these mice died. At autopsy, although the liver and other organs appeared normal, the kidneys were pale and the surface was granular. Histologic examination revealed a severe glomerulonephritis (tab. I) (8). The kidney pathology can be summarized as follows: In "early" lesions (30th day) the glomeruli appeared moderately enlarged with some thickening of the mesangium, some segmental foci of mesangial proliferation and some thickening of capillary loops. By immunofluorescence there were focal and segmental granular deposits of IgG, IgM and C3 in the mesangial spaces and along some glomerular basement membranes (GBM). Subsequently (>45 days) the lesions were more conspicuous with hyalinisation of glomerular tufts (with epithelial crescents and voluminous subendothelial deposits). In well advanced disease, virtually all glomeruli were sclerotic and there was extensive sclerosis and diffuse atrophy of tubules. By immunofluorescence there were coarse granular deposits of IgG, IgM and C3 along the GBM.

We should emphasize at this point, that under our experimental conditions – all Swiss and C3H mice treated with potent interferon preparations have shown the liver lesions described above, and all Swiss mice treated for the first week of life developed glomerulonephritis (tab. I). These syndromes have been seen in all strains of mice tested, including axenic mice and athymic nude mice.

These experiments were carried out with partially purified mouse interferon preparations having a specific activity of 10^7 reference units per mg protein. Mice were injected with approximately 50 μg of protein per day. Control preparations consisting of: mock mouse interferon preparations, heated or periodate inactivated mouse interferon preparations, or human interferon preparations did not induce disease in mice. All these preparations contained comparable amounts of protein.

Nevertheless, we wanted to determine the effect of even more purified mouse interferon preparations. Recently, we have tested electrophoretically pure mouse interferon with a specific activity of 0.5 to 1.0×10^9 units per mg of protein (2, 1). After electrophoresis in polyacrylamide gels in SDS, and staining with Coomassie Blue, only 3 bands corresponding to molecular weight of 35,000, 28,000 and 22,000 were observed. Biologic activity of interferon was found only in eluates of gel slices corresponding to these bands. We injected newborn mice with 25 μl of interferon (containing 500 nanograms of protein). We also tested the original partially purified interferon (having a comparable titer) and preparations containing the major contaminants collected during the purification procedures. Only partially and highly purified interferons inhibited mouse growth, delayed maturation of different organs, induced liver lesions, caused death, and induced glomerulonephritis. We have established that the minimal necessary daily dose is approximately 100 nanograms of interferon (6). We believe therefore, that there is no longer any doubt that interferon itself is the responsible factor in inducing these lesions.

Furthermore, this phenomenon is not confined to mice. When

newborn rats were injected with rat interferon preparations (but not mouse or human interferon preparations) there was also a clear cut inhibition of growth, delay in the maturation of several different organs, and the late development of glomerulonephritis (7). Liver lesions were, however, not observed.

The experiments discussed so far have concerned the use of exogenous interferon. Let us now consider the effects of endogenous interferon induced in newborn mice by lymphocytic choriomeningitis virus. When newborn mice of most strains were injected at birth with this virus, they grew poorly, and after the 8th day began to die. The only histologic lesions observed were in the liver, which showed steatosis and focal or diffuse necrosis. Mice surviving this acute episode, subsequently developed glomerulonephritis. In other words, the syndromes induced in mice with LCM virus were identical to those induced by treatment of mice with interferon (tab. II). To explain this similarity of syndromes we proposed two hypotheses (16): (a) interferon was lethal for suckling mice because it activated a latent virus (such as LCM virus) or (b) some of the manifestations of LCM virus disease in suckling mice were in fact due to endogenous interferon induced by the virus. We found no

Table II. Similarity in syndromes induced in suckling mice by interferon and LCM virus

Time	Clinical findings	Injected with		Comments
		mouse interferon	LCM virus	
Early 2—3 weeks	Decreased weight gain	+	+	Similar curves of weight gain
	Delay in maturation of several organs	+	+	
	Liver cell steatosis leading to necrosis	+	+	Similar histology
	Mortality	+	+	
Late >1 month	Glomerulonephritis	+	+	Similar histology Deposits of C3, IgG along GBM

evidence to support hypothesis (a) because neither LCM virus nor mouse hepatitis virus nor any other infectious agent was recovered from the liver and kidneys of interferon treated suckling mice (11). We therefore turned our attention to the second hypothesis. First, we found that despite previous reports to the contrary, LCM virus induced large amounts of interferon in suckling and adult mice (16). We then injected newborn mice with LCM virus and a potent sheep anti-globulin to mouse interferon (16). Injection of this immunoglobulin neutralized the endogenous interferon and resulted in a 100-fold increase in the serum LCM virus titer. Despite this marked increase in circulating virus, anti-interferon globulin treated mice grew normally, showed no liver lesions, and the incidence of death was much decreased (16). Furthermore, injection of this anti-interferon antibody also markedly inhibited the appearance of glomerulonephritis characteristic of late LCM virus disease (9). LCM virus was present in the blood and kidneys of these mice in amounts comparable to control virus infected mice developing glomerulonephritis (9).

The lethality of LCM virus for suckling mice varies markedly with the mouse strain. Thus virus infected BALB/c mice exhibit minimal liver lesions, and none die, whereas C3H mice have extensive liver lesions and all mice die. An intermediate pattern is observed with Swiss mice (36 % mortality). The results of our experiments suggest that the genetic control of susceptibility or resistance is determined not by the extent of virus multiplication but by the amount of interferon produced. Thus, although there was no difference in the titers of LCM virus in the plasma or liver between the three strains of mice, there was a marked difference in the amount of interferon produced and the duration of interferonemia. BALB/c mice produced small amounts of interferon detectable in the plasma mostly on the third day, Swiss mice produced more interferon detectable on the third and fourth day, whereas C3H mice produced larger amounts of interferon and interferonemia lasted between the third and sixth day (15). Our results suggest therefore, that the amplitude of the interferon response in C3H is in large part responsible for the severity of LCM disease. This interpretation is further supported by experiments showing a marked decrease in the incidence of mortality in virus-infected C3H mice when they were injected with a potent anti-mouse interferon globulin

(15). Furthermore, the relative resistance of BALB/c mice was not caused by insensitivity to interferon action because exogenous interferon did inhibit their growth and did induce comparable liver lesions. We suggest therefore, that the minimal disease occuring in BALB/c mice was related to their minimal interferon response.

We believe this ensemble of results shows that the presence of either exogenous or endogenous interferon in large quantities in an immature animal can be harmful. Furthermore, this brief exposure to interferon at a critical period of rapid growth and maturation can be an important factor in the development of disease that only becomes manifest later in life. Although space does not permit us to discuss the possible mechanisms of action of interferon in inducing disease, discussion of the various possibilities may be found in the reference cited (7, 9, 11, 13, 16). We should like instead to emphasize several points that we feel to be important and why we think that our results may be of relevance to human pathology. First, we may ask whether there are other instances in which interferon may be in part responsible for disease. Virelizier and Allison have shown that C3H and A2G strain mice can become carriers of mouse hepatitis virus (MHV-3) and develop a progressive neurologic disease lasting for weeks or months (17). These virus carrier mice also showed a chronic interferonemia (18), and thus the possibility exists that interferon may be exerting untoward effects in some of these mice.

Interferon does not cross the placental barrier in mice (3) and we believe for a very good reason. Pregnant animals can contract virus infections and produce circulating interferon. If interferon crossed the placenta, it might well affect the developing embryo. But what about a virus that crosses the placenta and infects the embryo? Is it possible that the embryo-toxic effects ascribed to rubella virus, for example, may be related to interferon induced in the embryo itself? In other words, is it possible that cellular lesions considered heretofore as being caused by the virus are in fact caused by host substances (such as interferon) induced by the infectious agent? Interferon has been detected in the sera of patients with a variety of diseases including systemic lupus erythematosus (SLE), rheumatoid arthritis, scleroderma and Sjögren's syndrome (12). Serial serum samples showed a good correlation between interferon titers and disease activity, and it has been suggested that the production of interferon might contribute to immunologic aberrations in auto-immune disease (12). It is of interest that microtubular structures similar to "lupus inclusions" found in the glomerular endothelium and the peripheral blood lymphocytes of patients with SLE, can be induced in several human lymphoid cell lines cultivated in the presence of human interferon (14).

Morphologically there is a considerable similarity between the "lupus inclusions" induced in interferon treated lymphoblastoid cells and the tubular aggregates associated with the granular endoplasmic reticulum in the cytoplasm of hepatocytes of suckling mice treated with mouse interferon (13). The biologic significance of the persistent presence of interferon in the sera of patients with chronic disease is unknown. It remains to be established whether this interferon is part of the host's response to chronic viral infection. It is also unknown whether this interferonemia exerts a beneficial effect, or contributes to some of the manifestations and pathology of the chronic viral disease, or is merely a marker of infection without much significance in the evolution of the disease.

In case these speculations might be misunderstood, we should like to emphasize, however, that we do not feel that the results presented herein constitute an argument against the use of interferon in patients. Over the years, we have pleaded long and hard the potential clinical usefulness of exogenous interferon (4). We have been too often witness in the past years to interferon's therapeutic efficacy in viral and neoplastic diseases of mice, not to believe that comparable results would be obtained in man, were it properly used. We have also been witness to its extraordinary activity – and to the varied effects interferon exerts on cells (5). In some instances, interferon inhibits specialized cellular functions and in other instances it enhances cellular functions (5, 10) Factors such as the amount of interferon, time of treatment, determine the biologic effects observed. Aside from the theoretical interest, these considerations are of utmost importance in determining how to use interferon in patients. Our results emphasize that we are dealing with a most potent substance – and as with hormones – use of interferon must be based on knowledge of its effects and, if possible, of its mode of action.

References

1 Aguet, M. (1980): Nature 284, 176.
2 De Maeyer-Guignard, J.; Tovey, M. G.; Gresser, I.; De Maeyer, E. (1978): Nature 271, 622.
3 Gresser, I. (unpublished observations).
4 Gresser, I. (1971): In "Third International Symposium of Medical and Applied Virology". Reprinted from Viruses affecting man and animals, ed. by M. Sanders and M. Schaeffer, p. 416. W. H. Green, Sr. Louis, Missouri, USA.
5 Gresser, I. (1977): Cell Immunol. 34, 406.
6 Gresser, I.; Aguet, M.; Morel-Maroger, L.; Woodrow, D.; Puvion-Dutilleul, F.; Guillon, J. C.; Maury, C. (1981): American J. Pathol. 102, 396.
7 Gresser, I.; Morel-Maroger, L.; Châtelet, F.; Maury, C.; Tovey, M. G.; Bandu, M.-T.; Buywid, J.; Delauche, M. (1979): American J. Pathol. 95, 329.
8 Gresser, I.; Morel-Maroger, L.; Maury, C.; Tovey, M. G.; Pontillon, F. (1976): Nature 263, 420.
9 Gresser, I.; Morel-Maroger, L.; Verroust, P.; Rivière, Y.; Guillon, J.-C. (1978): Proc. natn. Acad. Sci., USA 75, 3413.
10 Gresser, I.; Tovey, M. G. (1978): Biochem. Biophys. Acta 516, 231.
11 Gresser, I.; Tovey, M. G.; Maury, C.; Chouroulinkov, I. (1975): Nature 258, 76.
12 Hooks, J. J.; Moutsopoulos, H. M.; Geis, S. A.; Stahl, N. I.; Decker, J. L.; Notkins, A. L. (1979): New Eng. J. Med. 301, 5.
13 Moss, J.; Woodrow, D.; Sloper, J. C.; Rivière, Y; Guillon, J.-C.; Gresser, I. (1982): Brit. J. Europ. Pathol. 63, 43.
14 Rich, S. A. (1981): Science 213, 772.
15 Rivière, Y.; Gresser, I.; Guillon, J.-C.; Bandu, M.-T.; Ronco, P.; Morel-Maroger, L.; Verroust, P. (1980): J. Exp. Med. 152, 633.
16 Rivière, Y.; Gresser, I.; Guillon, J. C.; Tovey, M. G. (1977): Proc. natn. Acad. Sci., USA 74, 2135.
17 Virelizier, J.-L.; Dayan, A. D.; Allison, A. C. (1975): Infect. Immun. 12, 1127.
18 Virelizier, J.-L.; Virelizier, A.-M.; Allison, A. C. (1976): J. Immunol. 117, 748.

M. G. Tovey, M.D., Laboratory of Viral Oncology, B. P. No. 8, F-94802 Villejuif, Cedex (France)

Contr. Oncol., vol. 20, pp. 205—215 (Karger, Basel 1984)

Effect of α, β and γ Interferons on the Growth of Methylcholantrene Induced Tumor Cells in Mice

M. Lipková; L. Borecký; M. Novák; Z. Sekeyová; N. Fuchsberger

Institute of Virology, Slovak Academy of Sciences, Bratislava, CSSR

Introduction

Growing evidence indicates that interferon (IFN) inhibits the growth of several tumors in experimental animals and in patients. The available data suggest that the anti-tumorous effect of interferon is mediated by at least 2 mechanisms: They are a) stimulation of various cellular and humoral immunoregulatory mechanism, and b) modification of multiplication or phenotypic and enzymatic expression of .tumorous cells (1, 9, 19).

In this study, the effect of IFN on methylcholantren-transformed mouse embryonal cells was examined both in vivo and in vitro.

Materials and Methods

Viruses

The Newcastle Disease Virus (NDV), Hertfordshire strain, was used for IFN induction. The virus was maintained by serial passages in the allantoic cavity of 9 days old chicken embryos. The Indiana strain of Vesicular stomatitis virus (VSV) was used as challenge in antiviral IFN-assays. The virus was maintained by passages in monolayers of LB-cells (a subline of L-929 mouse fibroblasts).

Cells

Two sublines of L-929 mouse fibroblasts were used. The LG-subline originated in Imper. Cancer Res. Fund Laboratories (London). This line was used for induction of mouse interferon. Will be designated as β-IFN. The LB2 subline of L-929 mouse fibroblasts was used for IFN (β and γ) assay. Human diploid cells were used for assay of human IFN-α. Primary mouse fibroblasts from inbred C3H mice were used both for chemical transformation and in antiproliferative assays of IFN preparations together with spleen leukocytes from 10—12 weeks old BALB/c, C57BL/6 and C3H mice.

Animals

Newborn (48hrs old) C3H, BALB/c, C57BL/6, MRL/MpJ-ipr and BALB/c × C3H/F_1 mice were used for studying the growth of methylcholantren-transformed mouse embryonal cells in vivo as well as inhibitory effect of various IFN preparations on growth of tumor cells.

Growth Media

The following media were used: RPMI 1640, H-MEMd and MEM. They were completed with: 2 mM L-glutamin, Na-pyruvate (ImM), 2-mercaptoethanol (0,028 mM), streptomycine (50 μg/ml), penicillin (100 units/ml), 10 % heat-inactivated non-mitogeneic fetal calf serum and 15 mM HEPES (N-2-hydroxy-ethylpiperazine--N-2-ethanesulphonic acid).

Interferons

Mouse interferon designated as β, has been produced in LG cells after induction with NDV. It was purified by Zn-acetate precipitation and chromatography on SP-Sephadex C-25 according to (7). The human IFN used in these studies was of leukocyte origine (IFN-α) and was prepared essentially according to Cantell (3). The mouse IFN-γ has been prepared according to the method of Lipková et al. (16).

Transformation of primary mouse embryonic fibroblasts

C3H mouse embryonic fibroblasts in the 2nd passage were treated for 72 hrs with 0.1 μg of 20-methylcholantrene per 1 ml growth medium. From the transformed fibroblasts, the IFN-sensitive clone No. 36 has been derived. Its properties were described in detail (1).

Cell Proliferation Assay

Incorporation of ^{3}H-TdR has been used for measuring the synthesis of DNA in allogeneic and syngeneic leukocytes after stimulation with normal and/or transformed mouse fibroblasts. The normal or transformed fibroblasts in amount of 10^4 cells per 0,2 ml of complete H-MEM per well were distributed in 96-well flat bottom plates. The cell cultures were incubated for 24hrs at 37° C in a CO_2 incubator. After 24hrs, the fibroblasts (stimulator cells) were irradiated (20 Gy, ^{60}Co) and C57/BL/6, BALB/c and C3H spleen cells, respectively (responder cells), were added to the fibroblast suspensions (3×10^5 cells) well in 0,2 ml of RPMI 1640 medium. Two, 3, 4 and 5 days after sensitization ,the cultures were pulsed with ^{3}H-TdR (1 μCi per well) for 8 hrs. As controls, stimulator and responder cells incubated alone were used. After 8hrs, the cells were harvested on glassfiber filters with an automatic cell harvester. After addition of the scintillation fluid to dried discs, the incorporation of ^{3}H-TdR was measured in a β-counter.

This method was used also for measuring the proliferation of transformed cells. The 20-methylcholantrene transformed MCH-36 cells in amount of 10^4 cells per well were suspended in 0,2 ml of complete H-MEM medium in the absence or presence of 1000, 500, or, 100 units of IFN-α. The transformed cells were pulsed with ^{3}H-TdR always during the last 24hrs of cultivation.

The Growth of Tumors in Vivo

The growth of MCH-36 tumor cells in vivo was tested in non-irradiated syngenic, semisyngenic and allogenic recipients. Newborn (48hrs old) mice were injected with $0,5\times10^6$, 1×10^6, 1×10^7

transformed fibroblasts subcutaneously. Control mice received a comparable amount of normal fibroblasts or saline (PBS). The mice were under observation during 180 days.

One-Cell Progeny Test for Measuring the Inhibition of MCH-36 Cells by IFN In Vitro

The test was based on reported data (4, 6, 17). It makes possible to register the effect of IFN on one cell and its immediate progeny. The cloning medium RPMI 1640 completed with 15 % fetal calf serum and 0,25 agarose (Indubiose A37) was mixed with 1000 MCH-36 cells and various amounts of IFN-α (100, 500 and/or 1000 units). Subsequently, this mixture was layered over a sheet of feeder cells $1{,}4 \times 10^6$ syngenic spleen cells irradiated with 20 Gy) and imbedded in the cloning medium in 0,5 % agarose. The cloned cells were incubated in a humidified atmosphere with 5 % CO_2 at 37° C for 14 days.

Results

The Growth of Tumors in Vivo

The growth capacity of cloned MCH-36 tumor cells was studied in 5 inbred mouse strains differring in H-2 antigens. As follows from tab. I, MCH-36 cells grew as solid tumors in 100 % of inoculated C3H mice and in 30 % of BALB/c $\times$ C3H/F_1 mice. In other inbred strains no tumorous growth could be detected. Control mice were inoculated with normal embryonic C3H fibroblasts or phosphate buffered saline (PBS). No tumor could be detected in these mice during the observation periode (180 days).

Stimulation of proliferation of allogeneic and syngeneic spleen leukocytes by MCH-36 cells

The capability of MCH-36 cells and of C3H fibroblasts to stimulate the proliferation of allogenic and syngenic leukocytes was studied by incorporation of ^{3}H-TdR. As shown on fig. 1, while normal embryonic (C3H) fibroblasts stimulated both the syngeneic

Table I. Growth of MCH-36 tumor cells (C3H origin) in syngeneic, semi-syngeneic and allogeneic newborn mice

Host genotype	Number of tumor cells injected[1]	Mice with tumors[2]/mice injected	Mice with tumors/mice injected[3] — control	Mice with tumors/mice injected-PBS
C3H	$0{,}5 \cdot 10^6$	5/5	0/5	0/5
	$1 \cdot 10^6$	5/5	0/5	
C57BL/6	$1 \cdot 10^6$	0/5	0/5	0/5
	$1 \cdot 10^7$	0/5	0/5	
BALB/C	$1 \cdot 10^6$	0/5	0/5	0/5
	$1 \cdot 10^7$	0/5		
/BALB/c×C3H/F_1	$0.5 \cdot 10^6$	3/10	0/5	0/5
MRL/MpJ-ipr	$1 \cdot 10^6$	0/5	0/5	0/5
	$1 \cdot 10^7$	0/5	0/5	0/5

[1] S.c. application of cells.

[2] All tumors appeared within 3 months of injection; mice were observed for at least 6 months or until death.

[3] Mice received normal C3H fibroblasts in number equivalent to tumor cells.

(C3H) and allogeneic (C57BL6, BALB/c) leukocytes, the transformed C3H cells (MCH-36 clone) had a diminished stimulating capacity for spleen cells of either syngeneic or allogeneic origine. Moreover, the leukocytes stimulated with tumorous cells showed a depressed incorporation of ^{3}H-TdR when compared with the background incorporation of nonstimulated spleen leukocytes (less than 2×10^3 CPM). This finding suggests that activation of suppressor lymphocytes could occur during the stimulation with tumor cells or their products.

Inhibition by IFN of MCH-36 Cells-Growth In Vivo

Newborn (48hrs old) C3H mice were inoculated with 1×10^6 MCH-36 cells subcutaneously. 24hrs later, the mice were injected intraperitoneally with human α, and mice β or γ IFN, respectively.

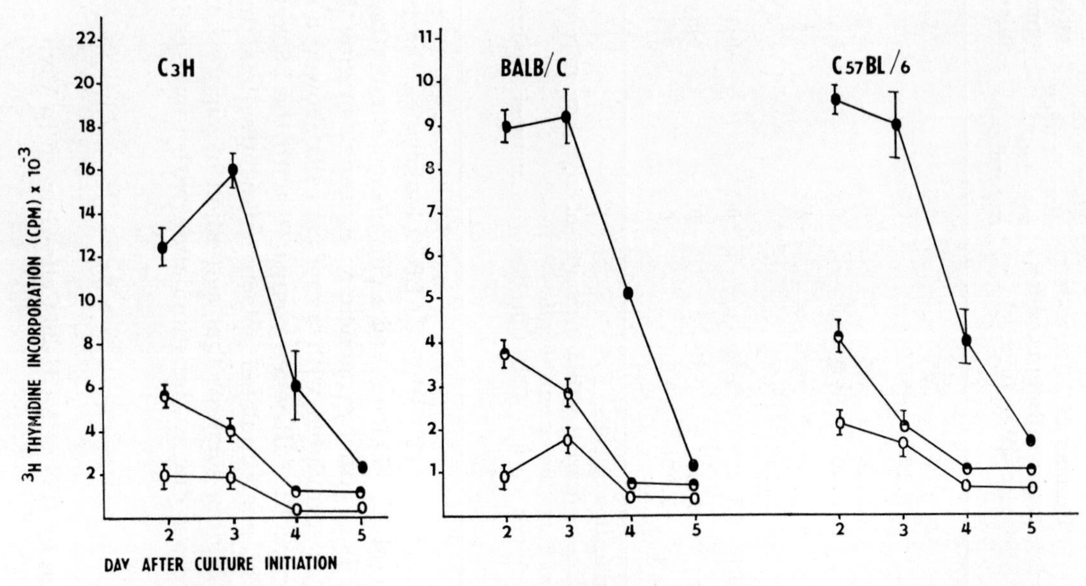

Fig. 1. Capability of normal and tumorous MCH-36 cells to stimulate proliferation of spleen leukocytes in vitro.
● Normal fibroblasts.
○ Tumor cells (MCH-36).
◐ Spleen leukocytes.

The IFN doses were repeatedly given to mice during 50 days in 48hrs intervals, altogether 25 doses per mouse. The mice were divided into 12 groups with 10 mice per group. The mice in the 1th group obtained repeatedly 50 units of IFN-α in the 2nd group — 50 units of IFN-β, in the 3rd group — 50 units of IFN-γ, in the 4th group — 25 units of IFN-γ, in the 5th group — 100 units of IFN-α, in the 6th group — 100 units of IFN-β, in the 7th group — 100 units of IFN-β and 25 units of IFN-γ, in the 8th group — 200 units of IFN-α and in the 9th group — 200 units of IFN-β. To each of these 9 groups, subgroups consisting of 5 mice served as IFN-controls i.e. received only IFN without tumor cells in the same intervals as the groups 1 to 9. The mice in the group No. 10 received only tumorous MCH-36 cells, the mice in the group No. 11 were injected only with PBS and the mice in the group No. 12, received neither tumor cells nor IFN or PBS. The first tumors were observed in the control

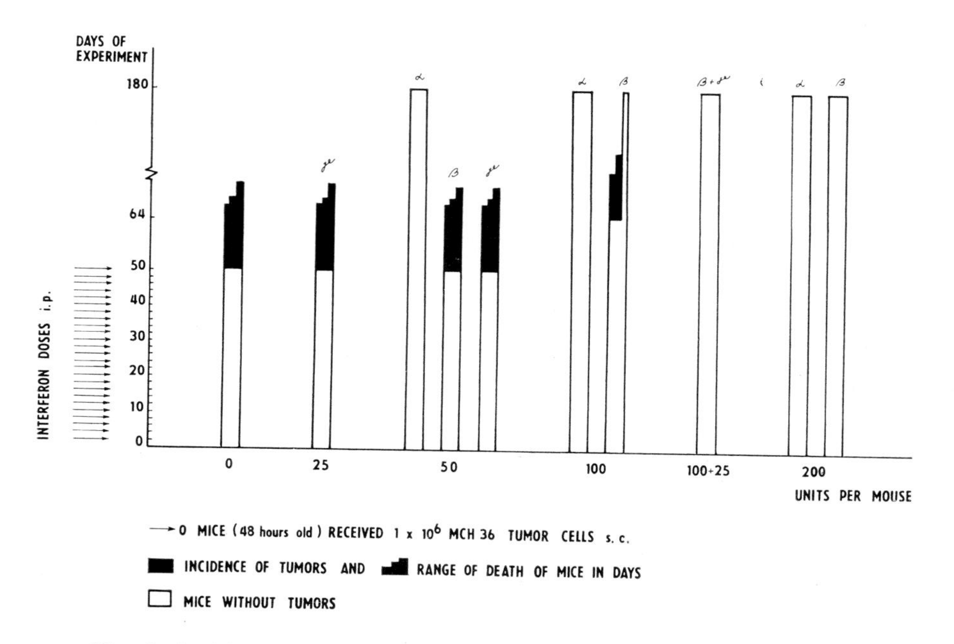

Fig. 2. Incidence of MCH-36 tumors in syngeneic C3H mice treated with interferons α, β and γ, respectively.
Arrows: Administration of IFN (i.p.). The newborn mice in days (10 per group) were injected on day 0 with 10^6 MCH-36 tumor cells subcutaneously.

group No. 10 50 days after inoculation of tumor cells. Concurrently, tumors were detected alsso in mice of the groups No. 2, 3 and 4. At this time, the administration of IFNs, was terminated in all groups. After further 14 days, tumors were detected in 6 of 10 mice in the group No. 6. No tumors could be detected in the mice in the group No 1, 5, 7, 8, 9, 11 and 12 during 180 days (fig. 2).

Inhibition of Tumor-Cells Proliferation by IFN-α In Vitro

The inhibitory effect of IFN-α on proliferation of transformed cells (MCH-36) in vitro was studied using the one-cell progeny test. In this test the number and the growth of clones from solitary MCH-36 cells treated with 100, 500 and 1000 units of IFN-α was evaluated according to the following criteria: a) small colonies with

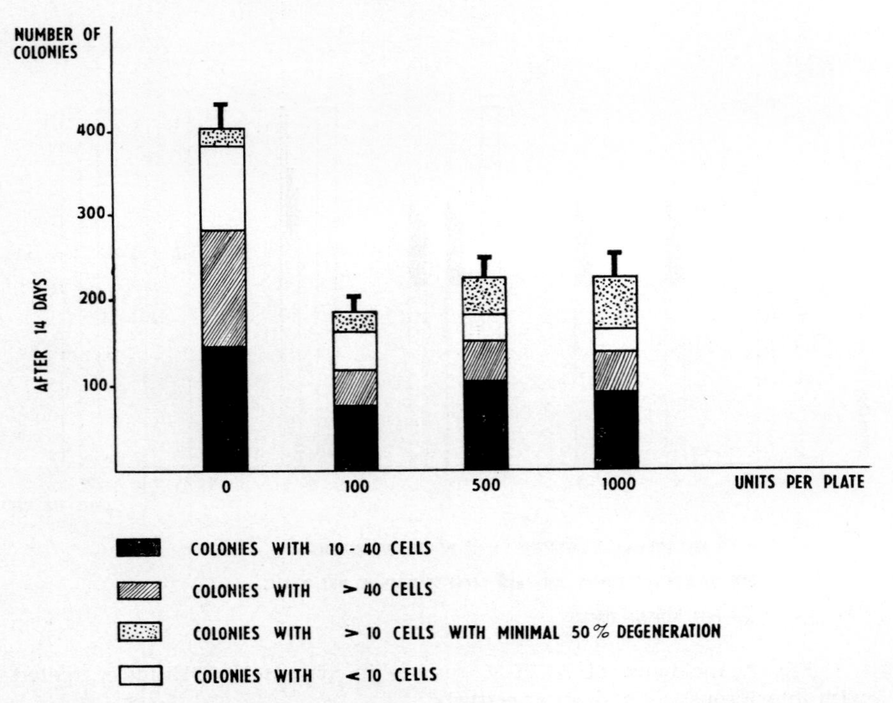

Fig. 3. Sensitivity of MCH-36 tumor cells to interferon in the "one-cell progeny test in vitro".

maximally 10 cells, b) medium-size colonies with 10 to 40 cells, c) large colonies with more than 40 cells, d) colonies consisting of more than 10 cells of which 50 % showed signs of degeneration. From a comparison with the control cells which were not-treated with IFN it follows that the clonning efficiency of this system is about 40 %. The results shown on fig. 3 are averages from 3 parallel cultures. In all IFN-treated cultures an inhibitory effect on colony-formation could be observed. The highest inhibitory effect on cell proliferation (50 %) was exerted by the 100 units dose. The degenerative effect in cells was strongest after treatment with 1000 units of IFN-α. All doses of IFN-α used for treatment had a depressive effect on the number of small and/or large colonies.

The results of the one-cell progeny test showed a significant correlation with the incorporation-inhibition test (Novák et al. – not published).

Discussion

The 20-methylcholantren transformed C3H mouse embryonic fibroblasts (MCH-36 clone) have been characterized in a previous paper (1). Their capability to produce tumors in syngenic C3H mice was 100 % while the cells did not multiply in allogenic. C57BL/6 or BALB/c mice. In accordance, the transformed cells produced tumors in 30 % of BALB/c $\times$ C3H/F_1 mice. In syngenic C3H mice, the 3 types of IFN (α, β and γ) administered in different doses (25 to 200 units per dose) were compared for their antiproliferative activity. Small doses (25 to 50 units) were unable to inhibit the growth of inoculated 10^6 tumor cells and the tumors in these mice appeared concurently with the tumor in control mice. When the doses of IFN were higher (100 units per dose), the tumors appeared with a delay of 14 days after termination of IFN treatment (group No. 6 treated with IFN-β). Interestingly, when IFN-γ in 25 units amount was added to 25 units of IFN-β, the tumors did not appear. This suggests a potentiating effect or synergism of 2 types of IFNs in a sense described by De Clercq et al. (5). Higher doses of IFN (such as 200 units) seemed to completely depress the growth of MCH-36 cells in syngenic mice and during the 180 days observation period a ressumption of tumor-cell multiplication could not be registered. This suggests that the transformed cells were eliminated from the organism.

The effect of human IFN-α in vitro was studied on the model of one cell progeny i.e. the clone growing in soft agarose on feeder cells. This test, in contradistinction to proliferation (^{3}H-TdR) test which reflects the sensitivity of the whole cell populations, indicates that also the clonned transformed cell line (MCH-36) is heterogenous with regard to the sensitivity to IFN (microheterogeneity). Our results showed that in one-cell progeny test IFN had a clear antiproliferative effect in small concentrations. It is unclear why were higher doses of IFN in this test less effective.

The mechanism of the inhibition of cell proliferation remains hypothetical despite of the fact that it has been described already in 1969 (10). Up to the present time, more than 100 reports brought evidence of the antiproliferative effect of α, β or γ IFNs (17). These reports showed that IFN alters both the structure and the behavior of the cell (9). However, still more data suggest that – in addition to the "direct" inhibitory effect on cell divison – IFN influences also the regulatory network of the immune system (9, 12). In our experiments in the presence of MCH-36 cells, the incorporation of ^{3}H-TdR into BALB/c, C57BL/6 and C3H spleen leukocytes was clearly suppressed. This finding suggests that also the cytotoxic T-lymphocytes might be suppressed in the presence of tumor cells since proliferation is a prerequisite also for differentiation of the precursor cell in the cytotoxic one (20).

It has been shown in several studies that during the tumor growth most of lymphocyte subpopulations (such as Ly1$^+$, Ly1,2$^+$, Ly1,2,3$^+$), and especially the suppressor T-lymphocytes, expand in number. In addition, it has been found that the Ly2,3$^+$ subpopulation of T-lymphocytes specifically enhanced the tumor growth probably via the Th-lymphocytes provided that the lymphocytes were transfered to the recipient concurently with tumor cells. In accordance, it has been reported that the immunoregulatory index (Th:Ts) decreases during the tumor growth (14). This suggests that IFN might in a dose dependent way influence the immunoregulatory index.

With regard to the paradox that in vitro resistant tumor cells proved sensitive to IFN in vivo, and that human IFN was also active in these tests, it seems that the antiproliferative effect may be indirect i.e. mediated through the activated Tc-lymphocytes (15) or NK cells (18, 8). We conclude that the evaluation of various effects of IFN requires a clarification of the role of interferon in the complicated network of immune regulation.

References

1 Borecký, L.; Hajnická, V.; Fuchsberger, N.; Kontsek, P.; Lackovič, V.; Russ, G.; Čapková, J. (1980): Ann. N. Y. Acad. Sci. 350, 188.
2 Brouty-Boyé, D. (1980): Lymphokine Rep. I, 99.
3 Cantell, K. (1970): In International Symp. on Interferon and Interferon Inducers. Symp. Ser. Immunobiol. Standards 14, 6. Karger, Basel, Switzerland.
4 Coffino, P.; Baumal, R.; Scharff, M. D. (1972): J. Cell. Physiol. 79, 429.
5 De Clerq, E.; Zhen-Xi Zhang; Huygen, K. (1982): Cancer Letters 15, 223.
6 Epstein, L. B.; Shen, J. T.; Abele, J. S.; Reese, C. C. (1980): Ann. N. Y. Acad. Sci. 350, 228.
7 Fuchsberger, N.; Hajnická, V.; Borecký, L. (1975): Acta virol. 19, 59.
8 Garnis, S.; Lala, P. K. (1978): Immunology 34, 487.
9 Gresser, I. (1982): Phil. Trans. R. Soc. Lond. B 299, 69.
10 Gresser, I.; Borali, C.; Levy, J.-P.; Fontaine-Brouty-Boyé, D.; Thomas, M. T. (1969): Proc. natn. Acad. Sci. USA 63, 51.
11 Gresser, I.; Maury, C. Brouty-Boyé, D. (1972): Nature 239, 167.
12 Handa,; K. Suzuki, R.; Matsui, H.; Shimizu, Y.; Kumagai, K. (1983): J. Immunol. 130, 988.
13 Lala, P. K.; Kaizer, L. (1977): J. natn. Cancer Inst. 59, 237.
14 Lala, P. K.; Mc Kenzie, I. F. C. (1982): Immunology 47, 663.
15 Lindahl, P.; Leary, P.; Gresser, I. (1972): Proc. natn. Acad. Sci. USA 69, 721.
16 Lipková, M.; Borecký, L.; Novák, M.; Sekeyová, Z. (1982): In 17th Regional Symposium of the Socialist Countries on Interferon, Varna, 8.—12. 11.
17 Novák, M.; Grešíková, M.; Sekeyová, Z.; Russ, G.; Zikán, J.; Pospíšil, M.; Čiampor, F. (1983): Acta virol. 27, 34.
18 Sakela, E. (1981): In Interferon 3 (ed. I. Gresser) 3, 45 New York and London: Academic Press.
19 Strander, H.; Einhorn, S. (1982): Phil. Trans. R. Soc. Lond. B 299, 113.
20 Wagner, H.; Röllinghoff, M. (1978): J. Exp. Med. 148, 1523.

M. Lipková, M.D., Institute of Virology, Slovak Academy of Sciences, Mlýnská Dolina 1, Bratislava (Czechoslovakia)

Contr. Oncol., 20, pp. 216—220 (Karger, Basel 1984)

A Role for Interferon in the Oncolytic Activity of Pichinde Virus

Norman Molomut[1], Morton Padnos[1], Daniel C. Pevear and Charles J. Pfau

Rensselaer Polytechnic Institute, Troy, New York, USA.

Historical Review

Since the advent of tumor immunology, a formidable problem has been to determine how proliferating tumors escape host rejection mechanisms. Early experimentation in this field was aimed at determining the antigenicity of neoplastic cells. Contamin (1) showed that protection against some transplantable tumor cell lines was possible when mice were preimmunized with the same strain of radiation-damaged cells. Similarly, Gross (7) found that small inocula of ground, viable sarcoma cells would often protect an inbred C3H mouse from subsequent challenge with a large number of sarcoma cells. Other methods of "immunization" proved equally as effective. Foley (6), using inbred C3H mice, showed that secondary isologous hosts were immune to reimplantation of the same methylcholanthrene-induced (MC) tumor after prior ligation of the implant. These results were supported by the findings of Prehn and Main (16) and Revesz (18), who reported that MC sarcomas possessed a distinct antigenicity since secondary hosts could readily reject a small MC sarcoma transplant.

In 1960 Klein et al. (9) confirmed the results of Prehn and Main, Foley, and Revesz, and demonstrated that the primary autochthonous host could also be rendered resistant to subsequent challenge with small doses of the same tumor line after surgical ex-

[1] Deceased.

cision of the tumor and repeated inoculations of irradiated sarcomas. Resistance was relative in that sufficiently large reimplantations could overwhelm the protective effect of immunization. This discovery confirmed that even in the autochthonous host, proliferating, invariably fatal tumor cells are recognized as distinctly foreign antigenic entities. Thus, it would appear that the immune system is altered during tumor progression so as to allow neoplastic growth. On the other hand, perhaps the cancers are only weakly antigenic and simply overwhelm naturally acquired immunity, as is suggested by the dose-dependent responses cited above. A third interpretation couples the first two, proposing that repression of the immune response is evoked by the rapidly accumulating tumor mass.

To combat a cancer, a logical line of action would be to attempt to alter either the weak immune response against the tumor or change the tumor cells so that they become more "immunogenic" from the host's standpoint. Many viruses have been exploited for their potential to change the immunogenicity of tumors, and though the volume of literature available on this subject is relatively small the results have been surprisingly encouraging (10). We will only point out a few examples: One is of historic interest and another is quite current.

The first report of a virus-induced, anti-neoplastic effect was by DePace in 1912 (5). He observed that patients receiving Pasteur's rabies treatment sometimes underwent regression of cervical carcinomas. One interpretation of DePace's observations surfaced in the early 1960's when Stuck (19) reported the production of viral antigens on the surface of infected leukemia cells. He inoculated virus-induced or spontaneous leukemias into the peritoneal cavity of BALB/c mice which had been previously infected with an unrelated leukemogenic virus (Rauscher virus). After harvesting the leukemia cells and treating them with anti-Rauscher leukemia antiserum and guinea pig complement, he found the formerly unrelated leukemia cells to be now fully susceptible to anti-Rauscher antibody mediated lysis.

In 1982 Austin et al. (1) showed that when cell membranes were prepared from cultured breast tumor lines infected with Vesicular stomatitis virus (VSV) and given to patients with breast cancer, lung carcinoma, or melanoma, 79 %, 13 %, and 15 % respectively displayed positive delayed-type hypersensitivity (DTH). Uninfected crude membranes were unable to elicit a DTH response in all thirteen breast cancer patients tested.

Anti-Tumor Properties of Arenaviruses

The ten viruses that compose the Arenaviridae family have been brought together on the basis of immunologic, morphologic, biochemical, and pathogenic properties. The prototype for this family is Lymphocytic choriomeningitis (LCM) virus. There is a long history of the antitumor properties of LCM virus. In 1956 Nadel and Haas (13) showed that LCM significantly increased the survival of guinea pigs bearing a transplantable leukemia cell line. Later studies with a particularly avirulent LCM virus, the M-P strain, showed that mice were protected to a marked degree against spontaneous leukemia (11, 14), transplantable leukemia (12, 14), and Rauscher virus-induced leukemia (20, 14). Furthermore, LCM murine infections afforded significant protection against spontaneous and transplantable mammary carcinoma (11, 14), Ehrlich ascites carcinoma (12), virus-induced osteosarcoma (17) and polyoma virus-induced tumors (8).

With the current resurgence of interest in the anti-tumor potential of viruses, as mentioned above, we decided to reinvestigate LCM and related viruses. We chose Pichinde virus, a member of the Tacaribe Complex of viruses in the Arenaviridae family. It appeared to be an ideal candidate for continuation of these studies for a number of reasons. First, it had not been identifed as the etiologic agent in any human disease (2). Second, it possessed the characteristics which in past animal trials had proven essential for success: i. e., maturation by budding from cell membranes and a strong immunoactivator. Third, unlike the prototype, Pichinde infection of adult mice was not lethal. Lastly, it was easily propagated and quantitated in tissue culture using a number of continuous cell lines.

Summary of Research at R.P.I.

We have found that C3H mice survived an invariably lethal injection of Sarcoma-180 cells if infected with Pichinde virus within 24 hours after tumor inoculation. Under these conditions if the tumor formed it regressed shortly thereafter. How did the immune responses differ in mice that would either succumb or survive? Within 12 hours and continuing for 5 days after infection with Pichinde, moderate levels of serum interferon were found. During this interval strong NK cell activity was noted against YAC-1 target cells. In mice receiving tumor cells alone serum interferon could not be detected and NK cell activity was minimal. Regardless of whether or not mice were infected with Pichinde virus, tumor-specific cytotoxic T cells could be isolated from spleens beginning on day 7. In untreated mice cytotoxic lymphocyte (CTL) activity remained constant for a few days and gradually disappeared by day 16. In virus-treated mice CTL activity gradually rose, peaked on day 13, and plateaued thereafter for 7—10 days. At all times 2—3 fold more lysis was observed in Pichinde-infected target cells than in non-infected cells-even using effector cells from mice never exposed to virus! The latter finding implied that Pinchide virus could have interfere with the repair mechanisms of sarcoma-180 cell membranes or caused rearrangement of tumor-specific antigens. Whatever the mechanism, the observation could have significance in the mouse since the sarcoma tumor was the main site of virus replication. Another potential advantage was that the mouse made a rapid and strong CTL response specifically against the virus. The response could have enhanced the development of the tumor-specific CTL response. The virus was cleared from the mouse by day 23, and a long lasting tumor specific CTL memory remained. This response undoubtedly enhanced the development of the tumor-specific CTL response since these mice became solidly immune to a second large implant of virus-free tumor cells.

References

1 Austin, F. C.; Boone, C. W.; Levin, D. L.; Cavins, J. A.; Djerassi, I.; Rosner, D.; Case, R.; Klein, E. (1982): Cancer 49: 2034—2042.

2 Buchmeier, M.; Adam, E.; Rawls, R. E. (1974): Infect. Immun. 9, 821—823.
3 Buchmeier, M. J.; Welsh, R. M.; Dutko, F. J.; Oldstone, M. B. A. (1980): Advances in Immunology 30, 275—331.
4 Contamin, A. (1910): Comptes Rend. Acad. Sc. 150: 128—129.
5 DePace, N. G. (1912): Ginecologia: Rivista Practica 9: 82—88.
6 Foley, E. J. (1953): Cancer Research 13: 835—837.
7 Gross, L. (1943): Cancer Research 3: 326—333.
8 Hotchin, J. (1962): Cold Spring Harbor Symposia on Quantitative Biology 27, 479—499.
9 Klein, G.; Sjogren, H. O.; Klein, E.; Hellström, K. E. (1960): Cancer Research 20: 1561—1572.
10 Kobayashi, H. (1979): Adv. Cancer. Res. 30: 279—299.
11 Molomut, N.; Padnos, M. (1965): Nature 208, 948—950.
12 Molomut, N.; Padnos, M.; Gross, L.; Satory, V. (1964): Nature 204, 1003—1004.
13 Nadel, E. M.; Haas, V. H. (1956): J. natl. Cancer Inst. 17, 221—231.
14 Padnos, M.; Molomut, N. (1973): in: Lymphocytic Choriomeningitis Virus and Other Arenaviruses (F. Lehmann-Grube ed.) Springer-Verlag, Berlin, pp. 151—163.
15 Pevear, D. C. (1982): Mechanisms of Pichinde Virus-Induced Tumor Immunity. M. S. thesis (70 pgs.) Rensselaer Polytechnic Institute, Troy, New York.
16 Prehn, R. T.; Main, J. M. (1957): J. natl. Cancer Inst. 18: 769—778.
17 Reilly, C. A.; Finkel, M. P. (1971): Abstracts Annual Meeting Amer. Soc. Microbiol., pp. 217.
18 Revesz, L. (1960): Cancer Research 20: 443—451.
19 Stuck, B.; Old, J. L.; Boyse, E. A. (1964): Nature 202: 1016—1018.
20 Youn, J. K.; Barski, G. (1966): J. natl. Cancer Inst. 37, 381—388.

C. J. Pfau, M.D., Department of Biology, Rensselaer Polytechnic Institute Troy, New York 12181 (USA)

Contr. Oncol., vol. 20, pp. 221—223 (Karger, Basel 1984)

The Anti-Inflammatory Effect of Human Interferons in Mice

I. Mécs; M. Koltai; S. Tóth, É. Ágoston

Institute of Microbiology and Pharmacology, Univ. Medical School, Szeged, Hungary

Recently we published that several inducers of interferon; i.e. poly I: poly C, statolone, tilorone and some viruses; as Semliki forest and Sendai virus, or homologous purified interferon (IFN) -beta decreased the inflammatory responses in mice provoked in the foot pad by carrageenin (4, 5). Furthermore, we noticed that there was a close correlation between the level of IFN in sera of animals and the extent of the inhibition of inflammatory responses (8, 1).

We report here that also heterologous human IFN types alpha and gamma cross the species specificity (3, 2, 9) and may reduce inflammatory reactions in mice.

Human interferons type alpha and gamma were produced in peripheral leukocytes stimulated with Sendai virus, or Concanavalin-A (7), respectively. Both IFNs were purified on Controlled Pore Glass Beads with a modified method of Whitman et al. (10) resulting in preparations with specific activities of 2×10^6 and 1.2×10^5 IU/mg protein, respectively. Groups of 10 CFLP mice weighing 26—30 grams were used for studying the inflammatory response, according to the method described by Levy (6). IFN preparations were injected subcutaneously immediately before the induction of paw edema provoked by 300 μg carrageenin (Viscarin 402) in a volume of 0.03 ml. Anti-human IFN alpha serum (produced in calves) with an IFN-neutralization capacity of 3×10^6 U/ml was given intravenously 3 hours prior to the IFN administration. The extent of the inflammatory response was measured 3 hours later, when the animals were exsanguinated and also the interferon levels in the sera were estimated.

Results

Both human IFNs types alpha and gamma markedly reduced the paw edema induced by carrageenin (tab. I). Human IFN type gamma was apparently more effective than type alpha, since ca. 1/100 dose produced approximately the same inhibition that observed when type alpha was used. One third dose of each human IFN injected simultaneously also evoked the same extent of reduction of inflammatory edema, indicating that there is an additive synergism between the two types of human IFNs.

The anti-inflammatory effect of human IFN alpha, but not that of human IFN gamma was almost completely suspended by the anti-human IFN alpha serum. This suggests that human IFN alpha is and possibly human IFN gamma might be responsible for the reduction of inflammation in mice.

The results show that mouse carregeenin paw edema test is a suitable model for studying in vivo cross-reactivty of pleiotropic

Table I. Inhibition of carrageenin paw edema in CFLP mice by human interferon alpha and gamma

Treatment	Dosis IU	% of inhibition	P
Human IFN alpha[1]	1.2×10^6	62	0.0005
Human IFN gamma[2]	1.0×10^4	68	0.0005
Human IFN alpha and Human IFN gamma	4.0×10^5 3.0×10^3	65	0.0005
Human IFN alpha and Anti-human IFN alpha serum[3]	1.2×10^6	12	N.S.
Human IFN gamma and Anti-human IFN alpha serum[3]	1.0×10^4	67	0.0005

[1] 2×10^6 IU/mg protein.

[2] 1.2×10^5 IU/mg protein.

[3] 1.2×10^6 neutralization units in 0.4 ml 3 hours prior to IFN, administered intravenously.

effect of human interferons. The pathophysiologic significance of the anti-inflammatory effect of various types of interferons administered parenterally, however, remains to be determined.

References

1 Béládi, I.; Tóth, M.; Mécs, I.; Tóth, S.; Taródi, B.; Pusztai, R.; Koltai, M. (1982): in The Clinical Potential of Interferons, p. 31—38. Ed. Reisaku Kono and J. Vilcek, University of Tokyo Press.
2 Bodo, G.; Palese, P.; Lindner, J. (1971). Proc. Soc. Exp. Biol. Med. 137, 1392—1395.
3 Havell, E. A. (1979): Virology 92, 324—330.
4 Koltai, M.; Mécs, I. (1973): Nature (Lond.) 242, 525—526.
5 Koltai, M.; Mécs, I.; Kása, M. (1981): Arch. Virol. 67, 91—95.
6 Levy, L. (1969): Life Sci. 8, 601—606.
7 Mécs, I.; Béládi, I. (1977): In: Proc. Symp. Preparation, Standardization and Clinical Use of Interferon. Yugoslav. Acad. of Sci. and Arts, Zagreb, p. 23—26.
8 Mécs, I.; Koltai, M. (1982): Acta Virol. 26, 346—352.
9 Samuel, C. E.; Farris, D. A. (1977): Virology 77, 556—565.
10 Whitman, Jr., J. E.; Crowely, G. M.; Tou, J.; Tredway, J. V.; Lung, C. L. (1981): J. Interferon Res. 1, 305—314.

I. Mécs, M.D., Institute of Microbiology, Univ. Medical School, Szeged (Hungary)

Contr. Oncol., vol. 20, pp. 224—232 (Karger, Basel 1984)

Age-Related Synthesis of Spontaneous Interferon in BALBtc Mice

Z. Błach-Olszewska; M. Cembrzyńska-Nowak; E. Kwaśniewska

Institute of Immunology and Experimental Therapy, Polish Academy of Sciences, Wroclaw, Poland

Introduction

The idea of the presence of the physiological interferon in an organism, has maturated for a long time in the minds of interferonists and genetics dealing with natural resistence of animals to viral infections. As reported by some authors, spleen cells or peritoneal cells freshly isolated from healthy animals produced spontaneous interferon in vitro (2, 4, 5, 6, 9, 11, 12). However, even if the spontaneous interferon production in vivo has not been reported yet, the connection between this interferon and natural resistance seems to be undeniable. Though the problem of spontaneous interferon has been signalized, there is still much to be elucidated.

It is known that the spontaneous interferon, like the induced interferons require mRNA and protein synthesis, in other words, transcription and translation processes are necessary to its synthesis. So, actually it is a synthesis de novo (4). Kinetics and the optimal conditions of its productions are also known (2, 4). Mouse spontaneous interferon was found to be type I (1, 2), probably it is IFN-β (our unpublished data). This IFN is produced by the Ia negative fraction of macrophages isolated from peritoneal cavity, spleen, bone marrow and lungs (6, 13, and our unpublished data). It was also found during the cultivation of freshly isolated liver cells (1, 10). But till now, we have no direct evidences for the function of this interferon in an organism and only indirect date are known. We do not know anything about the production of spontaneous interferon or mechanism unlocking its synthesis in organism. In vitro it is enough to lower the incubation temperature for the onset IFN-mRNA synthesis (4).

It is probable, that in the organism this process occurs in an other way. There are a lot of endogenous agents as: enzymes, hormones, lymphokines, some interactions between the cells, and a lot of exogenous agents which may take part in the onset of spontaneous interferon synthesis (3).

Unknown is the reason of the disturbance in the natural antiviral and antineoplastic resistance and the role of physiological interferon in these processes.

Particularly interesting seems the natural resistance or its lack during the maturation or senescent processes.

Since no such data were available a study was undertaken in mice of different age to provide insight in the age-dependent changes in spontaneous interferon production. Very young (less than 1 month) as well as very old (almost 2 year-old) mice were included in the experiments. For comparison, the age-related changes in interferon production induced by lipopolysaccharide of E. coli or by Newcastle disease virus were studied. These experiments were performed on explanted peritoneal cells in vitro as well as in vivo after intravenous inoculation of mice.

Materials and Methods

Animals

BALB/c mice 3-week to 2 year-old were used. They were maintained at the Inbred Mice Center of the Institute of Immunology and Experimental Therapy, Wrocław.

Preparation of Peritoneal Cells

Resident peritoneal cells were used for experiments. To obtain peritoneal cells, mice were killed by exsanguination and 5 ml of Eagle's medium supplemented with 10% calf serum was injected into the peritoneal cavity and immediately removed. After centrifugation (800 rpm, 10 min) cell suspension containing 3×10^6/ml was prepared.

Production of Interferons by Peritoneal Cells

a) Spontaneous interferon

The suspension of peritoneal cells was incubated at 26° C. After 24 h the supernatant was taken off and tested for interferon activity.

b) Interferon induced by LPS

Freshly isolated peritoneal cells (3×10^6 cells per ml) were incubated at 26° C with E. coli 0111:B4 lipopolysaccharide (DIFCO) in the concentration of 1 μg/ml. Twenty four hours later the medium was collected and tested for interferon activity.

c) Interferon induced by Newcastle disease virus (NDV)

This interferon was obtained by stimulation of peritoneal cells with $10^7 EID_{50}$ NDV per ml (NDV, the allogeneic strain Radom). After 30 min incubation, the virus was removed and the cells were suspended in Eagle's medium with 10% calf serum. Then the cells were incubated 24 h at 37° C. After this time the medium was collected and dialysed against glycine/HCl buffer (pH 2) for 72 h and redialysed to phosphate buffer (pH 7.4) for another 24 h.

Production of LPS and NDV-Induced Interferons In Vivo

Mice were inoculated intravenously with 1 μg of lipopolysaccharide E. coli or 2×10^8 EID_{50} NDV. After 2 and 5 hrs respectively, the mice were exsanguinated and their sera were checked for interferon activity.

Interferon assay

Interferon was assayed by the plaque reduction method using L-Borgen mouse cells and Col MM (EMC) virus as a challenge. One interferon unit was expressed as the reciprocal of the highest dilution of interferon resulting in 50% inhibition of plaque formation as compared with control cultures.

Results

Conventional BALB/c mice born at the same time and breeded under the same conditions were used in the experiments. These mice were periodically examined for some pathogens such as: Sendai, Rheo-3, Theiler GDVII, K, Ectromelia, PVM viruses and some bacteria.

In general, about 300 mice were separately tested for an age-dependent production of interferons.

The washouts of peritoneal cavity of individual mice at appropriate age were prepared. Each suspension was divided into 3 parts. The first part was left for spontaneous interferon production at 26° C, the second was intended for LPS-induced and the third for NDV induced interferon production (tab. I). As seen in table 1

Table I. Spontaneous interferon production by peritoneal cells of young individual mice

Age of mice (weeks)				
4	5	6	7	8
<5[1]	<5	20	40	320
5	<5	<5	10	320
<5	<5	5	5	40
<5	<5	10	10	80
<5	20	<5	10	160

[1] Titer of interferon U/ml.

the capacity of spontaneous interferon production developed in parallel with the maturation process. Very young, 4.5-week old mice were not found to produce this interferon. It was found not earlier than in peritoneal cells of 6—7 week old mice. The best spontaneous interferon producers were 8-week old mice (fig. 1). Usually this type of interferon was produced by animals until the age of 8 months. The older mice generally did not produce it. It should be added that about $5^{0}/_{0}$ of the mice in the age of 6-week to 8-months were negative when tested for the spontaneous interferon production and some of the old mice (about $5^{0}/_{0}$) produced small amounts of interferon.

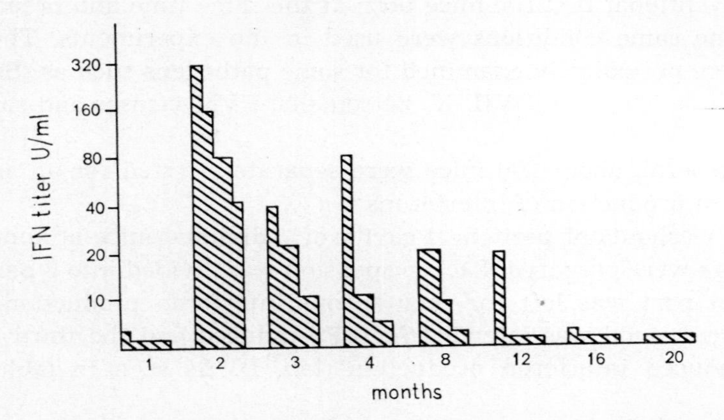

Fig. 1. Spontaneous interferon production by peritoneal cells of individual mice in different age.

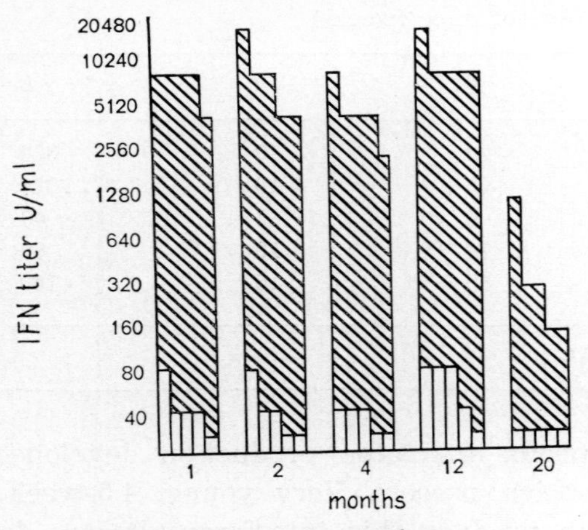

Fig. 2. LPS and NDV Induced interferons in sera of mice intravenously inoculated.

Table II. Comparison of interferon synthesis by noninduced and induced by NDV or LPS peritoneal cells of individual mice

Age of mice (months)	Titer of interferon in peritoneal cells		
	noninduced	induced with LPS	induced with NDV
1	<5	80	640
	5	80	640
	<5	80	320
	<5	40	640
	<5	40	640
2	320	640	2560
	320	320	1280
	40	80	640
	80	320	640
	160	640	640
12	<5	40	160
	<5	20	80
	<5	160	80
	<5	80	160
	<5	80	160
20	<5	20	160
	<5	20	160
	<5	20	320
	<5	20	320
	<5	10	160

The comparative investigations on the production of non induced, spontaneous and induced interferons showed that the peritoneal cells from the very young animals that do not produce the spontaneous interferon were able to synthetize a significant amount of interferons after induction with LPS or NDV (tab. II). Such interferons could be produced all over the life with downward tendency in older mice.

To obtain more complete picture of the ability of BALB/c mice of different age to produce interferon, the animals were inoculated intravenously with either NDV or LPS. The sera of the age-differentiated mice were than collected and examined for antiviral activity. The mice, were found to produce interferon independently of age. Only very old mice (over 1.5 year) had decreased titers. The results are presented in fig. 2.

In contrast to presented results with spontaneous interferon production, the synthesis of interferon in an organism after intravenal injection of inducers was observed much earlier (3 weeks) and lasted longer (12 month).

Discussion

Considering the results of our experiments dealing with the spontaneous interferon synthesis, 3 periods of time in BALB/c mice life could be distinguished: 1. the first five weeks of life in which peritoneal cells of mice do not produce spontaneous interferon. 2. 6th week to 8th month, when isolated peritoneal cells produced interferon, 3. one year old or older mice in which spontaneous interferon is not longer produced.

1. In the first period the intensive development and very high sensitivity to different viral infections are observed. At this time sensitive and simultaneously labile controlling system has to exist to take care of the proper development of the organism. The hormons, growth stimulating and other factors influence development of cells and their differentiation. On the other hand, the very precise control, gentle inhibition and protection of the young organism against infections have to exist.

Physiological interferon seems to be indispensable as a negative control of cell proliferation and cell maturation or as the defensive agent against different infections.

However, noteworthy is that we did not find spontaneous interferon production by peritoneal cells of young mice.

It might be that some agents produced by young organisms (inhibitors isolated from young animal organs, growth factors) interfere with the activity of spontaneous interferon, or perhaps particular intensity of metabolic processes lead to its destruction. On the other hand, functional immaturity of phagocytic system cells may be responsible for this phenomenon.

2. Peritoneal cells of 8-week-old BALB/c mice are the best

producers of spontaneous interferon. At that time the animals reach the best health stability.

Comparing the results of Mogensen's experiments (8) with ours a surprising convergence could be found. In our experiments 8-week-old BALB/c mice were the best producers of spontaneous interferon and Mogensen found 8-week-old BALB/c mice most resistant against Herpes simplex virus type 2. It is possible that this age convergence is not accidental.

A relative health stability lasted in BALB/c mice to reach 8 months of life make the best conditions for reproduction and for breeding of littermates.

3. The age over 8 months seems to be critical for BALB/c mice. The production of spontaneous interferon is broken. Mc Cubrey and Risser (7) observed the sudden rise of the spontaneous murine leukemia virus production in BALB/c mice over 8-month-old. And again the age-related convergence is striking, but the underlying mechanisms remains unclear. Animals over 8-months old begin slowly to be away, eliminating numerous functions of less and less efficient cells. The elimination of the spontaneous interferon production is an example of such dying down function.

May be that for some time this production could be restored by using interferon inducers but later this function decays as well making the animal defenceless against neoplastic and other diseases.

Summary

The capacity of BALB/c mice in different age to produce spontaneous interferon was studied. There was no activity of spontaneous interferon in peritoneal cells isolated from young 4—5 week old mice. The cells of mice 6 week to 8 month old but not the cells of older than 8 month animals were able to synthetize spontaneous interferon. On the other hand the peritoneal cells of BALB/c mice were able to synthetize interferons after induction with NDV or LPS all over the life with tendency to lower the titre in advanced old age animals. To obtain complete picture of IFN production, BALB/c mice were inoculated intravenously with NDV or LPS and sera were checked for antiviral activity. The interferon was produced independently on age. Only very old (20 months) mice were found to produce less IFN.

References

1 Arnheiter, H.; Haller, O.; Lindenman, J. (1980): Virology 103: 11—20.
2 Blach-Olszewska, Z.; Cembrzyńska-Nowak, M. (1979): Acta biol. med. Germ. 38: 765—773.
3 Bocci, V. (1981): Biol. Rev. 56: 49—85.
4 Cembrzyńska-Nowak, M.; Błach-Olszewska, Z. (1979): Acta Biol. Med. Germ. 38: 775—779.
5 De Maeyer, E.; Fauve, R. M.; De Maeyer-Guingard, J. (1971): Annales Institut Pasteur 120: 438—446.
6 Ito, Y.; Oaki, H.; Kimura, Y.; Takano, M.; Shimokata, K.; Macuo, K. (1981): Infect. Immunity 31: 519—523.
7 Me Cubrey, J.; Risser, R. (1982): J. Exp. Med. 156: 337—347.
8. Mogensen, S. C. (1978): Infect. Immunity 19: 46—50.
9 Nagano, Y.; Kojima, Y.; Arakawa, J.; Kanashiro, R. S. (1966): Japan J. Exp. Med. 36: 481—487.
10 Renton, K. W.; Deloria, L. B.; Mannering, G. J. (1978): Mol. Pharmacol. 14: 672—681.
11 Smith, T. J.; Wagner, R. R. (1967): J. Exp. Med. 125: 559—577.
12 Tálas, M.; Szolgay, E.; Rozsa, K. (1972): Archiv für die gesamte Virusforsch. 38: 149—158.
13 Wilson, M. R.; Cate, T. R. (1982): Infect. Immunity 38: 1249—1255.

Z. Blach-Olszewska, M.D., Institute of Immunology and Experimental Therapy, Polish Academy of Sciences, 53—114 Wrocław, Czerská 12 (Poland)

Contr. Oncol., vol. 20, pp. 233—243 (Karger, Basel 1984)

Effect of Experimental Water-Immersion Stress on the Induction of Interferon and Viral Infection

Y. Kojima; Y. Yamaguchi; H. Kuramoto; N. Shibukawa, S. Tamamura

The Kitasato Institute, Tokyo, Japan

Introduction

We have been performing experiments to test interferon induction by endotoxin in rabbit lymphoid cell cultures, repeating the test once a week for the past 6 years. Acording to our observation, interferon production after treatment with endotoxin was markedly depressed in winter from late December to early February. It is known empirically that the natural resistance of humans to infection of viruses changes when exposed to cold environment. Thus, people are liable to catch cold in winter. Especially, this has been the case frequently, when the weather changed suddenly from high to low temperature during the course of a day or when people put on rain-drown clothes for a long time.

Ruiz-Gomez and Sosa-Martinez (9) reported that adult mice infected with Coxsackie B1 virus and kept at 4° C died because they could not produce interferon at this temperature in contrast with animals kept at 25° C. However, Postic et al. (8) reported that cooling of rabbits at 4° C did not affect the titer of endotoxin-induced interferon.

In the present paper, we demonstrate that adult mice and young rabbits immersed in a water bath at 23° C for several hours cannot produce interferon after treatment with poly I:C and endotoxin in contrast with animals kept at 37° C.

Materials and Methods

Cell Culture

Rabbit RK-13 cells and mouse L cells were cultivated in Eagle's minimum essential medium (MEM) supplemented with 10% calf serum.

Viruses

New Jersey serotype of Vesicular stomatitis virus (VSV) was propagated in RK-13 cells and its yield was 0.6×10^7 plaque forming units (PFU)/0.2 ml as tested in RK-13 cells. D1 strain of Vaccinia virus was propagated in the back skin of rabbits and its yield was about 3.5×10^3 pock forming units/0.1 ml as tested by intravenous injection into the tail of mice.

Animals

Male specific pathogen-free (SPF) Japanese white rabbits weighing about 1 kg were used. Six-week old male ddy mice were purchased from the Shizuoka Agricultural Cooperative Association.

Interferon Inducers

Polyriboinosinic-polyribocytidilic acid (poly I:C) was purchased from Yamasa Shoyu Inc., Choshi, Japan. Endotoxin used was a lipopolysaccharide preparation (E. coli 0127:B8) purchased from Difco Lab.

Preparation of Rabbit Spleen Cells

Rabbits were sacrificed by cardiac puncture and the spleens were obtained. Cell suspensions were prepared by dispersing with the aid of a forceps and suspending in MEM solution. The cell suspensions were filtered through 90 mesh and diluted with MEM supplemented with 10% calf serum to a concentration of about 1×10^7 cells/ml.

Interferon Assay

The activity of rabbit interferon was assayed in RK-13 cells, while that of mouse interferon in L cells. The interferon titer was expressed as reciprocal of the dilution showing 50% reduction of the plaque count of challenge virus. About 100 PFU of VSV served at the challenge.

Water-Immersion Procedure

Male mice were immersed to the depth of the xiphoid in a water bath kept at the indicated temperature. Male rabbits were fixed in a narrow chamber immersed in a water bath whose temperature was kept at 23° C.

Results

Effect of Temperature on Water-Immersion Stress

It is known that temperatures of 23° C to 25° C may be the most convenient for uniform production of peptic ulcer (9, 11). In our initial experiments, mice were injected intravenously with 50 μg of poly I:C and immediately immersed in water bathes kept at 15, 23, 28 od 37° C. After 3 hours, the mice were bled and their sera collected. Usually sera from 3 mice were pooled and stored in an ice box until assayed. In water-immersion at 15° C mice were drowned after getting too cold. As shown in tab. I, inhibition of interferon production by water-immersion at 23° C was observed within several hours, although peptic ulcers occurred in most of the stressed mice in 15 to 20 hours at 23° C. In water-immersion at 28° C, interferon yield was about 10 to 20 % in contrast with

Table I. Effect of water-immersion temperature on the interferon induction with poly I:C in mice

Temperature of water bath (° C)	Interferon yield (IU/ml)		
	Exp. I	Exp. II	Exp. III
15	D[1]		
23	<15	<15	<15
28	155		70
37	400	360	390
Untreated	730	380	650

[1] Mice were drowned after being chilled in the water-immersion at 15° C.

Mice were injected intravenously with 50 μg of poly I:C and immersed immediately to the depth of the xiphoid in a water bath whose temperature was kept at 15, 23, 28 or 37° C. After 3 hours, the mice were bled and the sera collected. Usually, sera from mice were pooled.
Water-immersion

untreated mice. A slight impairment of interferon production was observed in water-immersion at 37° C. The effect of stress was due to the temperature rather than sensory shock at the water-immersion.

Effect of the Immersion Period

In the next experiment, we examined the effect of immersion period at 23° C (tab. II). As the first treatment, mice were exposed to water-immersion at the indicated times and then exposed or left unexposed to water-immersion for 3 hours after injection of poly I:C (50 μg). The gastric bleeding and contraction were apparently observed 1 hour after the water-immersion stress exposure. However splenic contraction could not be observed in response to the stress stimulus. When the mice were exposed to water-immersion for a longer period, recovery of interferon induction was retarded. When mice were freed from the water-immersion the general conditions of mice were quickly improved and showed production

Table II. Effect of immersion period on the interferon induction with poly I : C in mice

Exp. number	First treatment	Second treatment after induction	IFN yield (IU/ml)	
	Time of water-immersion (hr)	Water-immersion	Exp. I	Exp. II
1	Untreated	—	2360	570
2	Untreated	+	<15	<15
3	1	—	410	210
4	1	+	<15	<15
5	2	—	480	120
6	2	+	<15	<15
7	4	—	170	62
8	4	+	<15	<15
9	6	—	70	55
10	6	+	<15	<15

After the first treatment, the mice were injected with 50 μg of poly I : C. Three hours after the second treatment, the mice were bled and pooled sera assayed for IFN yield.

of interferon. Twenty-four hours after getting free from the stress the mice recovered completely to produce interferon normaly although this is not indicated in the table.

Effect of the Water-Immersion Stress on Endotoxin-Induced Interferon

Since the effect of ambient and body temperatures on interferon production in rabbits injected intravenously with virus differed from that after injection of endotoxin, we tested the effect of water-immersion stress on the endotoxin-induced interferon in mice and rabbits (tab. III). Mice and rabbits were exposed to water-immersion for 6 hours before stimulation with endotoxin to produce interferon. The results obtained by the endotoxin induction were similar to the above results obtained with poly I:C. In the first and second treatments, mice and rabbits exposed to water-immersion could not produce interferon.

Table III. Effect of water-immersion stress on the interferon induction with endotoxin in mice and rabbits

Experimental animals	First treatment Time of immersion	Second treatment Water-immersion	IFN yield (IU/ml)
Mice	Untreated	—	375
	Untreated	+	160
	6 hours	—	70
	6 hours	+	<15
Rabbits	Untreated	—	240, 170
	6 hours	+	<15 <15

Mice and rabbits were stressed for 6 hours, they were stimulated with 50 μg of endotoxin to produce interferon. The animals were bled 2 hours after the intravenous injection with endotoxin.

Effect of the Water-Immersion Stress to Virus Infection

It is well known that anti-interferon serum is capable of enhancing viral multiplication in vitro as well as in vivo by its neutralizing effect on interferon produced during the development of the virus (1, 2, 3). Water-immersion stress may also be capable of enhancing viral multiplication because interferon production is inhibited. In the following experiments, we examined the effect of water-immersion stress upon mouse tail lesions produced by Vaccinia virus. In a preliminary experiment, mice were exposed to water-immersion for 4 hours from 0 to 2 days before injection with lesions on the tail as revealed in the control unexposed group. Then, mice were immersed for 4 hours from 1 to 4 days after viral infection. As shown in tab. IV, when the mice were exposed to water-immersion on the first or second day after infection, the number of tail lesions were reduced. However, it was shown that the number of tail lesions increased when water-immersion was performed at the time of the development of the virus on the 4th day after infection.

Table IV. Effect of water-immersion stress of Vaccinia infection of mice

Stressed on day after infection	Number of tail lesions	'*t*' test
Not stressed	17 ± 5	
1	8 ± 5	$p < 0.05$
2	7 ± 5	$p < 0.05$
3	25 ± 15	N.S.
4	39 ± 19	$p < 0.05$

Groups of 10 mice were injected intravenously with 0.1 ml of Vaccinia virus and then exposed to water-immersion for 4 hours at the indicated intervals. Tail lesions were counted after 8 days.

Possible Mechanisms of the Water-Immersion Stress

Other experiments were conducted in an attempt to elucidate the mechanism of the stress-induced inhibition of interferon production.

In mice

We tested whether interferon production recovered or not when the water-immersion was stopped after induction with poly I:C. As seen in fig. 1, the mice lacked in the ability to produce interferon after freed from the water-immersion stress.

In Rabbits

A rabbit exposed to water-immersion for 6 hours at 23° C was bled. Serum from the stressed rabbit was added to spleen cell cultures prepared from a normal rabbit. Immediately after being incubated at 37° C for 2 or 5 hours, the spleen cell cultures were incubated with 10 ng of endotoxin at 25° C for 1 day. No significant alteration was noted in the ability of the spleen cells to produce interferon in vitro in contrast with untreated rabbit cells. This experiment indicated the absence of stress-induced humoral factors mediating the impairment of interferon production.

In the following experiment, rabbits were exposed to water-immersion for 6 hours and then injected intravenously with 50 μg of endotoxin. The rabbits were divided into two groups. 1. After a further immersion for 1 hour, a rabbit was sacrificed by cardiac puncture and spleen cells therefrom were further incubated with MEM at 25° C for 1 day in vitro. Interferon production no longer proceeded in the spleen cell culture. 2. After a further immersion for 2 hours, another rabbit was sacrified and the spleen homo-

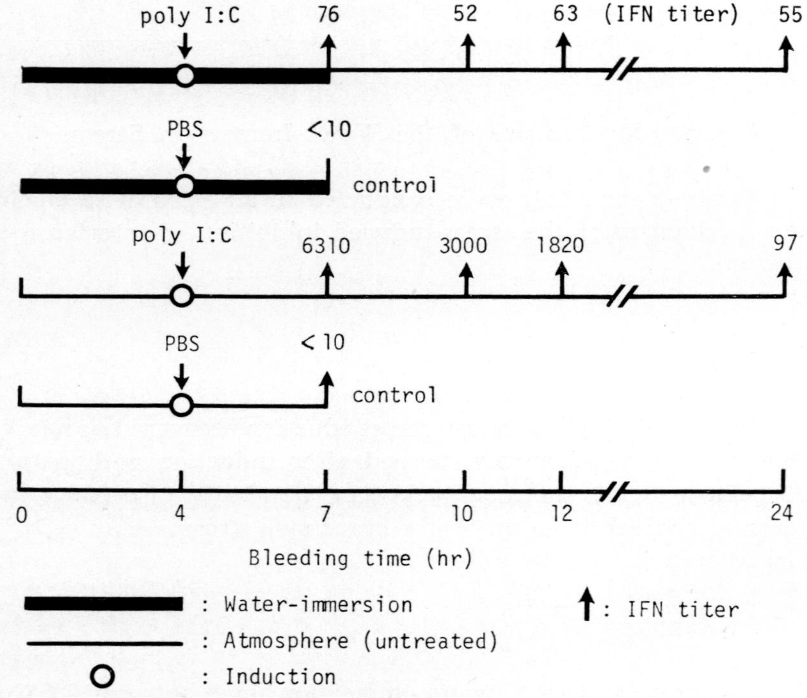

Mice were exposed to water-immersion for 4 hours at 23°C and injected intravenously with 100 ug of poly I:C. The mice were further exposed to the stress for 3 hours and then freed from the stress.

Fig. 1. **Lack of production of interferon after removing the water-immersion stress.**

genized to detect interferon production. Interferon production in the extracted fluid could not be detected.

As another approach, when spleen cells from a rabbit stressed for 6 hours at 23° C were incubated with 10 ng of endotoxin in vitro, interferon production was normaly observed as in the case of spleen cells from untreated rabbits.

These results may indicate the absence of penetration to spleen cells of endotoxin.

Discussion

There are many data concerning the effect of temperature on interferon production in vitro. These studies mostly aimed at analysis of interferon induction or at conditions for a large scale production of interferon in vitro (6, 7). Viewing from a different angle, influences of cold environments have interested us as a practical problem related to human diseases. According to our data, interferon production with endotoxin in cultured lymphoid cells from rabbits was markedly depressed in winter. The poor induction of interferon coincided with the prevalence of influenza in this season. For confirmation of this relationship, we examined the effect of cold treatment of mice and rabbits. For this purpose the technique of water-immersion known to produce peptic ulcers in these animals was adopted, since otherwise the hair covering the skin protected the animals from being directly exposed to atmospheric temperature. We could observe clearly a transitory inhibition of interferon production when the animals were exposed to the water-immersion stress at 23° C. These phenomena occurred with poly I:C as well as with endotoxin used as interferon inducers. The degree of inhibition was higher than that exerted by electrical stress which was reported by Jensen (4, 5). The effect was rapid in appearance in the case of water-immersion and readily disappeared following freeing from it. However, even when mice were freed the water-immersion stress after induction with poly I:C, production of interferon could not be recovered. Our experiments in vitro indicated the absence of any stress-induced humoral factors, and spleen cells from stressed rabbits showed no alteration in ability to produce interferon in contrast with those of untreated rabbits.

It may be that the water-immersion stress influences blood flow to other tissues to prevent interferon producing cells from contacting with the inducers. In any event, the water-immersion stress may influence the host-parasite relationship. It should be remarked that the duration of cold environments relative to the conditions of virus multiplication in the human body is an important factor determining the manifestation of disease.

Summary

Mice which were exposed to water-immersion stress at various temperatures were injected intravenously with poly I:C. When the mice were exposed to the stress at 23° C, interferon production was inhibited. The effect was rapid in appearence after water-immersion at 23° C and readily disappeared removing the stress. However, even when the mice were freed from stress immediately after induction with poly I:C, production of interferon could not be observed. When the temperature was raised the effect decreased. Similar results were obtained in rabbits and mice when endotoxin replaced poly I:C. It was indicated that no stress-induced humoral factors were present, and spleen cells from stressed rabbits showed no alteration in ability to produce interferon in contrast with those of untreated rabbits. The number of tail lesions appearing after infection of mice with Vaccinia virus significantly increased when the mice were treated by water-immersion at the time of the development of the virus on the 4th day of infection.

References

1 Fauconnier, B. (1971): Path. Biol. 19, 575.
2 Gresser, I.; Tovey, M. G.; Bandu, M. T.; Maury, C.; Brouty-Boye, D. (1976): J. Exp. Med. 144, 1305.
3 Gresser, I.; Tovey, M. G.; Maury, C.; Bandu, M. T. (1976): J. Exp. Med. 144, 1316.
4 Jensen, M. M. (1968): J. Inf. Dis. 118, 230.
5 Jensen, M. M. (1973): Proc. Soc. Exp. Biol. Med. 142, 820.
6 Kojima, Y.; Yoshida, F. (1974): Japan. J. Microbiol. 18, 217.

7 Lackovič, V.; Borecký, L.; Šikl, D.; Masler, L.; Bauer, S. (1967): Acta virol. 11, 500.
8 Postić, B.; De Angelis, C.; Breinig, M. K.; Ho, M. (1966): J. Bact. 91, 1277.
9 Ruiz-Gomez, J.; Sosa-Martinez, J. (1965): Arch. Ges. Virusforsch. 17, 295.
10 Takagi, K.; Kasuya, Y.; Watanabe, K. (1964): Chem. Pharm. Bull. 12, 465.
11 Takagi, K.; Okabe, S. (1968): Jap. J. Pharmac. 18, 9.

Y. Kojima, M.D., The Kitasato Institute, 5-9-1, Shirokane, Minato-ku, Tokyo 108 (Japan)

Contr. Oncol, vol. 20, pp. 244—250 (Karger, Basel 1984)

Interferon Modifies Host Defence Mechanisms against Bacterial Agents

M. Degré; H. Rollag; G. Bukholm

Dept. of Virology, National Institute of Public Health and Wilhelmsens Institute of Bacteriology, Rikshospitalet, University of Oslo, Oslo, Norway

Introduction

Interferon (IFN) was originally described as an antiviral agent and its role in the host defence against viral infections is wel established. More recently it is increasingly recognized that IFN also acts on normal cells and numerous physiological activities can be modified. These activities are now commonly called non-antiviral activities. Such activities include cell functions involved in host defence against different non-viral infectious agents, including bacteria, chlamydia, mycoplasma and protozoa.

Both clinical and experimental observations indicate, that IFN is often present in the organism during bacterial infection (1, 10). Production of α and β IFN may be induced either by a preceeding viral infection, by the bacterial agents while γ IFN is induced as a part of the immunological reaction to the agent.

It is well known that viral infection may pave the way for a secondary bacterial invasion, and that viral infection may greatly influence the outcome of bacterial infections. This may at least in part be mediated by IFN, as it has been shown that injection of exogenous IFN can modify the course of bacterial infections (3). It seems reasonable that this effect of IFN is achieved by means of influence on several host defence factor(s), both specific and non-specific ones. We have concentrated our studies of IFN effect on phagocytic cells. Recently we have initiated studies on IFN effect on primary contact of bacteria with their target cells and on invasiveness of bacteria into these cells.

Materials and Methods

Interferon Effect on Phagocytic Cells

The experimental design and the materials employed have been described in detail elsewhere (11, 12). The major part of the in vitro experiments were done with non-stimulated mouse peritoneal macrophages (MPM). Human peripheral monocytes (HPM) and polymorphonuclear leukocytes (HPML) were employed in some experiments. In vitro cultures of cells were exposed to suspensions of either ^{32}P labelled E. coli or sheep red blood cells (SRBC). Uptake of bacteria by the phagocytes was assayed either by microscopical examination or by measuring the radioactivity after uptake of radiolabelled particles. Degradation of ingested bacteria by the phagocytes was determined by measuring the release of radioactivity to the supernatant. The adherence and uptake of SRBC was determined by microscopical examination. The effect of MuIFN $\alpha + \beta$ or γ and HuIFN α and β was tested. Morphological alterations of the phagocytes during cultivation with IFN were observed by means of phase contrast light microscopy and by scanning electron microscopy.

Interferon Effect on Invasiveness of Bacteria in In-Vitro System

Invasive capability of several bacterial species in cell cultures is well correlated to the pathogenicity in the human organism. Human pathogens as Salmonella and Shigella species are invasive in Hep-2 cells (8, 9). Differentiation between attached and internalized bacteria can be achieved by a combination of Nomarski differential interference contrast microscopy and UV incident light microscopy applied on the same microscope (2). We have used this system to examine the invasiveness of bacteria in IFN treated cells.

Results and Discussion

Interferon Effect on Phagocytic Cells

Previous findings (5) indicated that i.p. injected MuIFN $\alpha+\beta$ reduced the mortality due to pulmonary infection caused by Haemophilus influenza. IFN treatment also enhanced the clearance of particles from blood circulation in mice (4). We believed that these effects were most probably mediated by the activity of phagocytic cells. Alveolar macrophages, obtained from mice given the same i.p. injection of MuIFN had more pronounced in vitro phagocytic activity than cells from control mice (3). Alveolar macrophages are, however, difficult to obtain in sufficient numbers, the following series were therefore performed with peritoneal macrophages. MPM treated with fibroblast IFN fibroblast IFN ingested more bacteria than control MPM (tab. I). This effect was dose and time dependent, most pronounced at 10^2—10^3 units per ml of IFN, while very high

Table I. Influence of various concentrations of interferon on the uptake of E. coli by MPM

Interferon units per ml	No. of bacteria per macrophage[a]	Per cent of macrophages containing bacteria[a]
0	1.5 (1.3—1.6)	49 (44—58)
10	1.7 (1.4—1.9)	57 (50—66)
100	1.9[b] (1.7—2.4)	58[c] (55—64)
1000	2.1[b] (1.8—2.5)	63[b] (59—70)
2500	2.1[b] (1.6—2.5)	64 (50—79)

[a] Median and range.
[b] $p < 0.014$.
[c] $p < 0.029$.

concentrations, more than 5×10^4 per ml were inhibitory. The stimulatory effect was similar whether the particles were opsonized or not, i.e. whether they were ingested by the C3b, Fc or non-specific receptors. Phagocytosis experiments, performed at 4° C, which allows only attachment of particles to the phagocytes showed that

Table II. **Effect of MuIFN-β and MuIFN-γ on attachment (4° C) and phagocytosis (37° C) of non-opsonized and** IgG-opsonized E. coli by mouse peritoneal macrophages[a]

Treatment	Uptake of non-opsonized E. coli[b]					Uptake of IgG-opsonized E. coli[b]				
	1	10^1	10^2	10^3	10^4 IFN (U/ml)	1	10^1	10^2	10^3	10^4 IFN (U/ml)
MuIFN-β 4° C	—	129 ± 14	157 ± 7[c]	154 ± 4[c]	106 ± 10	—	110 ± 7	155 ± 15[c]	155 ± 14[c]	131 ± 22
MuIFN-β 37° C	—	123 ± 16[c]	138 ± 5[c]	143 ± 15[c]	103 ± 11	—	104 ± 2	135 ± 16[c]	137 ± 2[c]	118 ± 11
MuFIN-γ 4° C	96 ± 13	69 ± 13[c]	68 ± 9[c]	—	—	93 ± 11	74 ± 13[c]	51 ± 8[c]	—	—
MuFIN-γ 37° C	89 ± 3	76 ± 3[c]	55 ± 10[c]	—	—	93 ± 5	74 ± 6[c]	66 ± 7[c]	—	—

[a] Five experiments each with 4 replicates.
[b] Attachment and phagocytosis expressed as per cent of controls, and given as mean and SEM.
[c] $\alpha < 0.05$ compared with controls.

Table III. **Specificity of interferon effect on phagocytosis by macrophages from different mouse strains**

Preparation	Interferon Activity (units per ml)	Uptake of E. coli by MPM from ICR	BALB/c	C57B1/6
Experiment 1				
Mock IFN	<10	70.5 ± 3.2[a]	58.0 ± 6.8	63.0 ± 5.1
Mu-IFN-α + β	500	106.0 ± 5.1 (150)[b]	102.5 ± 11.0 (176)[b]	70.0 ± 5.9 (111)[c]
Mu-IFN-α + β + anti IFN globulin	<10	59.5 ± 2.8 (85)[d]	65.0 ± 5.2 (110)[d]	55.5 ± 8.8 (90)[e]
Experiment 2				
Mock IFN	<10	84.0 ± 4.5		
Mu-IFN-γ	10	64.0 ± 5.5 (76)[b]		
Mu-IFN-γ (10u/ml) pH 2 treated	<10	83.0 ± 9.1 (98)[d]		

[a] CPMM, mean of four cultures ± SD (% of controls in parenthesis).
Significance of difference: [b] $p < 0.05$ from mock IFN, [c] $p > 0.1$ from mock IFN, [d] $p < 0.05$ from IFN, [e] $p > 0.05$ from IFN.

the phagocytosis-enhancing effect could partially be explained by an increased number of receptor structures during cultivation with IFN $\alpha+\beta$ (tab. II). However, stimulation of receptor expression did not fully explain the enhanced phagocytic activity which suggested that additional cellular mechanism(s) are also influenced. Comparable results were obtained with HPM and HPML treated with HuIFN $\alpha+\beta$ (8).

MuIFN γ treatment of MPM depressed phagocytosis of particles by MPM via all three types of receptors. The reduced phagocytic activity was partially a result of decreased numbers of receptors available and partially by reduced internalization of particles already attached. This difference in effect of MuIFN β and MuIFN γ on phagocytosis was confirmed using MPM from the C57B1/6 genotype of mice, known to be low responders for MuIFN. MuIFN had little effect on phagocytosis, whereas MuIFN had a marked depressive effect on the phagocytic activity of MPM from these mice (tab. III). Also in the human system HuIFN depressed the phagocytic activity.

Morphological and biochemical examinations indicate that the increased phagocytic activity of IFN α and β treated cells is most probably a part of a general activation of macrophages. Other parameters of activation, like intracellular cyclic AMP levels (10), spreading on glass surfaces and size of spreading macrophages are also enhanced by treatment with IFN $\alpha+\beta$ (11). The mechanism of this activation remains to be determined. Some other functions are, however, not stimulated. Effect of IFN on degradation of the ingested particles and on the activity of lysozomal enzymes was negligible when MuIFN $\alpha+\beta$ in concentrations which can be obtained in the organism were employed (12).

Interferon Effect on Invasiveness of Bacteria in In-Vitro System

Enteroinvasive bacteria as Salmonella, Shigella and Yershinia species can invade cells grown in vitro in cell culture systems. A high proportion of Hep-2 cells incubated with a suspension of S. typhimurium for three hours contain intracellular bacteria. If the cells are preincubated with human leukocyte IFN the number of internalized bacteria is reduced (tab. IV). This effect is time and dose dependent. Maximal suppression of invasiveness was observed on cell cultures treated with 100 units of IFN per ml. The effect was less pronounced after treatment with 500 and 1000 units per ml and even less after 10 000 units per ml. Both the proportion of cells that contained bacteria and the mean number of bacteria per cell was reduced. The interferon nature of the activity was proven by the finding that specific anti-human leukocyte-interferon globulin neutralized both the antiviral activity and the activity on bacterial invasiveness.

Treatment of Hep-2 cells with human leukocyte IFN also reduced the invasiveness of S. paratyphi B similarly to that of S. typhimurium. On the other hand, invasiveness of Shigella flexneri another enteroinvasive Gram negative rod was not influenced by treatment of cells with IFN.

The number of observed intracellular bacteria is a sum of several processes, i.e. adhesion and penetration of bacteria and to

Table IV. Number of intra-cellular S. typhimurium in HEp-2 cells following pre-treatment of cells with various concentrations of interferon

Interferon concentration (U/ml)	0	50	100	500	1000	10,000
Number of bacteria per cell						
0	271	297	390	369	375	323
1	2	9	2	7	2	20
2	10	16	2	5	2	14
3	32	22	4	6	2	18
4	25	22	1	9	5	9
5	17	0	—	0	2	9
6	13	4	—	2	1	8
7	15	—	—	—	2	8
8	0	—	—	—	—	6
9	7	—	—	—	—	2
11	—	—	—	—	1	—
15	1	2	—	—	—	—
18	1	—	—	—	—	—
21	—	1	—	—	—	—

some extent intracellular multiplication. Any of those parameters might be influenced by interferon.

Our data indicate that in addition to activation of specific host defense mechanism interferon may influence the entry of bacteria into cells and thereby directly interfere with establishment of bacterial infection.

References

1 Baron, S.; Howie, V.; Langford, M.; MacDonald, E. M.; Stanton, J.; Weigent, D. (1981—1982): Tex. Rep. Biol. Med. 4: 151—157.

2 Bukholm, G.; Johansen, B. V.; Namork, E.; Lassen, J. (1982): Acta path. microbiol. immunol. scand. B, 90: 403—408.

3 Degré, M.; Rollag, H. Jr. (1980): Acta pathol. microbiol. scand., Sect. B, 88: 177.

4 Degré, M.; Rollag, H. Jr. (1979): J. Reticuloendothel. Soc., Vol. 25, 5: 489.

5 Degré, M.; Rollag, H. (1981—1982): Tex. Rep. Biol. Med., 41: 388—394.

6 Degré, M., Rollag, H. (1982): J. Interferon Res., 2: 151—157.

7 Degré, M.; Rollag, H.; Kjeldsberg, E. (1982): Phil. Trans. R. Soc. Lond., B. 299: 131—133.

8 Gerber, D. F.; Wathins, H. H. S. (1961): J. Bacteriol., 82: 815—822.

9 Gianella, R. A.; Washington, O.; Gemski, P.; Formal, S. B. (1973): J. inf. Dis., 128: 69—75.

10 Haahr, S. (1968): Acta pathol. microbiol. scand., 73: 264.

11 Rollag, H. (1979): Acta pathol. microbiol. scand., Sect. C, 87: 99—105.

12 Rollag, H.; Degré, M. (1981): Acta pathol. microbiol. scand., Sect. B, 89: 153—159.

13 Rollag, H.; Mørland, B.; Degré, M. (1982): Acta pathol. microbiol. scand., Sect. B, 90: 113—118.

M. Degré, M.D., Dept. of Virology, National Institute of Public Health and Wilhelmsens Institute of Bacteriology, Rikshospitalet, University of Oslo, Oslo (Norway)

IV. Consequences of Interferon Administration to and Production in the Human Organism

Contr. Oncol., vol. 20, pp. 251—269 (Karger, Basel 1984)

The Clinical Application of Fibroblast Interferon — An Overview

A. Billiau

Rega Institute, University of Leuven, Leuven, Belgium

Introduction

Leukocyte interferon preparations, suitable for clinical trials became available at a relatively early time: already in the late sixties patients with various life-threatening diseases received such interferon in phase I/II trials. Fibroblast interferon was first injected systemically by our research group in 1976 (15). Already in 1978 we described observations on more than 30 patients with various clinical conditions: prophylaxis of viral disease following renal transplantations (48), chronic active liver disease (47), active and fulminant hepatitis and verruca vulgaris (6). Later on we also described observations in cancer patients (9, 28, 37): breast cancer, myeloma and head and neck epithelioma.

The main conclusions from our early studies was that the side-effects after local injection such as tissue reactions (16) or fever after systemic administration (3, 6) of interferon were not to be minimalized (6) as had been done in previous studies with leukocyte interferon, but that they represented a serious problem.

We also felt that fibroblast interferon was only minimally effective, if effective at all, in the diseases that we had chosen for treatment. A major problem, also recognized in these early studies, was the fact that fibroblast interferon had a different, and possibly a less favorable, pharmacokinetic behavior (3, 4, 18) in that blood levels were lower after intramuscular injection than those obtained with similar injections of leukocyte interferon. As a result German and Japanese workers who subsequently became engaged in trials with fibroblast interferon, have preferred to use intravenous infusions.

The production of HuIFN-β from fibroblasts is generally considered to be a much more expensive enterprise than the production of HuIFN-α from leukocytes. Around 1981 the prospects in this regard suddenly changed, as it seemed to become possible to produce HuIFN-β using bacteria or other microorganisms into which the HuIFN-β gene had been artificially introduced (14). Nevertheless the production of natural fibroblast interferon was pursued by at least two pharmaceutical companies: Toray Industries in Japan and Rentschler GmbH/Bioferon in West-Germany. In contrast, the pilot production plant of our Institute was dismantled because we felt that the costs were not outweighing the potential clinical benefits as they appeared from the early pilot trials.

In the present review I will only discuss the more recent clinical trials which were done mainly in Japan and in West-Germany. Some of the results are not yet available in published form and are quoted from abstracts of meetings. Partial data can be found in the Proceedings of meetings held in Bonn (34), Oiso (27) and Rotterdam (12, 13).

Antiviral Studies in Europe

Topical Administration

Scott et al. (39) have done a placebo-controlled double-blind trial, in which fibroblast interferon was instilled intranasally 4 times daily starting 1 day before intranasal challenge with rhinovirus 4. There was no measurable effect on virus shedding or clinical symptoms. This is in contrast to other studies of the same group in which they applied similar protocols using leukocyte interferon.

Very few if any other trials using topical administration of fibroblast interferon have been done recently in Europe.

Systemic Administration

1. Laryngeal papilloma. Schouten et al. (38) reported on two cases of juvenile laryngeal papillomatosis, treated with the two types of interferon: fibroblast and leukocyte. The dose was 3×10^6 units given 3 times weekly. The subcutaneous route was chosen for administering fibroblast interferon; leukocyte interferon was given intramuscularly. Fibroblast interferon was given first for 3 months, followed after an interruption of treatment for several months by a course of leukocyte interferon also given for 3 months. Under both types of treatment the papillomas clearly regressed. They recurred in the interim period when the treatment was discontinued and also after arrest of treatment.

Göbel et al. (19) described two cases of children with laryngeal papillomatosis in which they tried to compare the effects of fibroblast and leukocyte interferon. The former was given at a daily dose of 2×10^6 units by intravenous infusion; the latter was given subcutaneously at the same dose. Even after 6 weeks the fibroblast interferon treatment failed to affect the course of the disease, while leukocyte interferon therapy caused regression within about 4 weeks.

Ninane et al. (University of Louvain and Rega Institute, Belgium; not published) treated a single case of severe laryngeal papillomatosis in a 22 months old infant. The patient received intramuscular injections of 2.4×10^6 units 3 times a week for 5 to 6 weeks. Slight fever (38.5° C) developed after the injections, but the laryngeal vegetations continued to develop during therapy. Subsequently the schedule was switched to injections of leukocyte interferon (3×10^6 units, once a week for 8 weeks). Again fever developed and the vegetations continued to develop. Subsequent to this failure the patient was left without specific treatment, except for regular surgical removal of the vegetations. After 1 year, improvement occurred spontaneously.

2. Herpes zoster. In an uncontrolled study Heidemann et al. (22) obtained satisfactory results in generalized zoster in cancer patients. Fibroblast interferon was given by intravenous infusion (0.5×10^6 units per day). The results were comparable to those described in the controlled trial with leukocyte interferon, done in 1981 by Merigan et al. There was reduction of perivascular erythema

within 24 hr, disappearance of pain in 48 hr, arrest in development of new vesicles and progressive healing of existing vesicles.

3. Chronic hepatitis B. Following the relatively discouraging results reported with fibroblast interferon in early studies (47, 4, 33), a controlled randomized trial is now being performed in West-Germany in which interferon and adenine arabinoside are given in a combination treatment (36). Fibroblast interferon was given daily by intravenous injections of 2×10^6 units (1st week), 4×10^6 units (3rd and 4th week), 8×10^6 units (6th and 7th week) or 16×10^6 units (9th and 10th week). Arabinoside was given in the intervening weeks. During the treatment DNA polymerase decreased and remained low or undetectable for several months. The authors are anticipating that repeated courses of treatment might be able to prevent or arrest the progression of disease, but current available information does not yet allow to be affirmative in this respect.

Antiviral Studies in Japan

Topical Administration

In Japan, herpes simplex, dendritic keratitis and viral warts were evaluated as possible viral targets for human fibroblast interferon. Most studies in Japan are completed only recently or are still in progress. Published information (27) is only fragmentary. Recently, Yamasaki (49) discussed the Japanese trials, giving extensive detailed information. The present overview is an excerpt for this.

Following some preliminary trials in which the interferon was given as eye drops to patients with dendritic keratitis, a multicenter, randomized, double-blind study was done. A low (10^3 units/ml) and a high (10^6 units/ml) dosage regimens were compared. The study comprised 67 patients in total. The interferon was given as eye drops 4 times daily and the efficacy of treatment was compared by grading the clinical evolution on an arbitrary scale ranging from -1 (aggravation) to $+3$ (cured in 4 days). From this an efficacy score (% response) was calculated. If all patients were taken into consideration there was only a slight difference between low and high dosage regimen (18/36 vs. 11/32). However, if only those patients were considered whose eye-scrapings were proven to contain HSV antigen, the difference was more pronounced (13/21 vs. 6/20) and statistically significant.

In another controlled study fibroblast interferon was compared with IDU and placebo (albumin solution) as adjuvant to debridement. The study comprised 44 patients. The efficacy rates (%) of the treatments were 80.0 (20 patients with interferon), 70.6 (17 patients with IDU) and 58.8 (17 patients with placebo). Thus, the effect of interferon was not better than that of IDU, yet the side-effects (pain, allergic reactions) were less pronounced with fibroblast interferon.

Initial studies on viral warts were uncontrolled. Fibroblast interferon was injected intratumorally (potency 0.4×10^6 units/ml; 0.05 to 0.1 ml once a week). Verruca vulgaris (111 lesions in 43 patients) responded excellently (efficacy : 97 %). With verruca plana juvenilis (21 lesions in 8 patients) the efficacy was only 25 %. Good responses were also seen in 2 cases of condyloma accuminatum.

The low response rate of verruca plana juvenilis was considered possibly to be due to the difficulty of intratumoral injection. An uncontrolled trial on 50 patients using systemic, subcutaneous (3×10^6 units/week) rather than intratumoral injection yielded only a 30 % response rate. A controlled trial is now underway.

The high response rates of verruca vulgaris to local injection, on the other hand, was confirmed with a 83 % response rate on a total of 158 lesions. This in turn prompted the Japanese workers to perform a double-blind, placebo-controlled study. Out of 64 lesions, 52 (81 %) responded with cure, against only 11 (17 %) cures in the placebo group.

On this basis it was concluded that local administration of fibroblast interferon is remarkably effective for verruca vulgaris. Effectiveness in verruca plana juvenilis is considered not yet to be established.

Intravenous Administration

Suzuki et al. (42, 43) investigated the evolution of disease parameters in 65 patients with HBeAg-positive chronic active hepatitis.

The interferon was given by intravenous infusions (6×10^6 units per day for another 3 weeks). Polymerase levels decreased and in 59 % of cases became undetectable at the end of the treatment.

Antitumor Studies in West Europe and North America

Various trials using fibroblast interferon are done in Europe. Most of these are done in the Federal Republic of Germany where large amounts of fibroblast interferon are provided (under government contract) by a pharmaceutical company. However, some work is being done also in France and in Belgium. The design of these trials has recently been reviewed (11). They include patients with osteosarcoma (prophylaxis of metastasis), metastatic breast cancer, neuroblastoma, head and neck epithelioma, stomach cancer, myeloma, nasopharyngeal carcinoma and juvenile laryngeal papilloma. Most results are from preliminary (phase I/II) trials; large scale trials still being in execution.

Intratumoral injection has so far been explored in only one rather recent study (40). Systemic administration of fibroblast interferon by the intramuscular route is known to give only low blood levels. Nevertheless some groups have chosen the intramuscular or subcutaneous route for detecting possible effects on tumor. The reason for this is that NK activity can be shown to be enhanced after such injection, indicating that the interferon does penetrate the general circulation (37). Other groups, especially in West-Germany, have given preference to intravenous injections.

Intratumoral Injections

Seto et al. (40) compared preparations of HuIFN-α and HuIFN-β for effectiveness in reducing lesions of cervical intra-epithelial neoplasia. Either of the interferons was given twice weekly at a dosage of 2×10^6 units. Partial regression occured in all patients with both interferons. Complete regressions occurred in all patients (6/6) treated with HuIFN-α preparations; it occurred only in 2/5 patients treated with fibroblast interferon.

Intramuscular Injections

1. Breast cancer. At the Institute Curie in Paris limited Phase I and Phase II trials have been done with a view to obtain a preliminary evaluation of the efficacy of fibroblast interferon therapy in breast cancer. In an initial study on terminal patients with advanced cancer, single intramuscular injection of 8×10^6 units of fibroblast interferon were well tolerated, and they enhanced the natural killer activity of peripheral lymphocytes to about the same extent as similar doses of leukocyte interferon (28).

In a subsequent study (37), 11 patients with metastasized breast cancer received 8 intramuscular injections of 6×10^6 units of human fibroblast interferon over a period of 40 days. Although several types of metastases were monitored, only skin nodules consistently (10 out of 11 patients) exhibited changes that were suggestive of a therapeutic effect of the regimen, namely, a simple decrease in the size of some nodules, or central necrosis accompanied by an inflammatory reaction. The natural killer activity of peripheral blood leukocytes was significantly increased after administration of the first dose, but the effect of subsequent injections was less clear. Receptor activity for oestrogens and progesterogens was increased in the tumor biopsy specimens of, respectively, 2 out of 2 and 5 out of 6 patients tested.

2. Head and neck tumors. Van der Schueren and co-workers (Leuven University Hospital, Belgium; not publ.) treated 4 patients suffering from recurrent head and neck tumors (two spinocellular epitheliomas of the sinus piriformis, one carcinoma of the nasopharynx and one carcinoma of the salivary gland). The patients received intramuscular injections of fibroblast interferon (7×10^6 units, 4 times weekly for 4 to 5 weeks). The tumors continued to progress.

3. Myeloma. In a study on myeloma cases at the University of Leuven (9) fibroblast interferon (14 intramuscular injections of 8×10^6 units over a period of 4 weeks) failed to influence disease progression in a preterminal case of therapy-resistant light chain disease. In a second case that had not previously been treated, a first course of fibroblast interferon (3 weekly intramuscular injections of 10^7 units over a period of 14 weeks) also remained without effect. However, in this patient subsequent leukocyte interferon

therapy (3×10^6 units intramuscularly, 6 times weekly for 5 weeks) was associated with a decrease in urinary high light chain excretion. A third patient with light chain excretion. A third patient with light chain disease was resistant to chemotherapy ab origine. None of the disease parameters responded to either fibroblast or leukocyte interferon therapy (3.5×10^6 units intramuscularly, 6 times weekly for 6 weeks).

Intravenous Injection

1. Neuroblastoma. A multicenter trial was initiated in 1979 by the German Society of Pediatric Oncology. Interim reports have been published after one and two years (1, 2). All children with stage IV neuroblastoma, being older than one year at the time of diagnosis were entered in the trials. Treatment with fibroblast interferon (specific activity 1 to 2×10^6 units/mg) was started after an initial course of chemotherapy (including vincristine, adriamycine, cyclophosphamide and dacarbasine). The patients were randomly assigned to control and interferon treatment groups.

Beginning 9 weeks after the initiation of chemotherapy, fibroblast interferon was administered at a dose of 2×10^6 units in a 30 min infusion, daily for 3 weeks. This was followed by infusions three times a week for 21 weeks. Interferon was given 11 weeks longer since the chemotherapy protocol was finished after a total of 22 weeks.

At the time of this writing 34 children, having received interferon, could be compared to 31 children who had not received interferon. All children relapsed and eventually died. The medium survival time was 19.2 months in the interferon group and 16.3 in the control group. This difference was not statistically significant.

In conclusion, fibroblast interferon therapy according to the schedule used in this trial can be stated to have no striking effect on the course of the disease.

2. Nasopharyngeal carcinoma (NPC). After the initial report of the successful treatment of a boy suffering from the second relapse of a NPC with fibroblast interferon (44), 5 additional children have been treated in a similar way by the same group (45). In two children with a similar history stabilization of the disease could be

achieved for periods of more than 10 and 6 months, respectively. In both children the tumor started to grow rapidly after cessation of treatment. In two patients with initial extensive metastatic disease there was clear evidence of rapid tumor regression in spite of the administration of interferon. One child was treated for 9 days with interferon as a primary treatment but radiation had to be initiated because of rapid tumor growth.

3. Non-Hodgkin's lymphoma. Siegert et al. (41) treated 10 patients with advanced non-Hodgkin's lymphoma of low grade malignancy. They received daily infusions of 4.5×10^6 units per 4 weeks, followed by 9×10^6 units for 2 weeks. Progressive disease occurred in 6 out of 10 patients; 4 out of 10 had stable disease. Two of the latter developed progressive disease during subsequent consolidation therapy with cytostatics. The authors consider these results as rather disappointing.

Misset et al. (30) attempted to treat meningeal localisations of non-Hodgkin's lymphoma by giving intrathecal injections of fibroblast interferon (1.3×10^6 units, every other day). In 1 cause out of 6 a remission occured after 11 injections.

4. Myeloma. Misset et al. (31) treated 18 patients with malignant gammapathies (16 with myeloma and 2 with Waldström disease) with fibroblast interferon (intravenous infusions of 3 to 6×10^6 units weekly) during at least 3 months. In 3 patients the treatment was discontinued because of side-effects. Objective responses (reduction of M-component by $> 25\,\%$, disappearance of urinary Bence Jones protein, or reduction of bone marrow infiltration) was noted in 4 cases. Duration of treatment is considered by the authors to be a critical factor for activity.

Antineoplastic Studies in Japan

Intratumoral injection of fibroblast interferon into metastatic lesions was first reported to cause lymphoid infiltration and regression by Horoszewicz et al. (25). Similar observations were done in Japan on a large number of patients suffering from metastatic malignant melanoma.

Intratumoral delivery of fibroblast interferon into brain tumors was pioneered in Japan (35) as an approach to prove the therapeutic potential of interferon in cancer. Results were quoted to be rather encouraging.

Following these pilot studies with melanoma and brain tumors, trials of a larger scale were initiated. These were recently reviewed by Yamasaki (49). In a multicenter trial 43 patients with malignant skin lesions (various origins) received 0.3 to 3×10^6 units 2 to 3 times per week, by injection in the lesions. Complete regression occurred secondly (1 out of 20 cases with malignant melanoma; 1 out of 7 cases of metastatic breast or stomach cancer). If partial regression was also considered as positive response, the response rates were 50 % for malignant melanoma patients, and 43 % for patitents with other tumors. In total 285 lesions were treated and 70 % of those showed either a partial or complete regression.

A total of 30 patients with brain tumors received intrathecal or intratumoral injections of fibroblast interferon (0.4 to 6.0×10^6 units, 2 to 3 times a week, 2 months or longer). In 6 of the 30 patients, there was a positive response, including 1 case with complete regression. According to a recent interim report on the use of intravenous interferon injections, the positive response rates were 25 % for brain tumors, 25 % for myeloma and 11 % for lymphoma. No clinical improvement has been observed in patients with gastric carcinoma or liver carcinoma.

Application in Other Diseases

Fibroblast interferon has been given to patients with certain diseases of uncertain etiology and pathogenesis, in an attempt to see whether through its immunomodulatory effects, it might favorably affect such diseases. Thus, Jacobs et al. (26) gave intrathecal injections of partially purified fibroblast interferon to 10 patients with multiple sclerosis (10^6 units per m^2, 13 injections over a 6-months period). Severe side-effects were noted. The exacerbation rate during the treatment period was lower than during the pre-study period. In a concurrent control group the exacerbation rate remained unchanged. However, pretreatment exacerbation rates of controls and treated patients were different (non-matched, non-randomized study). The authors expressed cautious optimism that this form of treatment may be effective in altering the course of multiple sclerosis.

Side-Effects

The side-effects of fibroblast interferon therapy in man are comparable to those of therapy with leukocyte interferon. Fever and general malaise is the most salient feature. Although pyrogenicity was already reported in the very first patients to be injected with interferon (for review see Billiau, 1981), it was disregarded as being attributable to impurities and of a minor importance to the practicability of interferon injections. Authors working with fibroblast interferon were less prone to accept this viewpoint: fever was considered by them as a major problem (15, 4, 6). In fact, later studies using leukocyte interferon also attached much more importance to this phenomenon. As more and more pure preparations of both leukocyte and fibroblast interferon became available, it became also apparent that the pyrogenicity was at least partially due to the interferon itself. Current work with molecularly cloned leukocyte interferon is confirming this point of view (20). In the case of fibroblast interferon this point is still less clear.

The interpretation of literature data is confounded by the fact that some authors use intramuscular injections while others use intravenous infusions. If fibroblast interferon itself is pyrogenic, it is likely that this will be more apparent with intravenous than with intramuscular injections. In fact all studies in which intravenous fibroblast interferon was given, even if the interferon was completely pure, mention some degree of pyrogenicity in man (7). In contrast, intramuscular injections of completely pure fibroblast interferon seem to be devoid of pyrogenic effect (7), while under these circumstances, impure preparations are indeed pyrogenic. Another point of interest is the fact that fibroblast interferon prepared from the tumor line MG-63, is always nonpyrogenic by intramuscular injection (28, 37). This strongly suggests that certain fibroblasts, upon induction with interferon inducers, produce pyrogenic factors which are unrelated to interferon, but which may copurify to some extent.

A reasoneble but tentative conclusion from the available data is that HuIFN-β is slightly pyrogenic but that certain current preparations contain potent pyrogenic impurities.

Pharmacokinetics of Human Interferons in Experiments on Animals

Early experiments performed in rabbits failed to reveal any difference in pharmacokinetic behavior between the fibroblast and leukocyte interferons (17). The fibroblast interferon used for these experiments was obtained by superinduction in diploid skin muscle cells and had been partially purified by fractional ammonium sulphate precipitation, a method that yielded recoveries of around 10 %. Later experiments were performed with interferon that was also prepared from diploid cells but had been partially purified by adsorption/elution from controlled pore glass (CPG) beads, a method that routinely yields recoveries of over 50 % (5). With this interferon a clear-cut difference with leukocyte interferon was observed in serum titers obtained after intramuscular injection in rabbits. Unexpectedly, the same preparations injected intramuscularly in monkeys, failed to reveal a significant difference: high serum titers were obtained with either leukocyte or fibroblast interferon (10). Hence, the rabbit was considered to be a better model for interferon pharmacokinetics in man. In an independent study, Vilček et al. (46) confirmed that intramuscular injection of fibroblast interferon in rabbits resulted in lower blood titers than those obtained with similar injections of leukocyte interferon. In this study the difference between the two interferons was, however, less pronounced. The fibroblast interferon used by these authors was of diploid cell origin and had not undergone any purification.

Intravenous injections of leukocyte and fibroblast interferons in rabbits failed to reveal significant differences in clearance rate (4, 46). It is doubtful, however, whether these experiments were accurate and sensitive enough to reveal small differences that were sufficiently important to account for different blood values after intramuscular injection. Thus, these experimental results, while failing to support the idea of a more rapid clearance of fibroblast type interferon, did not refute the possibility either.

In our laboratory, experiments were done in mice to determine whether fibroblast interferon would be taken up more rapidly by tissues as compared to leukocyte interferon (10). Leukocyte or fibroblast interferon (0.1 to 1×10^6 units) were injected intraperitoneally and antiviral activity was determined in serum, lungs and spleen. The fibroblast interferon used was of diploid cell origin and had been purified by CPG adsorption. A 10 to 30-fold difference was apparent in the serum levels reached by leukocyte and fibroblast interferon. In contrast, antiviral activities measured in the spleen and the lungs were similar, irrespective of which interferon was used. These data go along with the idea that lower blood titers obtained with fibroblast interferon are not due to destruction at the site of injection but rather to more rapid uptake by some organs or organ systems.

However, evidence to support the contrary was reported by Hanley et al. (21). They report rapid destruction of human fibroblast interferon, and resistance to destruction of leukocyte interferon upon incubation with muscle homogenate. The finding was correlated with low blood titers of fibroblast interferon, as opposed to high titers of leukocyte interferon, after intraperitoneal or intramuscular injections in mice. The authors also found lower titers of fibroblast interferon in the tissues of injected mice.

Some of the discordant results in the literature may be accounted for by the molecular heterogeneity of fibroblast interferon. It has been known for quite some time that natural HuIFN-β, although consisting of a single type of peptide, is heterogeneous with respect to glycosylation (32). Evidence that this or other forms of molecular heterogeneity may affect the pharmacokinetics of fibroblast interferon was obtained in studies from our laboratory (23, 24). In particular, it was found that fibroblast interferon, when chromatographed on zinc chelate columns, separates into two subpopulations eluting from the column at pH 5.9 and pH 5.2, respectively. Each of these was prepared in sufficient quantities to perform pharmacokinetic experiments in rabbits. Intramuscular injection of the pH 5.2 variant yielded 30-fold lower blood titers than a comparable dose of leukocyte interferon. The pH 5.9 variant, on the other hand, yielded a blood curve that was superimposable to that obtained with leukocyte interferon.

It was also found (23) that fibroblast interferon prepared from most diploid cell strains contains more than 90 % of the pH 5.2 variant. In contrast, fibroblast interferon from MG-63 cells or from certain high passage diploid cell strains were found to contain the two components in about equal proportions. These data provide a possible explanation for the discordance in results obtained in earlier pharmacokinetic studies with fibroblast interferons. Most probably, the preparations used by the different research groups contained different proportions of the two molecular variants of fibroblast interferon.

Summary and Concluding Remarks

Preclinical as well as clinical studies with fibroblast interferons are still lagging behind of those with leukocyte interferon. For instance, no information whatsever is currently available on the behavior of recombinant DNA-derived HuIFN-β, while numerous such studies have already appeared using recombinant DNA-derived HuIFN-α.

Also very little is known on the side-effects of pure HuIFN-β. The little information that there is, do suggest that its side-effects are much less pronounced than those of HuIFN-α, be it that it may be slightly pyrogenic after intravenous injection. Pyrogenicity of current impure preparations might for the larger part be due to impurities.

It is now well established that considerably higher doses of HuIFN-β than of HuIFN-α are required to obtain measurable blood titers by intramuscular injections. Since there is concern about this being due to destruction of the interferon before it has reached its target organ(s), most current clinical studies use either local (e.g. intratumoral) treatment or intravenous infusions.

A study on topical treatment for acute rhinovirus infection has indicated that there is very little of any chance for fibroblast interferon to be a clinically useful substance to prevent or cure common cold. In herpetic dendritic keratitis eye drops of fibroblast interferon may be useful as such or in combination with debridement. The question is whether the new antiviral nucleoside analogs, acyclovir and bromovinyldeoxyuridine, will not be far superior to interferons. The main counter-argument is that Herpes virus strains will emerge that will be resistant to these drugs.

Topical treatment of warts (multiple intralesional injections) has been shown to yield a high success rate, especially in the case of verrucae vulgares, but less so in the case of verrucae planae juveniles. The question here is whether the discomfort of lengthy treatment will outweigh that of the current surgical removal. For certain localisation (e.g. under the finger or toe nails) interferon injections may be the treatment of choice. Studies on condyloma accuminatum are not so far advanced as to permit a documented conclusion.

Topical (intralesional) treatment by neoplastic diseases has been investigated, especially in Japan, to demonstrate that fibroblast interferon does have an antineoplastic effect in vivo. While there seems to be little doubt that local delivery does indeed cause tumor nodules to regress, the question in whether this procedure can offer a true clinical benefit to the patient. Perhaps in a few rare cases topical treatment may offer a chance of longer survival and/or less discomfort. Yet it is hard to imagine that the effects seen with intralesional interferon injections will be the basis for a broadly applicable anti-cancer therapy.

Systemic (intravenous) administration for chronic hepatitis B has been investigated further: given alone or in combination with adenine-arabinoside, fibroblast interferon seems to be able to reduce the level of viral activity. Whether this will lead to a generally accepted treatment of chronic active hepatitis is difficult to say at this moment. The natural evolution of the disease takes years and the number of patients available for study will decrease as a result of hygienic measures and vaccinations. Under these circumstances there is considerable doubt that chronic hepatitis B will provide a "market" for interferon.

Intramuscular injections have been used in studies on patients with various tumors. In breast cancer patients, metastases in the skin, but not in other organs, showed alterations suggestive of an effect on tumor progression. Yet there was no true clinical benefit for the patient. In other tumors, e.g. head and neck epithelioma, no effect was seen.

Intravenous infusions have met with some success in nasopharyngeal carcinoma: temporary stabilization or regression occurred, but no definitive cures were noted. A large neuroblastoma trial yielded negative results. Disappointing results were also obtained in non-Hodgkin's lymphoma. In initial studies on myeloma and brain tumors "response rates" of 25 % were recorded; the significance of this terms of clinical benefit to the patient is not clear yet.

The overall picture thus is that fibroblast interferon therapy will possibly have some limited application, e.g. in certain cases of warts or in selected cases of localized metastases of various tumors. However, there is at present no sound basis for expecting fibroblast interferon to become a drug with wide applicability in various diseases or even in a single frequently occurring disease. This conclusion which was already reached several years ago can now not be challenged any more with the argument that the dosages were too small. Increases in dosages have not improved the results obtained with leukocyte interferon. Therefore, there is little hope that higher doses of fibroblast interferon, e.g. through the use of recombinant DNA-derived HuIFN-β, will alter the current situation.

References

1 Berthold, F.; Treuner, J.; Niethammer, D.; Lampert, F. (1981): Klin. Pädiat., 193, 198—201.
2 Berthold, F.; Treuner, J.; Brandeis, W. E.; Evers, G.; Haas, R. J.; Harms, D.; Jürgens, H.; Kaatsch, P.; Michaelis, J.; Niethammer, D.; Prindull, G.; Riehm, H.; Winkler, K.; Lampert, F. (1982): Klin. Pädiat., 194, 262—269.
3 Billiau, A.; De Somer, P.; Edy, V. G.; De Clercq, E.; Heremans, H. (1979a): Antimicrob. Agents Chemother., 16, 53—63.
4 Billiau, A.; Edy, V. G.; De Somer, P. (1979b): in "Antiviral Mechanisms in the Control of Neoplasia", ed. by P. Chandra, Plenum Publishing Corporation, New York and London, pp. 675—696.
5 Billiau, A.; Van Damme, J.; Van Leuven, F.; Edy, V. G.; De Ley, M.; Cassiman, J. J.; Van den Berghe, H.; De Somer, P. (1979c): Antimicrob. Agents Chemother., 16: 49—55.
6 Billiau, A.; De Somer, P. (1980): in "Clinical Use of Interferons in Viral Infections", ed. by D. A. Stringfellow, Marcel Dekker, Inc., New York, pp. 113—144.
7 Billiau, A.; Heine, J. W.; Van Damme, J.; Heremans, H.; De Somer, P. (1980): Ann. N. Y. Acad. Sci., 350, 374—375.
8 Billiau, A. (1981): Arch. Virol., 67, 121—133.
9 Billiau, A.; Bloemmen, J.; Bogaerts, M.; Claeys, H.; Van Damme, J.; De Ley, M.; De Somer, P.; Drochmans, A.; Heremans, H.; Kriel, A.; Schetz, J.; Tricot, G.; Vermylen, C.; Waer, M. (1981a): Eur. J. Cancer Clin. Oncol., 17, 875—882.
10 Billiau, A.; Heremans, H.; Ververken, H.; Van Damme, J.; Carton, H.; De Somer, P. (1981b): Arch. Virol., 68, 19—25.

11 Billiau, A.; Niethammer, D.; Strander, H. (1983): in "Clinical Application of Interferons and Their Inducers", ed. by D. Stringfellow, Marcel Dekker, Inc., New York, in press.
12 De Maeyer, E.; Galasso, G.; Schellekens, H. (1981): "The Biology of the Interferon System", Elsevier/North-Holland Biomedical Press, Amsterdam.
13 De Maeyer, E.; Galasso, G.; Schellekens, H. (1983): "The Biology of the Interferon System", Vol. II, Elsevier/North-Holland Biomedical Press, Amsterdam.
14 Derynck, R.; Remaut, E.; Saman, E.; Stanssens, P.; De Clercq, E.; Content, J.; Fiers, W. (1980): Nature, 287, 193—197.
15 Desmyter, J.; Ray, M. B.; De Groote, J.; Bradburne, A. F.; Desmet, V. J.; Edy, V. G.; Billiau, A.; De Somer, P.; Mortelmans, J. (1976): Lancet, ii, 645—647.
16 De Somer, P.; Edy, V. G.; Billiau, A. (1977): Lancet, ii, 47—48.
17 Edy, V. G.; Billiau, A.; De Somer, P. (1976): J. Infect. Dis., 133, A18—A21.
18 Edy, V. G.; Billiau, A.; D Somer, P. (1978): Lancet, i, 451—452.
19 Göbel, H.; Arnold, W.; Wahn, V.; Treuner, J.; Jürgens, H.; Cantell, K. (1981): Eur. J. Pediatr., 137, 175—176.
20 Guttermann, J. U.; Fine, S.; Quesada, J.; Horning, S. J.; Levine, J. F.; Alexanian, R.; Bernhardt, L.; Kramer, M.; Spiegel, H.; Colburn, W.; Trown, P.; Merigan, T.; Dziewanowski, Z. (1982): Ann. Internal Med., 96, 549—556.
21 Hanley, D. F.; Wiranowska-Stewart, M.; Stewart II, W. E. (1979): Intern. J. Immunopharmacol., 1, 219—226.
22 Heidemann, E.; Wilms, K.; Treuner, J.; Niethammer, D. (1982): Deutsch. Med. Wochenschr., 107, 695—697.
23 Heine, J. W.; Van Damme, J.; De Ley, M.; Billiau, A.; De Somer, P. (1981): J. gen. Virol., 54, 47—56.
24 Heine, J.; Billiau, A.; Van Damme, J.; De Somer, P. (1982): in "The Clinical Potential of Interferons", ed. by R. Kono and J. Vilcek, University of Tokyo Press, pp. 69—74.
25 Horoszewicz, J. S.; Leong, S. S.; Ito, M.; Buffett, R. F.; Karakousis, C.; Holyoke, E.; Job, L.; Dölen, J. G.; Carter, W. A. (1978): Cancer Treat. Rep., 62, 1899—1906.
26 Jacobs, L.; O'Malley, J.; Freeman, A.; Eks, R. (1981): Science, 214, 1026—1028.
27 Kono, R.; Vilcek, J. (1982): "The Clinical Potential of Interferons", University of Tokyo Press.
28 Lucero, M. A.; Magdelenat, H.; Fridman, W. H.; Pouillart, P.; Billardon, C.; Billiau, A.; Cantell, K.; Falcoff, E. (1982): Eur. J. Cancer Clin. Oncol., 18, 243—251.
29 Merigan, T. C.; Randn, K. H.; Pollard, R. B.; Abdallah, P. S.; Jordan, G. W.; Fried, R. P. (1978): New Engl. J. Med., 298, 981—987.
30 Misset, J. L.; Mathé, G.; Gastiaburu, J.; Goutner, A.; Dorval, T. I.; Gouveia, J.; Hayat, M.; Jasmin, C.; Schwarzenberg, L.; Machover, D.;

Ribaud, P.; de Vassal, F.; Horoszewicz, J. P. (1982a): Biomedicine, 36, 167—170.

31 Misset, J. L.; Mathé, G.; Gastiaburu, J.; Goutner, A.; Dorval, T.; Gouveia, J.; Hayat, M.; Jasmin, C.; Schwarzenberg, L.; Machover, D.; Ribaud, P.; de Vassal, F.; Horoszewicz, J. S. (1982b): Anticancer Res., 2, 63—66.

32 Morser, J.; Kabayo, J. P.; Hutchinson, D. W. (1978): J. gen. Virol., 41, 175—178.

33 Müller, R. W.; Siegert, W.; Hofschneider, H. P.; Deinhardt, F.; Frösner, G., Vido, I.; Schmidt, F. W. (1981): in "The Biology of the Interferon System", ed. by E. De Maeyer, G. Galasso and H. Schellekens, Elsevier//North-Holland Biomedical Press, Amsterdam, pp. 355—359.

34 Munk, K.; Kirchner, H. (1982): "Interferon. Properties, Mode of Action, Production, Clinical Application", Contributions to Oncology, Vol. 11, S. Karger AG, Basel.

35 Nagai, M.; Arai, T.; Kohno, S.; Kohase, M. (1982): in "The Clinical Potential of Interferons", ed. by R. Kono and J. Vilcek, University of Tokyo Press, pp. 257—273.

36 Obert, H. J. (1983): in "The Biology of the Interferon System", Vol. II, ed. by E. De Maeyer, G. Galasso and H. Schellekens, Elsevier/North-Holland Biomedical Press, Amsterdam, in press.

37 Pouillart, P.; Palangie, T.; Jouve, M.; Garcia-Giralt, E.; Fridman, W. H.; Magdelenat, H.; Falcoff, E.; Billiau, A. (1982): Eur. J. Cancer Clin. Oncol., 18, 929—935.

38 Schouten, T. J.; Weimar, W.; Bos, J. H.; Bos, C. E.; Cremers, C. W. R. J; Schellekens, H. (1982): Laryngoscope, 92, 686—688.

39 Scott, G. M.; Reed, S.; Cartwright, T.; Tyrrell, D. (1980): Arch. Virol., 65, 135—140.

40 Seto, W. H.; Choo, Y. C.; Hsu, C.; Merigan, T. C.; Tan, Y. H.; Ma, H. K.; Ng, M. H. (1983): in "The Biology of the Interferon System", Vol. II, ed. by E. De Maeyer, G. Galasso and H. Schllekens, Elsevier/North-Holland Biomedical Press, Amsterdam, in press.

41 Siegert, W.; Theml, H.; Fink, U.; Emmerich, B.; Kaudewitz, P.; Huhn, D.; Nöning, L.; Abb, J.; Joester, K. E.; Bartl, R.; Riethmüller, G.; Wilmanns, W. (1982): AntiCancer Res., 2, 193—198.

42 Suzuki, H. (1983a): Antiviral Res., Special Abstract Issue, P99.

43 Suzuki, H. (1983b): Antiviral Res., Special Abstract Issue, P100.

44 Treuner, J.; Niethammer, D.; Dannecker, G.; Hagmann, R.; Neef, V.; Hofschneider, P. H. (1980): Lancet, i, 817.

45 Treuner, J.; Niethammer, D.; Dannecker, G.; Jobke, A.; Aldenhoff, P.; Kremens, B.; Nessler, G.; Bömer, H. (1981): in "Cancer Campaign", Vol. 5, "Nasopharyngeal Carcinoma", Gustav Fischer Verlag, pp. 309.

46 Vilcek, J.; Sulea, I. T.; Zerebeckij, I. L.; Yip, Y. K. (1980): J. Clin. Microbiol., 11, 102—105.

47 Weimar, W.; Heijtink, R. A.; Schalm, S. W.; Van Blankenstein, M.;

Schellekens, H.; Masurel, N.; Edy, V. G.; Billiau, A.; De Somer, P. (1977): Lancet, ii, 1282.
48 Weimar, W.; Schellekens, H.; Lameijer, L. D. F.; Masurel, N.; Edy, V. G.; Billiau, A.; De Somer, P. (1978): Eur. J. Clin. Invest., 8, 255—258.
49 Yamazaki, S. (1983): in "The Biology of the Interferon System", Vol. II, ed. by E. De Maeyer, G. Galasso and H. Schellekens, Elsevier/North-Holland Biomedical Press, Amsterdam, in press.

A. Billiau, M.D., Rega Institute, University of Leuven, B-3000 Leuven (Belgium)

Contr. Oncol., vol. 20, pp. 270—277 (Karger, Basel 1984)

A Hydrophilic Interferon Gel in the Treatment of Herpes Virus Infections in Man

Birger Møller[a]; *Kurt Berg*[b]

[a] Department of Dermatology and Venerology Marselisborg Hospital, and [b] Institute of Medical Microbiology, Bartholin Building, University of Aarhus, Denmark

Introduction

During the last few decades a dramatic rise in the incidence of infections with herpes simplex virus (HSV) has occurred. Especially, genital herpes has emerged from relative obscurity and exploded into a full-fledged epidemic, first of all in the United States. Thus, an estimated 20 million Americans now have genital herpes, and furthermore, half a million new cases are expected this year (5, 9).

No currative treatment of herpes infections in man has been reported (2, 4, 5, 6, 7, 9). Acyclovir, a drug known to interfere with the viral replication, has proved to be effective, but only during the first episode of a herpes attack. Thus, patients suffering from recurrent herpes infections are still left without any effective treatment (5).

In this preliminary report we have treated a limited number of patients with cold sores, genital herpes and herpes zoster via topical application of a hydrophilic human leukocyte interferon gel to the affected area. The gel was found to alleviate the symptoms significantly by speeding up the healing processes in patients with primal as well as recurrent herpes eruptions. Preliminary results suggest that also herpes zoster patients may be treated beneficially.

Material and Methods

Interferon was a generous gift from Prof. Kari Cantell (Helsinki). It consisted of partially purified leukocyte interferon (PIF, spec. act.: 2×10^6 units/mg protein). All interferon units are expressed in international reference units (69/19B reference, M. R. C., U. K.) (1).

Interferon gel was made by mixing the appropriate carboxymethylcellulose with PIF yielding a gel containing 60,000 units/g gel. The gel was ampouled into 2.5 ml portions, sealed and stored at $-20°$ C until used. When used, the gel was kept at $+4°$ C for no longer than 15 days.

Stability tests were carried out by weighting out about 0.2 g of the interferon gel into a glass tube. After adding 2.0 ml of 2% medium the tube was sealed with a rubber stopper and turned upside down for at least 2 h (or until all the gel had dissolved). This solution was then titrated as previously described (8). Ampoules containing the interferon gel were kept at the indicated temperatures (cf. fig. 1), and the interferon content was determined with the intervals as indicated in fig. 1.

Patients. In total, 38 patients with herpes eruptions were treated with interferon gel as the only treatment. Two groups of patients with recurrent herpes eruptions were studied: group I consisted of 15 patients with recurrent HSV infection of the lips and/or chin; group II consisted of 12 patients with recurrent HSV infection of the genital organs. Patients with symptoms lasting less than two years or patients with fewer than two eruptions during the last year were excluded from the study.

During the very first visit the patients were interviewed with regard to the duration of symptoms lasting from the first sensation of local itching, "burning", pain and/or erythema, the occurrence of vesicles, and to the development of crust leading to the usual epithelialisation. The duration of symptoms was defined as the time-span between the first sensation and the crustation of the whole affected area.

Specimens were also taken from the eruptive vesicles and cultivation attempts for HSV were performed. Only patients who were culture-positive at the first visit were included in this study. The patients were asked to return to the hospital for treatment at the beginning of the next prodrome. Patients with herpes labialis received the interferon gel five to seven times during the first 24 h followed by 2—3 times a day until the symptoms had disappeared (tab. 1). Patients with genital herpes were treated only once a day with the gel except patient no. 26 (tab. II) who received the gel-treatment 4 times during the first 24 hrs and 6 times during the next 24 hrs.

Table I. Topical treatment with human leukocyte interferon gel of 15 patients with recurrent Herpes simplex virus infection of the lips and/or chin

Patient no.	Duration of disease (years)	No. of eruptions per year	Duration of symptoms without treatment (days)	Duration of symptoms during treatment (days)
1	26	3—4	10—14	4—5
2	10	10—12	5—6	1—2
3	6	6—8	5—6	1—2
4	20	3—4	4—6	1
5	8	12	6—7	1—2
6	15	4—6	5—6	1
7	2	5—6	3—5	1—2
8	10	3—4	5—7	2
9	20	5—10	4—8	1—2
10	3	6—7	5—6	2
11	5	2—3	5	1
12	10	4—8	4	1
13	2	12	5	1—2
14	5	10	4—5	1
15	15	2—3	6—8	1—2

Sign. test: $p < 0.0001$.

Seven female patients with primary genital HSV and four patients with herpes zoster were treated with the interferon gel during hospitalisation. The treatments were in all cases started within the first day after the appearance of the symptoms. In general, the gel was applied two times a day until the symptoms had disappeared.

Table II. Topical treatment with human leukocyte interferon gel of 12 patients with recurrent Herpes simplex virus infection of the genital organs

Patient no.	Duration of disease (years)	No. of eruptions per year	Duration of symptoms without treatment (days)	Duration of symptoms during treatment (days)
16	2	4—6	6—8	4—5
17	5	1—2	10—14	3—5
18	20	3—5	5—8	3—4
19	2	5—6	5	3—4
20	5	4	4—8	2—3
21	8	1—2	7	2—3
22	1	5—6	12—14	4—6
23	5	5	4—6	2
24	2	6—7	5—6	2—3
25	2	1—2	14	3—4
26	2	4—8	7—8	1—2[1]
27	1	5	10—12	6

[1] Treated 4—6 times daily.
Sign. test: $p < 0{,}0005$.

Results

From fig. 1 it can be seen that the interferon gel is stable for at least a year at –20° C; at +4° C the gel is stable for more than 5 weeks. The gel became unstable at +37° C and after 4 days more than 50 % of the activity was lost. The gel tolerated repeated thawing/freezing cycles without loosing its activity.

As can be seen from tab. I and II the interferon containing gel was able to yield a marked decline in the duration of symptoms in all patients belonging to group I and II. This shortening of duration was highly significant compared with the corresponding untreated groups (Sign. test; $p < 0.0001$ and $p < 0.0005$, respectively).

Characteristically, the treatment produced relief of symptoms and resolution of the lesions after a few hours or 1–2 days, if the gel was applied 5–7 times a day (group I). In group II the symptoms persisted for up to six days. It is noteworthy that in the patient treated more than once a day (patient no. 26, tab. I), the relief of symptoms occurred in less than 48 hours compared to seven to eight days without treatment.

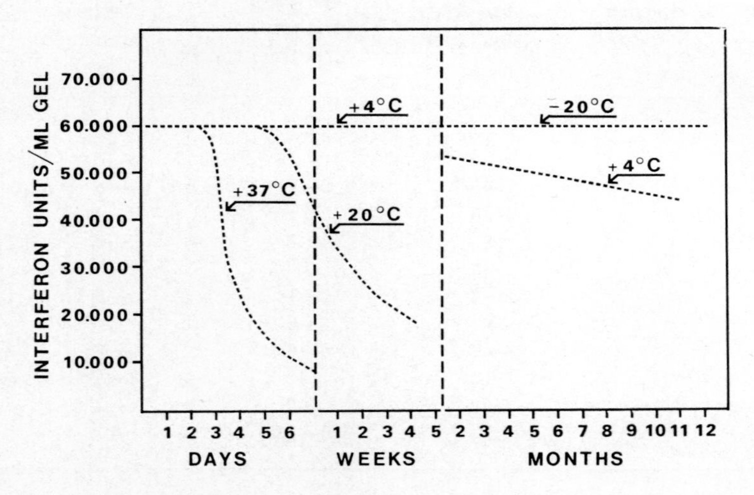

Fig. 1. Stability of the interferon gel.
Tubes containing 2 ml of the interferon gel were kept at +37° C, +20° C, +4° C, and at —20° C, respectively. At intervals, about 0.2—0.4 g of the gel was dissolved into 2 ml of 5 % medium and titrated for interferon activity.

In the group of seven female patients, having their first episode of genital herpes, eruption symptoms disappeared and lesions became crustated within five days after admission to the hospital. With regard to the four patients with herpes zoster the symptoms disappeared after five to six days of treatment, if the interferon gel was applied twice a day. One of these patients developed host-hepatic pain.

Most of the patients in this study have been followed for more than six months. The treatment with the interferon gel as such did not seem to reduce the frequency of recurrency of the attacks; rather, it seems only to reduce the duration of symptoms and the time of healing. Apart from a mild transitory itching — lasting 15—30 min subsequent to the gel application — no side effects were seen. No allergic reactions, whatsoever, occurred during treatment.

Discussion

The interferon gel appeared to be stable if used within the 12 months period subsequent to its preparation. At –20° C the gel may turn out to be stable for more than a year (experiments are in progress). The more limited stability at +4° C, comprising fifth weeks, did not cause any practical problems during the period of administration since the patients, generally speaking, used at least one tube of interferon per herpes attack. With regard to stability the present gel formulation turned out to be superior compared to the previously described (8) for reasons not fully understood.

The topical use of the interferon gel, per se, does not seem to have a curative effect on herpes infections. Thus, we found that recurrence of the eruptions was just as frequent before as after the treatment. However, treatment reduced the duration of symptoms markedly and hastened the healing of lesions in patients with perioral herpes and genital infection, although the efficacy of the treament of the latter group was not so convincing. In this context it is interesting to note that when one patient, suffering from genital herpes, was treated 4–6 times a day with the interferon gel (cf. table 2), a remarkable reduction of the duration of the symptoms was recorded (from 7–8 days to 1–2) which is more in line with the results shown in table 1 with regard to patients suffering from cold sores.

The question could be raised if the interferon gel also could be of benefit if used before the outbreak of the lesions. On a single occasion, one patient, suffering from recurrent herpes on the lips, was treated (twice daily) with the interferon gel prior to the usual outbreak of the lesions, but at the onset of the usual pain-itching period (which is known to proceed the outbreak of the lesions). This time no lesion appeared, at all, and the pain-itching disappeared very fast (compared to the untreated instances).

Subsequent to the first gel administration a mild transitory itching preceeding the relief of pain was noted in most of the patients. It may be the result of the actions of interferon since it is known that interferon, per se, will activate a wide variety of immunological systems including, for example, the NK-cells, macrophages, and other cell-mediated immunological functions (3). The time interval (15—30 min) corresponds very well with the minimal time necessary for stimulation of for example the NK-systems (3). It will be interesting further to analyze mechanisms that are the most important when using the interferon locally.

The results in tab. I and II showed a highly significant shortening in the duration of symptoms in both groups of patients as measured by the sign. test. By comparing tables 1 and 2 it is tempting to suggest that the patients themselves, somehow, could be considered to "act" as controls, especially comparing the columns called: "Duration of symptoms without treatment" with "Duration of symptoms with treatment".

The rapid disappearance of the pain/itching was associated with no more outbreaks of vesicles. The rapidity of the disappearance of the symptoms far exceed the pattern seen, for example, by Acyclovir: here the pain/itching will not disappear before 10 h, at the earliest, subsequent to the introduction of Acyclovir (E. de Clercq, 1983, personal communication). The marked difference in rapidity, could be explained by the fact that interferon, apart from being an antiviral substance, also is able to modulate the immune system (as discussed above).

The present work was performed as an open study and all the patients were fully informed of the treatment in advance. However, in general, the patients were very sceptical to a possible effect of such treatment, mostly because all patients with recurrent herpes previously had been treated with a number of other regimens such as zinc sulphate solution, idoxuridine, or corticosteroides (topically). Although very subjective, all patients with recurrent herpes stated that the effect of the interferon gel was convincing.

The preliminary results thus seem to indicate that the interferon gel shortens the duration of symptoms and speeding up the healing of primal and of recurrent herpes in patients with cold sores or genital herpes. However, a double-blind study is highly needed before any final conclusions can be made (experiments are in progress).

Acknowledgements:

The study was supported by the Danish Medical Research Council.

References

1 Berg, K. (1982): Acta path. microbiol. immunol. scand., Sec. C, No. 279, p. 1—136.
2 Crane, L. R.; Levy, H. B.; Lerner, A. M. (1982): Antim. Ag. Ch. 21, 481—485.
3 Heron, I.; Berg, K. (1979): Scand. J. Immunol. 9, 517—526.
4 Levin, M. J.; Leary, P. L. (1981): Infect. Immun. 32, 995—999.
5 Luby, J. (1982): N. Engl. J. Med. 306, 1356—1357.
6 Milman, N.; Jessen, O.; Scheibel, J. (1979): Ugeskr. Læg. 141, 2960—2962.
7 Milman, N.; Scheibel, J.; Jessen, O. (1980): Ugeskr. Læg 142, 1202—1203.
8 Møller, B. R.; Johannesen, P.; Osther, K.; Ulmsteen, U.; Hastrup, J.; Berg, K. (1983: J. Gyn. & Obstr. (in press).
9 Raab, L. Al. (1981): J. Am. Acad. Derm. 5, 249—263.

K. Berg, M.D., Institute of Medical Microbiology, Bartholin Building, University of Aarhus, DK-8000 Aarhus C (Denmark)

Contr. Oncol., vol. 20, pp. 278—284 (Karger, Basel 1984)

Dipyridamole As an Interferon Inducer in Man

A. S. Galabov; M. Mastikova

Department of Virology, Institute of Infectious and Parasitic Diseases, Medical Academy, Sofia, Bulgaria.

Introduction

In previous papers (1, 2, 3) we characterized dipyridamole [2,6--bis(diethanolamino-4,8-dipiperidinopyrimido-[5,4-d]-pyrimidine] as an interferon (IFN) inducer both in vitro, in murine and human lymphoid and non-lymphoid cells, and in vivo – in mice by different ways of administration (intravenous, intraperitoneal or oral). Oral application was found to be the most efficient. Dipyridamole in doses ranging from 3.12 to 100 mg/kg body weight elicited IFN blood levels at the 6th hr after administration. Following application of the optimum IFN inducing doses of 12.5 and 25 mg/kg, peak IFN levels of 4096—8192 IU/ml were reached at the 48th hr. Twenty-four hours later the blood IFN content was sharply reduced, but nevertheless IFN titres remained higher than those in the control group until the 120th hr. The IFN inducing capacity of dipyridamole showed marked selectivity, the selectivity index being found to be 86—172.

The pharmacological and toxicological aspects of dipyridamole are very well studied and sufficient clinical experience has been accumulated with view of its wide application as coronary vasodilatator and antiaggregant (4). This gave us the possibility to undertake immediate trials on the use of dipyridamole as an interferon inducer in humans, which is the aim of the present paper.

Table I. Interferon blood levels in healthy volunteers in response to oral administration of dipyridamole (single dose of 100 mg)

Trial Nr.	Volunteers	Serum interferon titre (IU/ml)		
		0 hr	24 hrs	48 hrs
I	1 A.G.	4	1024	
	2 L.G.	4	32	
	3 P.R.	8	8192	
	4 M.Tz.	4	8192	
	5 Y.G.	16	512	
	6 A.S.	4	≦ 4	
	7 K.G.	4	4096	
	8 A.K.	4	16	
	9 S.D.	4	512	
	10 G.G.	8	2048	
	11 M.M.	4		≦ 4
	12 S.P.	8		≦ 4
	13 T.V.	4		1024
	14 E.T.	4		8192
	15 G.S.	32		4096
	16 E.B.	<4		1024
	17 M.D.	≦ 4		1024
	18 K.R.	≦ 4		≦ 4
	19 I.B.	4		1024
	20 M.N.	≦ 4		1024
II	21 Y.M.	8	32	
	22 V.M.	8	512	
	23 S.U.	≦ 4	128	
	24 Ya.I.	≦ 4	8192	
	25 V.S.	≦ 4	512	
	26 M.S.	≦ 2	4096	
	27 V.H.	8	4096	
	28 A.Tz.	2	1024	
	29 N.N.	32	1024	
	30 V.G.	8	1024	
	31 N.R.	4	8192	
	32 E.Tz.	8	4096	
	33 P.M.	8	4096	
	34 N.I.	4	256	
	35 E.P.	8	4096	
	36 V.S.	8	64	
	37 H.Ch.	≦ 4	64	
	38 L.V.	4	256	
	39 V.T.	4	2048	
	40 G.G	2	512	

Material and Methods

The compound was applied orally in a single dose of 100 mg (4 tablets in 2 hrs). Two separate trials were carried out on a group of 20 healthy male and female volunteers of age 30 to 50. Individual serum samples for IFN assay were taken immediately before and 24 and 48 hrs after dipyridamole administration. IFN was assayed by the CPE inhibition method in diploid human embryo lung fibroblasts using VSV as a challenge virus.

Results

In the first trial (tab. I) 16 volunteers (80 %) responded with a markedly elevated blood IFN content at the 24th and 48th hr after dipyridamole administration. The mean geometrical values of the serum IFN titres in responders showed a 200-fold increase, from 5.2 ± 1.7 IU/ml before treatment to 1100 ± 16.2 IU/ml afterwards. In 4 volunteers (20 %) no interferon induction was found. The difference between IFN levels 24 and 48 hrs after dipyridamole administration was not statistically significant.

Addendum to Table I. Mean values of the serum interferon titres in responders

Trial Nr.	Time post dipyridamole administration	IU/ml
I	0 hrs	5.2 ± 1.7
(1—20)	24—48 hrs	1100.0 ± 16.2
II	0 hrs	4.6 ± 1.9
(21—40)	24 hrs	835.0 ± 17.5
Σ	0 hrs	4.9 ± 2.2
(1—40)	24—48 hrs	958.0 ± 23.5

In the second trial (tab. I) the percentage of responders reached 100 %, the mean IFN titre being 835 ± 17.5 IU/ml, i.e. 180 times higher than that taken before dipyridamole application.

The summarized mean value for IFN content in the responders

from the two trials (36 out of 40 volunteers) was 958 $\pm$ 23.5 IU/ml.

In 14 of the responders the IFN inducing activity of dipyridamole was subjected to a repeated check. In 9 out of 14 a full coincidence was found in the response and in the IFN levels, in 4 – a partial coincidence (some differences in the IFN titres) and only in one case an opposite response was observed.

Meanwhile we studied (in collaboration with Dr. V. Stoyanov) the IFN inducing effect of dipyridamole in 35 patients with chronic pulmonary diseases from the Sofia District Pneumophthisiatric Clinic (27 male and 8 female, age 50 to 70). In this trial positive IFN responses to the inducer were found in 60 % of the patients (48.15 % of the men) and the mean IFN blood titre was 314 $\pm$ 11.3 IU/ml, i.e. three times lower than the IFN blood level found in dipyridamole treated healthy volunteers. Moreover, as the initial level in the patients was 9.2 $\pm$ 4.0 IU/ml, the serum IFN increase in this case was only 34-fold.

A following step in our study was the analysis of the hyporeactivity (tolerance) phenomenon after dipyridamole administration. For this purpose we selected a number of responders and distributed them in several test groups. They were treated twice with the inducer with a time interval between the first and second application ranging from 3 to 20 days. Serum samples were taken before and 24 hr after the first and second dipyridamole application.

The initial results of this study, summarized in tab. II, showed that the period of tolerance was within the time interval of 4th–6th day after dipyridamole application, and it was most strongly expressed on the 4th day.

Conclusions

The results of the first dipyridamole trial in human volunteers reveal this agent as a potent IFN inducer after oral application. These data are in line with the results of the experimental study in vivo (albino mice). Ninety percent of the healthy volunteers are responders. Dipyridamole elicits a 195-fold increase in the blood serum IFN values. Of special interest are the differences among

Table II. Interferon blood levels in healthy volunteers in response to single and repeated oral administration of dipyridamole (100 mg)

Time interval between 1st and 2nd dipyridamole administration (days)	Volunteers	Serum interferon titre (IU/ml)			Hyporeactivity state (+)
		Initial level	24 hrs after 1st application	24 hrs after 2nd application	
3	A.G.	2	≧ 2048	4096	—
	P.R.	8	≧ 8192	≧ 8192	—
	V.M.	4	4096	2048	—
4	G.G.	2	512	≦ 8	+
	S.D.	4	512	≦ 8	+
	P.R.	16	1024	32	+
	L.G.	4	32	≦ 8	+
	V.T.	4	2048	≦ 8	+
5	N.N.	32	1024	128	+
	P.M.	8	4096	256	+
	V.N.	2	512	128	+
6	N.I.	4	4096	512	+
	N.R.	4	8192	≦ 32	+
	Zh.Z.	4	4096	128	+
7	E.Tz.	8	4096	8192	—
	M.N.	16	≧ 8192	≧ 8192	—
	Ya.D.	16	8192	4096	—
	V.G.	8	1024	2048	—
	E.P.	8	64	64	—
8	G.G.	4	4096	4096	—
	E.B.	4	64	64	—
	V.G.	8	1024	1024	—
	N.R.	4	8192	8192	—
	E.Tz.	8	4096	4096	—

Table II. (continued)

Time interval between 1st and 2nd dipyridamole administration (days)	Volunteers	Serum interferon titre (IU/ml)			Hyporeactivity state (+)
		Initial level	24 hrs after 1st application	24 hrs after 2nd application	
12	A.Tz.	2	1024	1024	—
	A.G.	2	2048	4096	—
	V.M.	2	1024	1024	—
	A.K.	4	2048	2048	—
	N.N.	32	1024	1024	—
	S.U.	4	1024	1024	—
16	I.B.	2	2048	2048	—
	P.M.	8	4096	4096	—
	N.I.	4	256	256	—
	Zh.Z.	4	4096	4096	—
	V.N.	2	512	512	—
20	V.H.	8	128	128	—
	K.G.	≦ 4	1024	1024	—
	E.P.	8	4096	4096	—
	M.S.	≦ 2	4096	4096	—
	M.Tz.	8	1024	1024	—

the individual responses to the inducer — negative (10 %) or positive, ranging from 16 to above 8192 IU/ml.

A tolerance state in man was established from the 4th to the 6th day after dipyridamole application. Its duration in mice was found (2) to be longer — within the time interval of 3rd—16th day after inducer administration. These data are of practical use in the elaboration of an optimal scheme for the application of dipyridamole as interferon inducer.

We think there is sufficient evidence to recommend the use of dipyridamole as IFN inducer in man.

References

1 Galabov, A. S.; Mastikova, M. (1982a): Acta Virol., 26, 137.
2 Galabov, A. S.; Mastikova, M. (1982b): Seventh Regional Symposium of the Socialist Countries of Interferon, Varna, November 8—12, 1982.
3 Galabov, A. S.; Mastikova, M. (1983): Acta Virol. 27 (in press).
4 Simon, H. (1972): in "Herzwirksame Pharmaka-Wirkungsweise und klinische Anwendung", Urban und Schwarzenberg.

A. S. Galabov, M.D., Department of Virology, Medical Academy, Belo More 8, Sofie — 1527 (Bulgaria)

Contr. Oncol., vol. 20, pp. 285—289 (Karger, Basel 1984)

Influence of Thymosin and Thymopoietin Pentapeptide on the Production of Interferon in Lymphocytes

E. Rentz

Academy of Sciences of the GDR, Central Institute of Cancer Research, Dept. of Chemical Carcinogenesis, Berlin-Buch.

Since some years evidences have been accumulated which indicate that the thymus gland plays an important role in the development of immunological competence in animals and man.

Although there is little knowledge about the molecular events by which the thymus exerts the control over immunological events, it appears that the vital part of the process occurs via a hormonal mechanism.

The thymus produces several polypeptides that play an important role in the maturation, differentiation, and function of T cells. Some authors (7, 5, 9) have also described an effect of thymosin and its derivates, respectively, on introduction and production of interferon.

We examined the influence of thymosin fraction 5 and thymopoietin pentapeptide (32—36) on cellular interferon production in presence of Concanavalin A.

The mean component of the thymosin fraction 5 is thymosin α_1, one of the best characterized thymic hormones. It is a polypeptide with a molecular weight of 3108 D and effects, for instance, the maturation of helper and suppressor T cells and can modulate the functional capacity of differentiated T cells which influence immunoregulatory control over T—T and T—B lymphocyte interaction. It increases phenotypic T cell markers on mouse spleen and bone marrow cells. The used thymosin fraction 5 was prepared according to Hooper et al. (4).

The second substance we used was thymopoietin pentapeptide

(TP 5). The pentapeptide arginyl-lysyl-aspartyl-valyl-tyrosine is a synthetic fragment of the hormone thymopoietin and it posesses the biological activities of the native molecule. It has a molecular weight of 680 D and it is rapidly degraded by proteolytic enzymes (half-life approximately 30 sec.). The TP 5 initiates biological changes for days or weeks and has immunoregulatory actions that appear to restore immune perturbations towards normal balance. The used pentapeptide was prepared according to Abiko et al. (1).

Material and Methods

In our experiments (fig. 1) we treated mice with thymosin fraction 5 or thymopoietin pentapeptide 3 times. On the 4th day the mice were sacrified and the removed spleens were cultivated

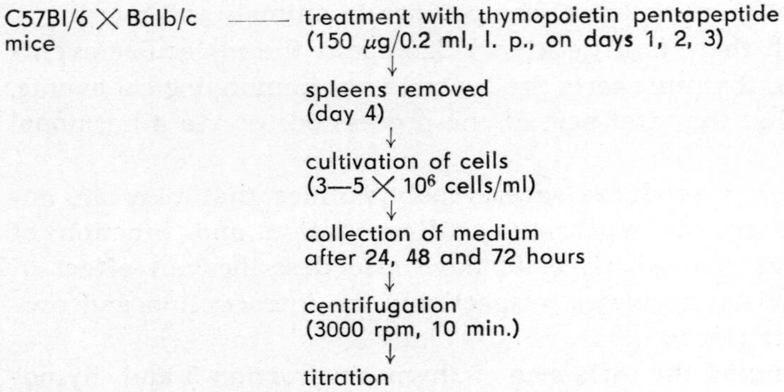

C57Bl/6 × Balb/c mice —— treatment with thymopoietin pentapeptide (150 μg/0.2 ml, I. p., on days 1, 2, 3)

↓

spleens removed (day 4)

↓

cultivation of cells (3—5 × 10^6 cells/ml)

↓

collection of medium after 24, 48 and 72 hours

↓

centrifugation (3000 rpm, 10 min.)

↓

titration

Medium: RPMI 1640 with 3 % neonatal calf serum, 1 % L—glutamine, 10 mM HEPES — buffer, 10^{-5}M 2—mercaptoethanol, antibiotics

Mitogen: Concanavalin A (3 μg/ml) (Pharmacia Fine Chemicals)

Pentapeptide: thymopoietin pentapeptide (32 — 36) prepared and synthetized according to T. Abiko et al.: Chem. Pharm. Bull. 28 (1980), 2507

as suspension for 24, 48, and 72 hours. The supernatants were harvested and centrifugated and the interferon activity was determined by cytopathic inhibition assay employing VSV on L_{929} cells. Results are given in terms of the International Standard (IU, NIH Mouse Fibroblast Reference Standard G 002-904-511). For the stimulation of cells we used Con A.

After pretreatment of mice with thymosin fraction 5 we found an enhancement of interferon production in Con A-stimulated spleen cells (tab. I). Each titer represents the mean of interferon activity of triplicate cultures from 9 separate prepared mice. The difference between controls and thymosin-treated mice after 24 and 48 hours was significant.

Table I. Production of mitogen-induced IFN by mouse spleen cells

Pretreatment[1] in vivo	Mean interferon titer		(IU/ml)
	24 hrs	48 hrs	72 hrs
PBS (control)	168	157	50
Thymosin	416	472	100

[1] Injections on days 1, 2, 3; PBS 0.2 ml, thymosin 100 μg/0.2 ml.

For partial characterization of interferon samples were incubated, respectively, with anti-mouse type 1-interferon globulin and at pH 2 until the determination of interferon activity. Because of the lack of neutralization by antiIFN-β and the sensitivity to pH 2 we suppose that the produced interferon is mainly IFN-γ.

The pretreatment of mice with TP 5 affected also an significant difference in comparison with the normal stimulated controls after 24 or 48 hours of cultivation (tab. II).

In the first experiment we used XVII x AKR hybrids and found between 7 and 10 times more interferon than in controls. In the other experiments we treated C57B1/6 × Balb/c mice and found about 3 times more interferon than in untreated controls.

By use of thymosin fraction 5 as well as TP 5 we found after 72 hours of cultivation no augmentation of interferon production. Our findings correlate with results of Shoham and coworkers (8) who found an higher interferon production after pretreatment of human lymphocytes with thymosin followed by incubation of cells with Con A.

In our mouse model we are unable to find any augmentation of interferon production after simultaneous incubation of thymic peptides and Con A in comparison with Con A-treated controls.

In the mouse cells thymosin as well as TP 5 alone had no effect

Table II. Influence of treatment of mice with thymopoietin pentapeptide (TP5) on the Con A-induced interferon production in mouse spleen cells

Pretreat-ment	Exp.	Num. of animals	Mean interferon titer (IU/ml)					
			24 hrs.	Comp. with control	48 hrs.	Comp. with control	72 hrs.	Comp. with control
Without (control)	1	6	250		250		300	
	2	6	270		420		249	
	3	9	300		200		213	
TP5	1	6	1750	$p < 0.02$	2500	$p < 0.01$	450	n.s.
	2	6	700	$p < 0.05$	1200	$p < 0.01$	256	n.s.
	3	9	950	$p < 0.05$	500	n.s.	250	n.s.

[1] Results indicate the mean interferon titer (IU/ml) of separately prepared and cultured spleens.

[2] U-test (Mann and Whitney) values and significance levels (p) for the indicated comparisions.

n.s. = no significance.

Table III. Influence of thymopoietin pentapeptide on cAMP/cGMP levels and ^{3}H-thymidine incorporation in leukocyte cultures

Treatment	cAMP (pMol)	cGMP (pMol)	Thymidine[1] incorp. (%)
Without (control)	1.40 ± 0.2	0.10 ± 0.002	100
Pentapeptide[2]	3.10 ± 0.9	0.60 ± 0.01	79

[1] Mean values and standard devations of 12 separately prepared mice/group were determined by Student's t-test.

[2] Cultures were incubated with 100 μg TP-5/ml for 30 hrs at 37° C.

on the IFN production. In human peripheral lymphocytes thymosin enhances significant the interferon production without additional inducer (2).

TP 5 influenced the cAMP/cGMP levels and the ^{3}H-thymidine incorporation. The cAMP level was determined according to Gilman (3) and we found a significant difference between the control and pretreated mice with $p < 0.001$ (tab. III). The cGMP level was determined according to Kuo et al (6) and the difference was significant with $p < 0.01$. The ^{3}H-thymidine incorporation was significant lower in pretreated cells.

In line with Pugliese (7) and Hyang (5) we suppose an in vivo relationship between action of thymic peptides on spleen cells and interferon production.

But the mechanism by which this peptides affects the interferon production is still unclear.

All experiments were done in corporation with Dr. Diezel from the Department of Dermatology of the Charité, Berlin, and with Dr. Forner from the Institute of Drug Research Berlin.

References

1 Abiko, T. et al. (1980): Chem. Pharm. Bull. 28, 2507.
2 Diezel, W. et el.: Exp. Clin. Endocrinol., in press.
3 Gilman, A. (1970): Proc. Nat. Acad. Sci. USA 67, 305.
4 Hooper, J. A. et al. (1975): Ann. N. Y. Acad. Sci. 249, 125.
5 Hyang, K.-Y. et al. (1981): J. Interferon Res. 1, 411.
6 Kuo, J. F. et al (1976): J. Biol. Chem. 251, 1759.
7 Pugliese, A.; Tovo, P. A. (1980): Thymus 1, 305.
8 Shoham, J. et al. (1980): J. Immunol. 125, 54.
9 Svedersky, L. P. et al. (1982): Eur. J. Immunol. 12, 244.

E. Rentz, M. D., Central Institute of Cancer Research, Dept. of Chemical Carcinogenesis, Berlin-Buch (DDR)

Contr. Oncol., vol. 20, pp. 280—297 (Karger, Basel 1984)

Interferon Production by Leukocytes Obtained from Patients with Systemic Lupus Etythematosus

J. Rovenský[a]; *V. Lackovič*[b]; *L. Borecký*[b]; *D. Žitňan*[a]; *J. Lukáč*[a]

[a] Research Institute of Rheumatic Disease, Piešťany,
[b] Institute of Virology, Slovak Academy of Sciences, Bratislava, Czechoslovakia

Introduction

Antiviral activity with high frequency was found in the serum samples of patients with systemic lupus erythematosus (SLE) and other related autoimmune disease (6). Recent studies suggest that interferon (IFN) present in a large number of sera from patients with SLE had the antigenic and biologic characteristics of the human IFN-α (7, 8, 10). The source of this IFN is not known.

Neighbour and Grayzel (9) reported that peripheral blood leukocytes from many of the SLE petients failed to produce detectable levels of IFN in response to various IFN-inducers in vitro. On the other hand, these autors observed than leukocytes from less than 5 % of the SLE patients released IFN spontaneously when cultured for 48 hrs in the absence of known inducers. They postulated that IFN found in the serum is not produced by leukocytes in the circulation but originated from stimulated lymphocytes located either within lymphoid organs, or more likely, at localized sites of lymphocyte infiltration and activation. The same has been suggested by Preble et al. (8). These autors were not able to detect any significant spontaneous release of IFN from non-stimulated purified mononuclear cells obtained from IFN-positive patients after their cultivation in vitro.

In the present study, the IFN and anti-IFN activities of serum samples from patients with SLE were studied. Attention has been paid to IFN-levels in the blood of patients repeatedly examined during the course of the disease and the cellular origin of the IFN in the circulation.

Materials and Methods

Patients population. All patients fulfilled four or more of preliminary criteria for the diagnosis of SLE of the American Rheumatism Association (1). The patients were in various clinical stages of disease activity and were receiving the following treatment either singly or in combination: non-steroidal antirheumatic drugs, prednisone, cyclophosphamide and Levamisole. They suffered from no obvious viral illness during the collection of serum samples. Three patients with SLE were concurrently treated against active tuberculosis of the lungs. The control group consisted of healthy donors.

Leukocyte preparation and interferon production. The mononuclear cells and polymorphonuclears (PMNs) were separated from heparinized blood by the method of Bøyum with the use of Dextran T-250 sedimentation and/or Ficoll-Isopaque isopycnic centrifugation (2). Mononuclears were further fractioned into E-rosette-positive (ER^+) population by rosetting with sheep erythrocytes as described by Fitzharris et al. (3).

For the induction of IFN, 4×10^6 cells suspended in 1 ml of Eagle's basal medium supplemented with 10 % fetal calf serum and antibiotics were incubated either alone or with Newcastle disease virus (NDV-B_1) at a multiplicity of 2—3 infectious virus particles per leukocyte and BCG (10 μg per ml), respectively. Mycobacterium bovis, strain BCG, was obtained from the Institute of Sera and Vaccines, Prague. The BCG was suspended in sterile normal saline at concentration 5 mg/ml just prior to use. The cells treated with NDV-B_1 or BCG were left in contact with inducer for 24 hours. The supernates from these cells were kept at pH2 for 72 hours (NDV-B_1), or used directly (BCG) for IFN titration.

Interferon assay. IFN assays were done by modification of Vesicular stomatitis virus cytopathic effect method (4). Series of twofold dilutions of the tested samples were incubated with human diploid cells. As positive samples were defined those containing $\geqq$ 16 units per ml of serum and/or $\geqq$ 4 units per ml of cell culture supernates. The results were standardized using reference human leukocyte IFN preparation obtained from the National Institutes of Health, Bethesda, USA.

Interferon neutralizing activity. To determine the IFN neutralizing activity of serum samples, a constant dose of human IFN-α (16 units) was mixed with 0.1 ml of patient's serum heated previously at 56° C for 30 min. After 30 min incubation at 36°C the presence of the remaining IFN activity in the mixtures was determined as above.

Results

When sera of 70 patients with SLE and 26 healthy volunteers were examined in IFN assay, significant IFN activity (≧ 16 units per ml) was detected in the serum samples of 14.3 % (10 of 70) patients with SLE (tab. I). Higher levels of IFN, ranging from 32 to 128 units, were detected in 5 patients. These data are consistently lower than those published by Hooks et al. (5). This discrepancy might be explained by the fact that the serum samples examined in our study were obtained from patients in various clinical stages of their disease.

Table I. Interferon and anti-IFN levels in sera of patients with SLE

Activities	SLE patients	Normal donors
IFN		
No of donors	70	25
No of IFN-positive donors	10	0
%	14,3	
Geometric mean of IFN/units/ml/	36,5	
Anti-IFN		
No of IFN-negative donors	60	25
No of anti-IFN positive donors	9	0
%	15	

The neutralizing activity for IFN-α was present in the sera of 15 % (9 of 60) of IFN-negative patients. The question whether this neutralizing activity is associated with immunoglobulins was not yet examined. However, Panem et al (7) recently reported on the presence of IgG antibodies to IFN-α in the serum of a patient with SLE. No IFN or anti-IFN activities could be detected in the serum samples of 26 healthy donors.

The immunological characterization as well as the pH 2 lability of IFN found in the sera of IFN-positive patients (5) suggests that this IFN is probably a subtype of the IFN-α (tab. II).

Table II. Characterization of interferon by antibody sensitivity and pH 2 lability

Source of IFN	Interferon (units per ml)		
		Treatment	
	None	Anti-IFN alfa	pH2
Serum from SLE patients:			
1	64	<8	<8
2	64	8	16
3	128	<8	32
4	16	<8	<8
5	16	<8	<8
6	32	<8	<8
Standard (IFN alfa)	128	<8	128

Sheep anti-IFN alfa serum was added in concentration sufficient to neutralize all homologous IFN activity (500 units). The IFN-antiserum mixtures were incubated for 30 min at 37° C and the residual IFN activity was then quantitated in human diploid cell system. Acid (pH 2) treatment was performed by mixing equal parts of serum and 0.2 M HCl and leaving the mixtures in refrigerators (+4° C) for 24 hrs after which the neutral pH wasrrestored and the residual IFN was assayed.

The analysis of the dynamics of appearance of IFN and anti-IFN activities in the serum samples of SLE-patients repeatedly tested during several years periodes revealed 3 phases: a) IFN-positive phase, b) IFN-negative, and, c) anti-IFN-positive phase (tab. III). It is of interest that the anti-IFN-positive phase seems to be neither a permanent nor a final stage because, after several months, such patients were found to release IFN again.

Additional experiments were aimed to explain the origin of the IFN in the blood of IFN-positive patients with SLE. For this purpose, T-lymphocytes and PMNs from peripheral blood were isolated and assayed for IFN release. We were unable to detect any significant spontaneous release of IFN from the unstimuled ER^+ T-lymphocytes. Surprisingly, all the unstimulated PMNs obtained from 5 IFN-positive SLE patients secreted detectable levels of IFN in vitro (fig. 1). When leukocytes from IFN-negative SLE patients were examined, the secretion of spontaneous IFN in vitro was detected in 1 of 12 (8.3 %). The same cell fractions from 10 healthy donors did not release detectable IFN amounts spontaneously in vitro. Neutralization of the antiviral activity by antibody to IFN-α human supports the view that the IFN spontaneously secreted

Table III. Interferon and anti-IFN activities in the sera of patients repeteadly tested during the course of SLE

Patient	Age/Duration of illness (years)	Serum received (Date)	IFN (Units/ml)	Anti-IFN
M.E.	48/18	31. 8. 1980	<16	+
		2. 2. 1981	<16	+
		17. 11. 1981	<16	—
		6. 5. 1982	32	NT
		6. 10. 1982	16	NT
G.A.	29/10	15. 12. 1976	NT	—
		15. 10. 1979	32	NT
		26. 6. 1980	16	NT
		29. 6. 1981	<16	+
		10. 2. 1982	<16	—
		2. 9. 1982	32	NT
M.M.	28/8	13. 2. 1978	<16	+
		25. 6. 1980	<16	—
		5. 10. 1982	64	NT
B.A.	26/3	13. 4. 1977	<16	+
		29. 6. 1981	32	NT
		26. 7. 1982	<16	—
		5. 10. 1982	16	NT

Serum samples were stored from the time of collection until use at —20° C. The sera were first examined for the presense of IFN, then the IFN-negative samples were tested for presence of anti-IFN activity.

+ = IFN neutralizing activity present.

— = IFN neutralizing activity absent.

NT = non tested.

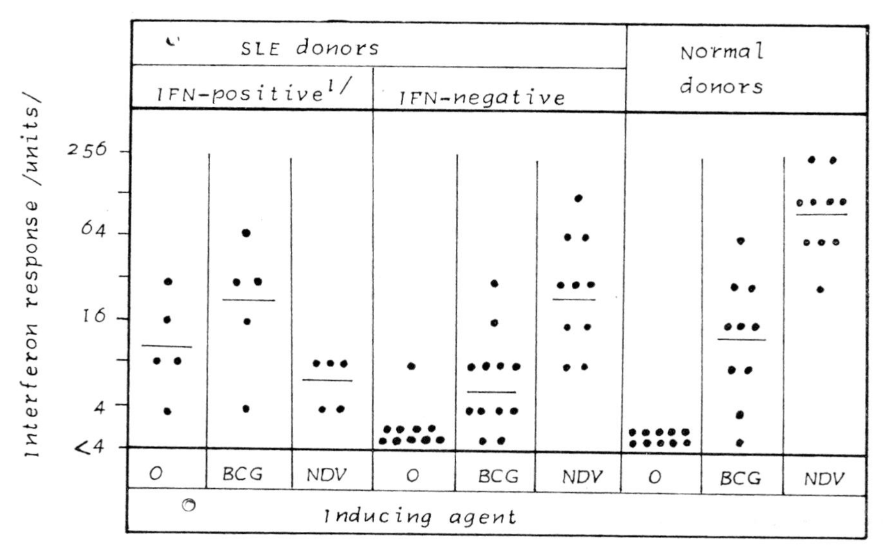

Fig. 1. Individual interferon responses of leukocytes to BCG and NDV in vitro from normal and SLE donors.
1. Geometric mean of IFN titer in sera: 48.5. The horizonal bars indicate the geometric mean of IFN responses.

from peripheral leukocytes obtained from IFN-positive SLE patients is also of IFN-α type.

The NDV-induced and BCG-induced IFN response of leukocytes from IFN-positive and IFN-negative SLE patients were studied also (fig. I). Our study lends support to the data previously reported by Neighbour and Grayzel (9) that the leukocytes of SLE patients exhibit a defect in their IFN response to NDV. In addition, we analysed the capability of IFN production in leukocytes obtained from IFN-positive and IFN-negative SLE patients. Our results strongly suggest that the leukocytes from IFN-negative patients are better producers of IFN-α than the cells obtained from IFN-positive patients.

The BCG induced IFN-response of leukocytes from both IFN-positive and IFN-negative patients did not differ significantly from that of normal donors. However, a depressed IFN response to BCG was observed in those SLE patients which had an active tuberculosis of lungs. This observation deserves further studies.

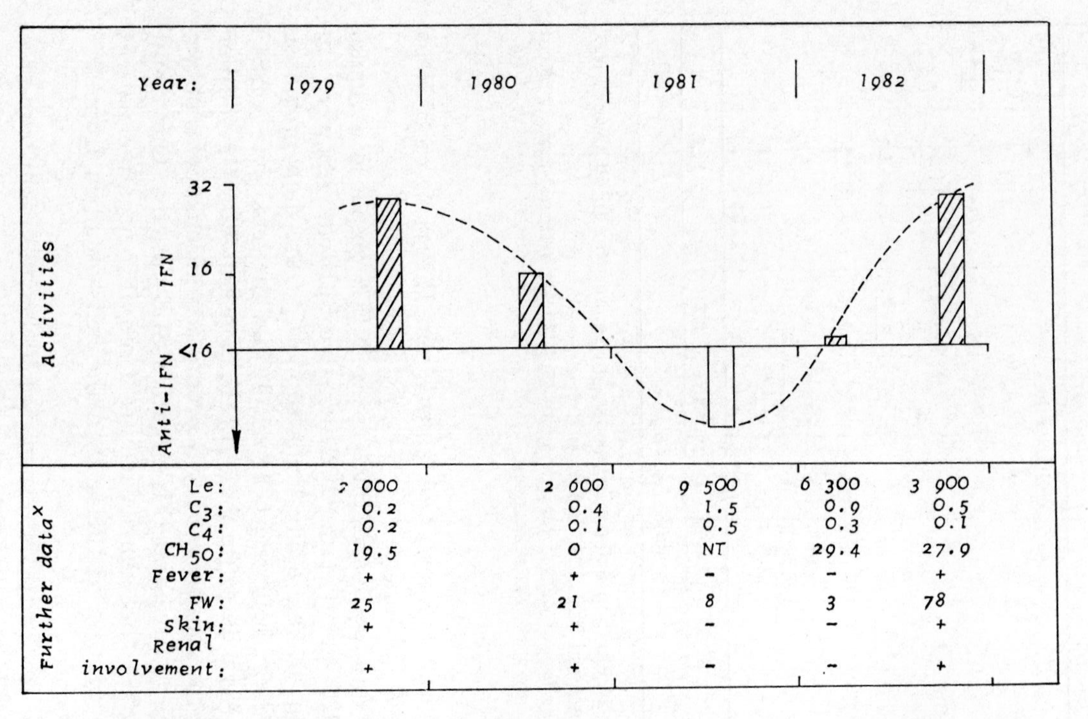

Fig. 2. Interferon and anti-IFN activities found in the serum during the clinical course of a patient with SLE.
Le: Leukocyte number, C_3, C_4, C_{50}: Values for total complement (CH_{50}) or components C_3 and C_4 (in units).
FW: Sedimentation of erythrocytes according to Westergreen, NT: not tested.

On the basis of our findings we conclude that:

1. A periodicity of IFN- or anti-IFN production can be observed during the course of SLE disease (fig. 2). The appearance of IFN in the serum seems to show a relation to the clinical manifestation of SLE. The anti-IFN-positive phase is not a final stage because several months later the same patient may produce IFN again. The reason of these changes are unclear.

2. The source of IFN found in the serum is unclear. It might be produced by leukocytes in the circulation. However, ER^+ T-lymphocytes from IFN-positive patients do not spontaneously secrete IFN in vitro. The IFN secreted from the PMNs was characterized as IFN-α.

References

1 Cohen, A. S.; Reynolds, W. E.; Franklin, E. C.; Kulka, J. P.; Ropes, M. V.; Shulman, L. E. (1982): Bull. Rheum. Dis. 21, 643.
2 Ferenčík, M.; Rovenský, J.; Štefanovič, J. (1982): Agent Action. 12, 478.
3 Fitzharris, P.; Alcocer, J.; Stephens, H. A.; Knight, R. A.; Snaith, M. L. (1982): Clin. exp. Immunol. 47, 110.
4 Havell, E. A.; Vilček, J. (1972): Antimicrob. Agent, Chemother. 2, 476.
5 Hooks, J. J.; Jordan, G. W.; Cupps, T.; Moutsopoulos, H. M.; Fauci, A. S. Notkins, A. L. (1982): Arthritis Rheum. 25, 396.
6 Hooks, J. J.; Moutsopoulos, H. M.; Geiss, S. A.; Stahl, N. I.; Decker, J. L.; Notkins, A. L. (1979): N. Engl. J. Med. 301, 5.
7 Panem, S.; Check, I. J.; Henriksen, D.; Vilček, J. (1982): J. Immunol. 129, 1.
8 Preble, O. T.; Black, R. S.; Friedman, R. M.; Klippel, J. H.; Vilček, J. (1982): Science 216, 429.
9 Neigbour, P. A.; Grayzel, A. I. (1981): Clin. exp. Immunol. 45, 576.
10 Ytterberg, S. R.; Schnitzer, T. J. (1982): Arthritis Rheum. 25, 401.

J. Rovenský, M.D., Res. Inst. Rheum. Dis., Piešťany, (Czechoslovakia)

Contr. Oncol., vol. 20, pp. 298—303 (Karger, Basel 1984)

Interferon in Patients with Psoriasis

S. R. Waschke; W. Diezel

Central Institute of Molecular Biology, Academy of Sciences of the GDR, and [a]Dept. of Dermatology, Medical School, Charité, Humboldt University, Berlin, GDR

Psoriasis is a chronic skin disease characterized by extended skin lesions and, in severe cases, arthritis. Several lines of evidence point to the possibility that psoriasis may be a systemic rather than a localized disorder caused by a genetically determined abnormality in the keratinocytes and/or Langerhans cells, a constituent of the monocyte-macrophage series influencing the keratinocyte growth and, in this connection, resulting in increased proliferation which is characteristic for the psoriatic lesions (5).

The aetiology of this disease is unknown as is the mode of heredity. Some recent publications report on the detection of interferon (IFN) in the circulation of patients with autoimmune diseases (for review see 2). This prompted us to look for IFN activity in the sera of psoriatic patients.

Blood samples of thirty-seven patients attending the hospital to undergo photochemotherapy were drawn before the beginning of therapy and serum samples were analyzed.

The antiviral activity was determined by IFN microtest on human diploid embryonic fibroblasts using Vesicular stomatitis virus as the challenge virus. Virus-induced cytopathic effect was evaluated microscopically, and the IFN titer was defined as the reciprocal of the highest dilution of sample to protect 50 percent of the cells. Standard alpha- and gamma-IFN were included in each assay. Results were standardized to 023-901-527 reference human leukocyte IFN (NIH, Bethesda, USA).

As shown in tab. I, all of the samples tested contained measurable antiviral activity the titers ranging from 8 to 512 units/ml. There was no difference in the IFN levels of patients with psoriasis vulgaris to patients suffering from psoriasis arthropathica (not shown).

Table I. Comparison of IFN from psoriatic patients with human α- and γ-IFN

IFN	No. of Samples	Titer on HEF (IU/ml)	Titer on HEF/ Titer on MDBK	Neutralization factor (Anti-body to α-IFN)
α-IFN	1	100	1 to 2	>60
γ-IFN	1	100	63	0 to 2
Psor. Pat. Sera	2	8 to 16	N.T.	N.T.
	2	32	1 to 2	N.T.
	18	64	0.5 to 2	0 to 4
	13	128	1 to 2	0 to 8
	1	256	2	2
	1	512	2	8

For method and materials see text.

All of the samples were tested in duplicate on human and bovine cells, line MDBK. The titer on bovine cells was the same as on human cells within the twofold error inherent in the assay. Thirty-three of the samples were also analyzed with sheep antibody to human alpha-IFN kindly provided by Prof. Borecký, Institute of Virology of the Slovac Academy of Sciences, Bratislava. 50 μl of IFN-positive serum (64 to 512 I.U./ml) were mixed with 50 μl of medium or 50 μl of antibody.

Dilutions of antibody capable of specifically neutralizing 100 I. U. of homologous IFN per milliliter were used (1:500 to 1:1000). After incubation at 37° C for 60 minutes, residual IFN was assayed. The neutralization factor is the IFN titer of the medium controls divided by the titer of IFN remaining after treatment with the antibody.

Most of the samples were not significantly affected by this antibody even at the higher concentration, only about 30 % of them showed marginal neutralization of 50 to 75 %.

We now evaluated the acid sensitivity and thermal stability of the serum IFN of our psoriatic patients (tab. II). IFN in serum samples of 15 patients included in tab. I were incubated at pH 2 for 24 hours at 4° C or for 60 minutes at 56° C. Standard alpha- and gamma-IFN samples were acidified, incubated, and neutralized in an identical manner.[1] The serum IFN was inactivated fourfold or more by dialyzis for 24 hours at pH 2, whereas standard human alpha-IFN was completely stable, and human gamma-IFN produced by Con A-stimulated human lymphocytes (10) was inactivated more than 70fold under these conditions. In addition, eight of the samples were inactivated fourfold or more heating at 56° C for one hour.

Taken together, these results make it likely to regard the IFN in psoriatic sera as an acid-labile alpha subtype like the IFN in

Table II. Properties of IFN in serum of 16 psoratic patients

Source	Titer on HEF	(IU/ml) on MDBK	Neutralization factor (Antibody to α-IFN)	Inactivation factor 56° C	pH 2
Rie.	64	128	4	4	8
Höf.	32	8	N. T.	N. T.	N. T.
Hen.	128	64	0	4	>8
Leh.	64	64	4	2	4
Wie.	64	128	2	2	16
Gie.	64	128	2	4	8
Wur.	128	128	2	8	>16
Zan.	128	64	0	4	>16
Rut	64	32	0	2	8
Kuh.	64	64	2	2	4
Wib.	128	128	2	2	4
Wik	64	16	2	4	>8
Hai.	128	128	2	4	16
Sch.	128	64	4	2	4
Nag.	512	128	8	16	>8
Dre.	64	64	2	2	8
α-IFN	512	512	128	16	0
γ-IFN	1000	>10	0	>30	>70

[1] The inactivation factor is the IFN titer of the pH 7.2 (PBS) — threated controls for acid treatment and of untreated samples (heating), respectively.

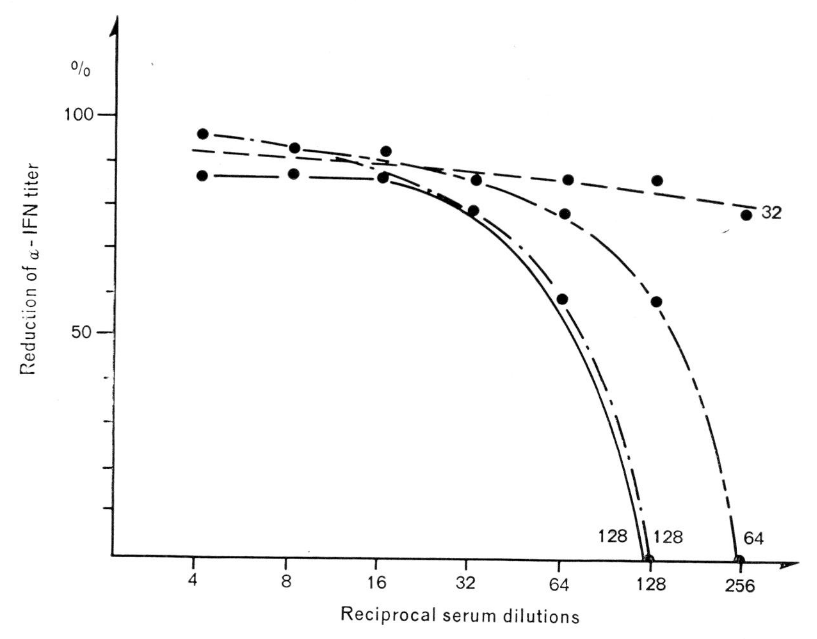

Fig. 1. Inhibition of antiviral activity of human α-IFN by serum of psoriatic patients.

lupus erythematosus patients characterized by Preble et al. (8), although, in contrast to these autors, we failed to get any significant neutralization by a conventional alpha-IFN antibody raised with virus-induced human leukocyte IFN which was treated at pH 2 before concentration and application to the animal. This treatment may well have been deleterious to any acid-labile subtype of the original human alpha-IFN preparation.

In view to the known side effects of circulating IFN we have been wondering how the psoriatic patients manage to live with their IFN levels. We, thus, looked for anti-IFN activity in their sera (fig. 1, 2). Standard human alpha- (5000 I.U./ml) and gamma-IFN (3000 I.U./ml) were incubated with serial twofold dilutions of serum samples of four patients included in tab I for one hour at 37 °C. The remaining IFN activity was then titrated as usually.

All of the tested sera showed strong anti-IFN activity against both alpha- and gamma-IFN inversely correlated to the IFN titer of the tested serum sample. Hence, the anti-IFN activity may be an unspecific IFN inhibitor rather than an IFN-type specific antibody. Healthy controls showed only marginal inhibitory activity of up to 1:8 at the highest. Studies are now under way to isolate and characterize this IFN inhibitor present in psoriatic sera.

The nature of the IFN inducer(s) involved in psoriasis remains obscure as does, up to now, the type of producer cell(s). In very preliminary studies we got high IFN titers of up to 3000 I.U./ml released by Ficoll/Visotrast-isolated peripheral blood lymphocytes of three female psoriatic patients within the first two hours of cultivation in medium laking any classical type of inducer.

Furthermore, it remains an open question whether the IFN in psoriasis is of any pathogenic significance to the disease. Recently, photochemotherapy consisting in oral application of methoxsalen (0.6 mg/kg) and subsequent whole-body exposure to UV-A radiation two hours later (PUVA therapy) became increasingly popular in treatment of psoriasis and several other skin diseases (6, 9). Se-

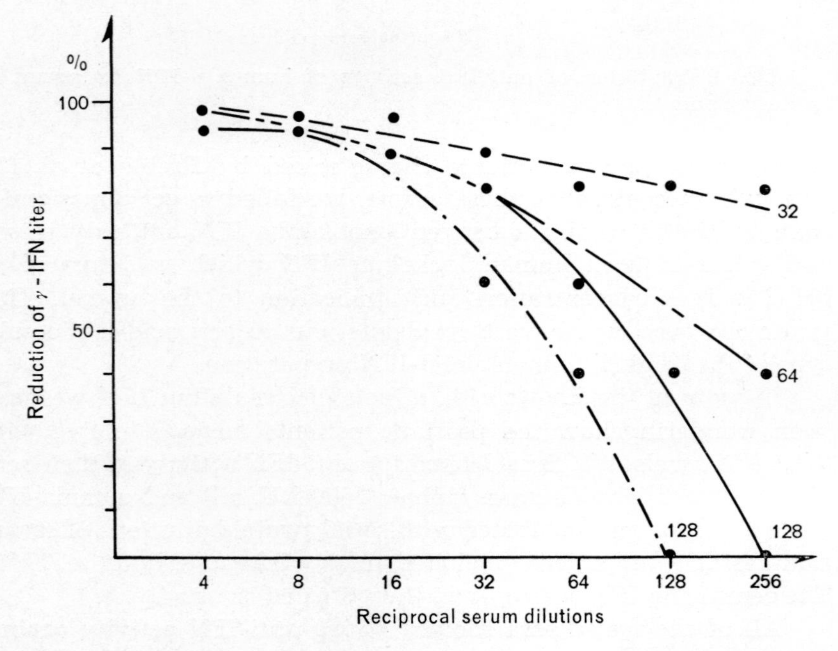

Fig. 2. Inhibition of antiviral activity of human γ-IFN by serum of psoriatic patients.

veral previous publications, moreover, state that UV radiation should be capable of potentiating an induced IFN production (3, 7). Therefore, we followed the behaviour of serum IFN levels in our psoriatic patients under PUVA therapy.

Serum samples of 19 patients subjected to PUVA therapy three times a week for four weeks (4) were collected before and two weeks (6 treatments) and 4 weeks (12 treatments) after initiation of the therapy and analyzed (1). IFN levels significantly increased in all cases but one over the whole period of treatment.

Provided this beneficial effect of PUVA is mediated by IFN it may open a future field for an applied IFN research including the possibility of clinical application of suitable IFN inducers in psoriasis, mykosis fungoides, and certain other skin disorders where PUVA treatment proved to be of therapeutical value.

References

1 Diezel, W.; Waschke, S. R. (1983): Brit. J. Dermatol., 109, 549.
2 Hooks, J. J.; Moutsopoulos, H. M.; Notkins, A. L. (1982): Texas Rep. Biol. Med. 41, 164.
3 Lindner-Frimmel, S. J. (1974): J. gen. Virol. 25, 147.
4 Meffert, H.; Metz, D.; Sönnichsen, N. (1978): Dermatol. Monatsschrift 164, 481.
5 Mier, P. D.; Gommans, J. M.; Roelfzema, H. (1980): Brit. J. Dermatol. 103, 457.
6 Melski, J. W.; Tanenbaum, L.; Parrish, J. A.; Fitzpatrick, T. B.; Bleich, H. L. (1977: J. invest. Dermatol. 63, 328.
7 Mozes, L. W.; Vilcek, J. (1974): J. Virol. 13, 646.
8 Preble, O. T.; Black, R. J.; Friedmann, R. M.; Klippel, J. H. Vilček, J. (1982): Science 216, 429.
9 Vella-Briffa, D.; Rogers, S.; Greaves, M. W.; Markers, J., Shuster, S.; Warin, A. P. (1978): Clin. exp. Dermatol. 3, 339.
10 Venker, P.; Waschke, S. R., in press.

S. R. Waschke, M. D., Zentralinstitut für Molekularbiologie, Akademie der Wissenschaften der DDR, Berlin (GDR)

Contr. Oncol., vol. 20, pp. 304—309 (Karger, Basel 1984)

Production of Interferon in Children with Chronic Respiratory Diseases

K. Vaněček; A. Lehovcová

Research Institute of Child Development, Prague, ČSSR

Introduction

The production of interferon is one of the basic protective mechanisms of the organism against virus infection. It can be influenced by various pathological processes occurring in the organism (5, 1, 3, 7, 2, 6) and probably also by environmental factors. Deficient or anomalous production of interferon may contribute to the pathogenesis of virus infections. Our investigation was directed to the production of interferon in children suffering from respiratory tract diseases, repeated or protracted in particular. The ability of interferon production was tested in 104 children hospitalised or followed-up at the 2nd Pediatric Clinic in Prague with different diagnoses, predominantly with respiratory tract infections.

Methods

Eight ml of heparinised blood was taken from patients and a lymphocyte tissue culture was prepared in the amount of 10 million in 1 ml of culture medium. Lymphocytes were incubated with 5000 haemagglutination units of Newcastle disease virus for 24 hours. The obtained sample was titrated to determine the amount of interferon-alpha in cells of human embryonal lungs infected by the virus of vesicular stomatitis.

The results were analysed by a Hewlett-Packard computer by a one-way variance analysis using the logarithmical transformation of data and by the "t"-test.

The examined children, aged 6 months to 18 years were divided into 5 groups according to the clinical diagnosis. The first group included children with acute respiratory diseases (acute pharyngitis, laryngotracheitis, acute and asthmatic bronchitis, possibly associated with light bronchopneumonia). The second group comprised children with repeated or long-term respiratory infections (recurrent bronchopneumonia or sinusitis, chronic bronchitis and cystic fibrosis). The third group included children with allergic or autoimmune diseases (hemorrhagic purpura, toxic exanthema, bronchial asthma, collagenosis and lung fibrosis). The fourth group comprised children with bacterial infections (lacunar angina, urinary tract infections, etc.). The fifth group included children who showed no clinical symptoms of a disease and who had come to a control examination or those suffering from non-infectious diseases (for example children with genetically conditioned congenital developmental deffects or with metabolic disorders, etc.). Almost in all children normal immunoglobulin levels were observed. Only in one child with acute bronchopneumonia striking hypogammaglobulinemia was found.

Results

In the majority of the children examined, the amount of interferon produced by leukocytes varied within 32 and 256 units/ml (fig. 1). For normal values we considered the mean amount of 64 units of interferon-alpha in 1 ml which corresponds to the findings obtained in control children and is in agreement with data found in the literature (4).

The graph shows that children with acute respiratory disease or bacterial infections do not differ significantly from the controls. In the group including repeated and chronnic respiratory diseases the percentage of children with deficient or no production of interferon was higher than in other groups. If we compare the geometrical mean values of interferon-alpha in individual groups then we see that this difference was significant (tab. I).

In the group of allergic and autoimmune diseases strikingly low or negative titers were found in the children treated with high doses of Prednison (20 to 40 mg daily) at the time of examination

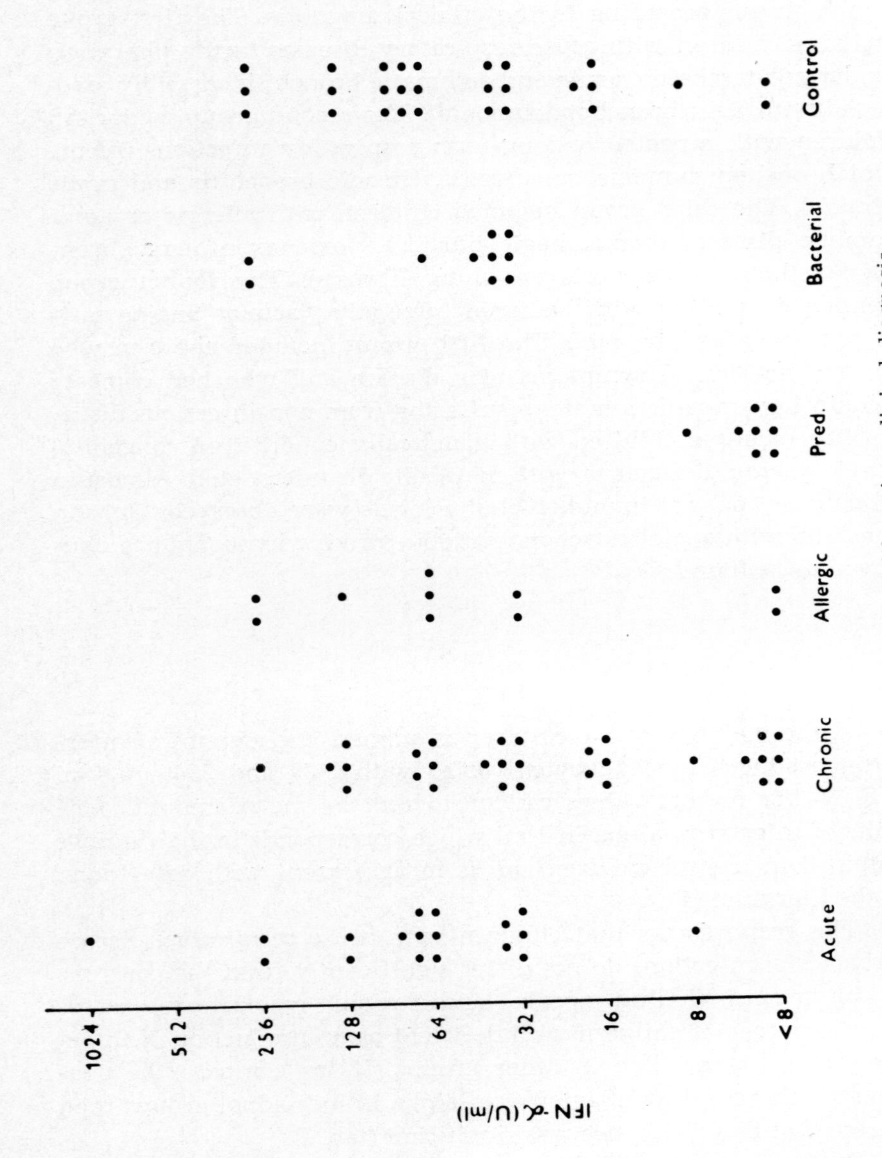

Fig. 1. Production of IFN — alpha in patients with various clinical diagnosis.

Table I. Geometric mean values of IFN-alpha in groups according to the diagnosis

Group	Number of patients	Geometric mean
Acute respiratory infections	18	69.12
Recurrent and chronic respiratory diseases	30	17.55
Allergic and autoimmune diseases	10	34.30
Prednisone treatment	8	1.30
Bacterials infections	10	51.90
Controls	28	39.00

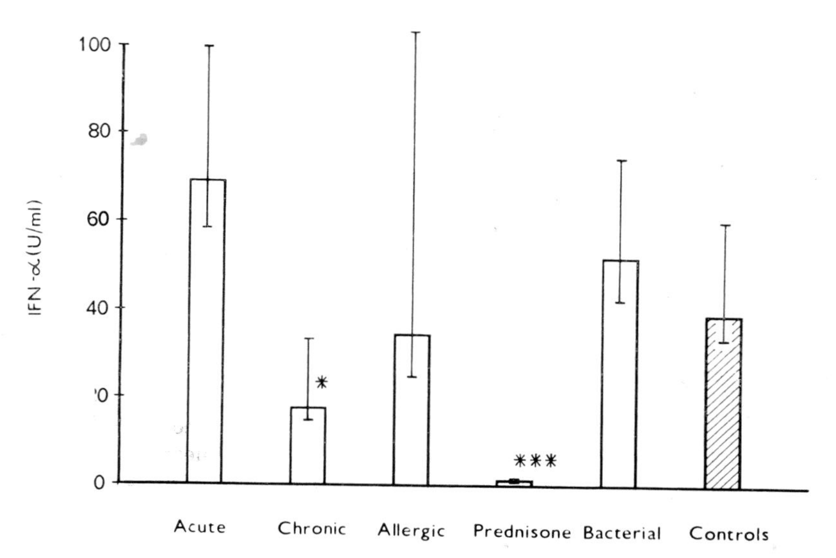

Fig. 2. Comparison of geometric mean values of IFN — alpha titres.
* $P < 0.05$
*** $P < 0.01$

(fig. 2). If we compare these children with the children suffering from the same diseases but who were not treated with corticoids during the investigations, the difference is highly significant. The suppression of the production of interferon-alpha by the application of high doses of Prednison might explain negative effects of corticoid therapy on the course of viral infections. Therefore the application of corticoids in such states should be properly considered.

The production of interferon-alpha was not influenced either by the age or sex of the patient (tab. II and III).

Interferon production in the child with hypogammaglobulinemia was not affected and the amount of interferon produced by leukocytes was relatively high – that is 1024 units in 1 ml. Our further investigation will be orientated to the long-term follow-up of children with a decreased production of immunoglobulins.

Table II. Dependence of IFN-alpha production on age

Age (years)	Number	Geometric mean
0—5	25	37.8
6—10	19	22.2
11—15	27	29.6
16—20	3	50.8
Total	104	—

Table III. Dependence of IFN-alpha production on sex

Sex	Number	Geometric mean
Male	53	24.0
Female	51	32.9
Total	104	—

References

1 Daniels, J. C.; Sakai, H.; Cobb, E. K. et al. (1970): Am. J. Med. Sci. 259, 214—227.
2 Emödi, G.; Just, M. (1974): Acta Paediat. Scand., 63, 183—187.
3 Hadházy, G.; Gergely, L.; Toth, F. D.; Szegedi, G. (1967): Acta microbiol. Acad. Sci. Hung., 14, 391—397.
4 Isaacs, D.; Tyrrell, D. A. J.; Clarke, J. R.; Webster, A. D. B.; Valman, H. B. (1981): Lancet ii, 950—952.
5 Sanders, C. V.; Luby, J. P.; Hull, A. R. (1977): J. Lab. Clin. Med. 77, 786—796.
6 Starr, S. E.; Tolpin, M. D.; Friedman, H. M.; Plotkin, S. A.; Paucker, K. (1977): Lancet II, 1357.
7 Strander, H.; Cantell, K.; Leisti, J.; Nikkila, E. (1970): Clin. exp. Immunol. 6, 263—272.

K. Vaněček, M.D., Research Institute of Child Development, V úvalu 84, 150 06 Prague 5-Motol (Czechoslovakia)

Contr. Oncol., vol. 20, pp. 310—317 (Karger, Basel 1984)

Synthesis of Gamma Interferon by Blood and Bone Marrow Cells in Children with Hemoblastoses

T. G. Orlova; S. V. Pavlushina; I. E. Gavrilova; G. L. Mentkevich

N. F. Gamaleya Institute of Epidemiology and Microbiology, Academy of Medical Sciences, Moscow, USSR

Introduction

Many reports in the literature consider that the capacity of blood cells to produce virus-induced (alpha) interferon is the index of host responsiveness in various pathologic conditions. There is evidence, however, that in a number of severe diseases of the hemopoietic system (acute lymphocytic leukemias, etc.) the capacity of blood and bone marrow cells to produce alpha interferon remains unimpaired (4, 5). The suggestion has been made (1) that the capacity of lymphocytes to produce immune (gamma) interferon should be considered the principal index of the effector function. The present work was mainly aimed at the study of the gamma-interferon (IFN-γ) response of mononuclear cells in three of the most prevalent child hemoblastoses: acute lymphocytic leukemia (ALL), lymphosarcomas (LS), and lymphogranulomatosis (LGM).

As a rule, blood mononuclears are the source of cells for such studies, as only one report (2) is available on the capacity of bone marrow mononuclears to produce IFN-γ. It was therefore important to determine the interferon-producing activity of bone marrow mononuclears in sick children, and to compare the IFN-γ response of blood and bone-marrow mononuclears to stimulation with phytohemaglutinin.

Materials and Methods

The study covered 35 children with ALL, 16 with LS and 18 with LGM. Nineteen of these children had been given no antitumor therapy before the study. The results of examinations of the blood from 11 adult donors was also included. Blood was obtained by venipuncture, and bone marrow by diagnostic puncture of the iliac bone.

Mononuclear blood cells recovered by one-step Ficoll-Verographin density gradient centrifugation. For the first time this method has been extended to recovery of bone marrow mononuclears. The results of morphological examinations showed mononuclears to comprise at least 81 % of the isolated fractions (6).

To induce IFN-γ, 10^6 viable mononuclears were suspended in medium 199 containing 20 % human AB serum, Monomycin (100 μg/ml), and PHA in a concentration of 10 μg/ml. The stimulated mononuclears were cultivated for 8—9 days. During this period 7.5 % sodium bicarbonate solution was added drop-wise to the cultures to prevent the destruction of the acid-labile IFN-γ. Based on previous data on the kinetics of IFN-γ production (7), supernatants of most blood cultures were collected after 7 days.

Interferon titres in the supernatants were determined by 50 % inhibition of the cytopathic effect of 100 TCD_{50} of Vesicular stomatitis virus in human embryo skin-muscle diploid cell monolayers. The resulting IFN was classified as "immune" according to its lability at pH 2. The results were statistically treated by calculation of Fischer exact criteria.

Results

IFN-γ Response of Bone Marrow Mononuclears

Fig. 1 presents the results of the studies on the kinetics of IFN-γ production by bone marrow mononuclears in children with ALL in the acute period and remission, as well as in children with lymphosarcoma and lymphogranulomatosis. In the latter two instances the patients bone marrow showed no evidence of malignant transformation.

It will be seen from the data presented that mononuclears of the intact bone marrow (ALL in remission, LS, and LGM) were capable of producing IFN-γ in response to PHA stimulation. Maximum titres of IFN-γ (up to 300 U/ml) were observed after 7 days of cultivation.

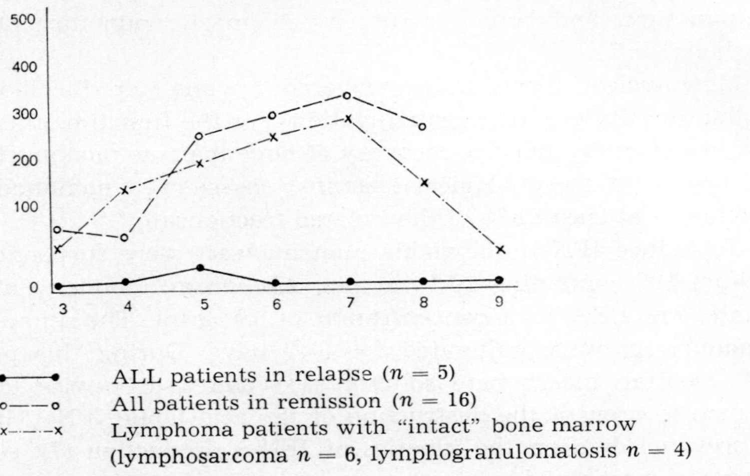

Fig. 1. Kinetics of immune interferon production by bone marrow mononuclears.
Abscissa — days of observation
Ordinate — here and in all the other figures: IFN—γ in U/ml

This kinetics of IFN-γ production in bone marrow cells differs from that observed with blood mononuclears. In this instance, two peaks of interferon production are known (1, 2): at 3 and at 6 days. The first peak has been shown to be associated with interferon production by T-lymphocytes, the second by B-lymphocytes. Previously (7) we also had demonstrated two peaks of IFN-γ response in cultures of blood mononuclears from these children. The lack of a peak at 3 days in bone marrow cell cultures agrees well with bone marrow cell composition. While most lymphocytes of bone marrow are known to be B-lymphocytes, T-lymphocytes (giving maximum IFN-γ production at 3 days) are practically absent. It should be strongly emphasized that bone marrow mononuclears of children in the acute period of the disease are incapable of immune interferon production.

Thus, the results indicate that mononuclears of intact bone marrow taken from children malignant hemoblastoses produce IFN-γ. Maximum titres at 7 days which most likely attests to the participation of B-lymphocytes in this process. Since, bone marrow blastogenesis leads to the loss of this capacity, it seems likely that immature blast cells are incapable of IFN-γ response.

Comparative Study of IFN-γ Response by Blood and Bone Marrow Mononuclears

In the first series of experiments, responses of 33 pairs of parallel cultures ond blood and bone marrow mononuclears (obtained from children with all forms and stages of hemoblastoses) were compared. Statistically significant agreement in the titres of IFN-γ produced was found. Arithmetic mean titres in all the groups examined were 178 U/ml for blood cells and 188 U/ml for bone marrow cells. Complete coincidence of the intensity of interferon production was observed in 18 pairs.

In the next series of experiments blood and bone marrow cell cultures from children were compared with reference to the form of the disease. The results of calculated and statistically significant

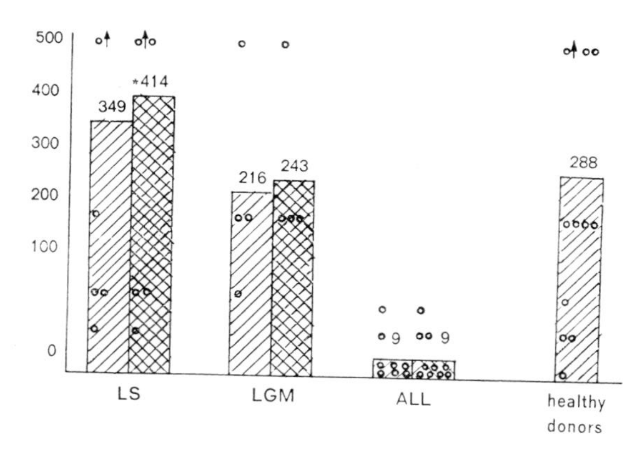

Fig. 2. IFN—γ response by blood and bone marrow mononuclears in patients with hemoblastosis before chemotherapy.
columns — here and in all the other figures: arithmetic mean
o — here and in all the other figures: IFN in individual patients.

arithmetic mean titres of IFN-γ are shown in fig. 2 along with the data on IFN-γ response of blood cells from normal donors.

IFN-γ production by two kinds of nuclears from children with LS and LGM as well as mononuclears of normal donors was found to be approximately similar, varying from 200 to 400 U/ml. A significant decline in titres was observed with blood and bone marrow mononuclears from children with ALL in the acute period.

Thus, similar titres of IFN-γ produced by blood and bone marrow mononuclears of children with different hemoblastoses have been obtained. The intensity of response of these two kinds of cells is determined only by the stage of the disease. The high IFN-γ producing activity of blood and bone marrow mononuclears from children with LS and LGM appears to be of primary importance.

IFN-γ Response of Blood and Bone Marrow Cells in Relation to Chemotherapy

The foregoing data suggested that immature blast cells were incapable of IFN-γ production. Therefore we studied the intensity of IFN-γ production in relation to the per cent of blast cells in the bone marrow, the content of which could be decreased by appropriate course of chemotherapy (fig. 3). It was found that with the elimination of blast cells from the bone marrow and restoration of hemopoiesis IFN-γ response of bone marrow mononuclears in-

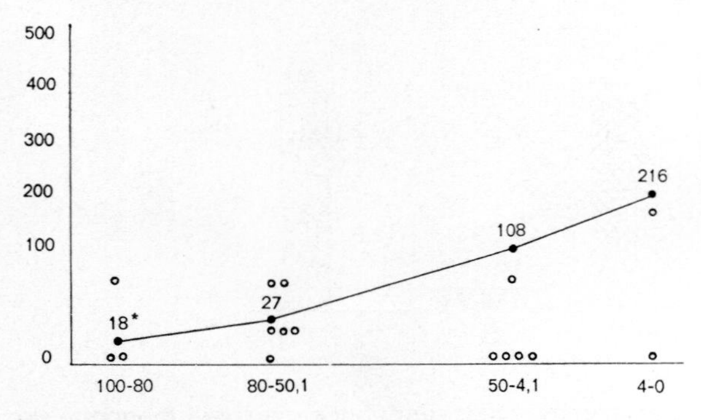

Fig. 3. IFN—γ response by bone marrow mononuclears in children with ALL in relation to bone marrow blastogenesis.
Abscissa —% of blast cells in bonne marrow.

creased. Thus, at 80—100 % blastosis the titre of IFN-γ was 8 U/ml to 27 U/ml. When blastosis decreased to 4—50 % the titre increased to 108 U/ml. When blastosis was 4 % which corresponds to myelogram of a normal bone marrow, the titre of IFN-γ rose to 200 U/ml. This correlates with commonly observed interferon production by intact bone marrow.

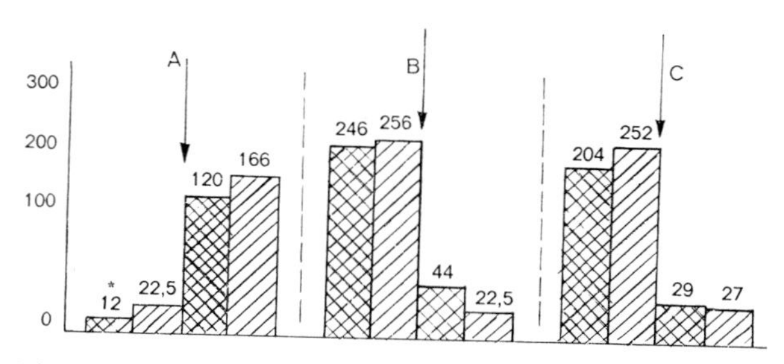

Fig. 4. The effect of cytostatic therapy on IFN—γ response by blood and bone marrow mononuclears in patients with hemoblastoses in relation to the stage of the disease.
A — acute period, induction of remission
Remission: B — consolidation of remission
C — reinduction of remission
Arrow — here and in fig. 5 — beginning of massive cytostatic therapy.

The therapy of acute lymphocytic leukemia in children includes several courses of massive polychemotherapy. The first course is given in the acute period of the disease to destroy blast cells and induce remission.

The study of IFN-γ response of mononuclears at different stages of chemotherapy showed (fig. 4) that during the induction of remission characterized by a decrease in the number and disappearance of blast cells a titre of produced IFN-γ increased from 12—22.5 U/ml before chemotherapy to 120—166 U/ml. The next course given to consolidate the remission resulted in a drop of IFN-γ response from 240 U/ml to 20—40 U/ml. This is probably due to damage of cells of normal hemopoiesis. The same effect of massive chemotherapy was observed in courses of reinduction (a drop in titres from 204 to 250 U/ml to 29—27 U/ml).

Chemotherapy was also used in treatment of LS and LGM. Again, the study of IFN-γ response of mononuclears of the intact bone marrow also showed an inhibition of this function of lymphocytes (fig. 5).

Thus it follows from the foregoing that leukemic blast cells are incapable of producing IFN-γ in response to stimulation with PHA. This is in agreement with the available reports (3, 4). This is an important observation because leukemic cells are completely

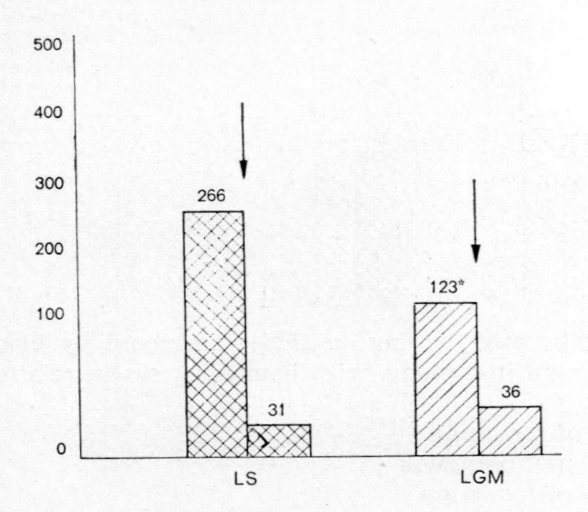

Fig. 5. The effect of cytostatic therapy on IFN—γ response by bone marrow mononuclears in patients with lymphomas.
GM — lymphogranulomatosis
LS — lymphosarcoma

capable of producing virus-induced IFN-α (4, 5). Hence, IFN-γ response of bone marrow mononuclears from leukemic patients may serve as a prognostic test indicating the degree of malignant transformation of bone marrow.

Courses of massive chemotherapy exert different effects on the IFN-γ response of blood and bone marrow cells. Thus, if the course is directed at the destruction of blast cells and normalization of hemopoiesis, this function of the bone marrow is recovered. Antitumor therapy given to patients with intact bone marrow inhibits IFN-γ response and it also is recovered after the termination of the course.

Summary

Production of immune interferon by bone marrow and blood cells was studied in children with acute lymphocytic leukemia, lymphosarcoma, and lymphogranulomatosis. The IFN-γ production by bone marrow and blood cells was found to be similar. A marked reduction of IFN-γ response by mononuclear bone marrow cells of children with ALL was demonstrated. A normal response was noted in those with LS and LGM without bone marrow involvement. In ALL, a close correlation was established between the degree of compensation of the leukemic process and IFN-γ response of bone marrow and blood mononuclears. Being extremely low before the onset of treatment with cytostatics, it increases considerably by the time hematologic remission was achieved and when bone marrow became normal (probably indicating the inability of leukemic blast cells to produce IFN-γ).

In periods of clinico-hematologic remission, course of massive chemotherapy sharply reduce the IFN-γ response by mononuclears. This suggests an inhibiting effect of chemical drugs on the potencies of normal lymphocytes. Similar effects of massive chemotherapy were demonstrated in children with LS and LGM.

References

1 Epstein, L. B.; Kreth, H. W.; Herzenberg, L. A. (1974): Cell Immunol. 12, 404—421.
2 Epstein, L. B.; Salmon, S. E. (1974): J. Immunol. 112, 1131—1138.
3 Freeman, A. I.; Grossmayer, B. J.; O'Malley, I. A. (1981): Second Annual International Congress for Interferon. Res. San Francisco, 21—23, Oct.
4 Lazar, A.; Jacson, R.; Grossmayer, B.; Mizrani, A.; Freeman, A. (1981): J. Interferon Res. 1, 433—450.
5 Orlova, T. G.; Berulava, I. I.; Gavrilova, I. E. (1982): Voprosy virusol. 5, 15—19.
6 Pavlushina, S. V.; Orlova, T. G.; Tabagary, D. Z. (1983): Exp. Oncol., in press.
7 Pavlushina, S. V.; Orlova, T. G.; Gavrilova, I. E. (1983): Voprosy virusol., in press.

T. G. Orlova, M.D., N. F. Gamaleya Institute of Epidemiology and Microbiology, Moscow (USSR)

V. The Perspectives of Interferon and Inducer Therapy

Contr. Oncol., vol. 20, pp. 318—332 (Karger, Basel 1984)

Interferon: The Lesson Learned from Endocrine and Immunotherapy

L. Borecký

Institute of Virology, Slovak Academy of Sciences, Bratislava, Czechoslovakia

The Systemic Approach to Interferon

There is a growing support for the concept that the neural, hormonal and immune systems are in vivo interconnected in a multisystem of as yet unsharp contours (20, 24, 30). Its long-term goal seems to be the coordinated metabolism of the multicellular organism through the normal functioning of its cellular components. Several more or less defined subsystems, such as prostaglandins, complement, chalons, mediators of inflammation, etc., are further candidates for inclusion into this multisystem.

The multisystem is characterized by: 1. predominantly peptid or amino acid derivative character of effector molecules (polypeptide and amino-acid (tyrosin) derivative hormones and neurotransmitters, antibodies and cytokines, interferons, chalones, complement components, etc.). However, an integrated and, probably, phylogenetically older part of the multisystem is represented by steroid hormones and fatty acid derivatives known as prostaglandins, etc.

2. The multisystem shows a wide distribution of cells producing the effector molecules. The producing cells may form specialized organs (endocrine and lymphoid glands) or may be dispersed in different organs (CNS, gastro-entero-pancreatic system, etc.).

3. The activity of the system can be stimulated by a variety of agents. The effector molecules may either be constantly (intermittently?) present in the body fluids (hormones), or appear there

only after appropriate stimulation (immune effector molecules, interferon?).

4. The effector molecules may exert both direct (via specific receptors) and indirect effects on target organs that are mediated by specialized cells such as T-lymphocytes, cytokines, prostaglandins etc.

5. The effector molecules exert often multiple yet inseparable effects (catecholamines may function both as neurotransmitters and hormones, the "bone resorption factor" (a cytokine) and prostaglandin E may both cause bone resorption, histamine as well as corticosteroids may depress cell-mediated immunity etc. They often operate according to the "ying-yang" principle, i.e., small and large doses may have opposite effects. Also, different subsets of effector molecules may act antagonistically (prostaglandins of the E and F or A series).

6. Deficiencies or disturbances of the multisystem may lead to diseases (absence of thymus profoundly influences the immune system, growth factors may function as tumor promoters, antibodies may cause autoimmune disease, etc.).

The interferons (IFNs) fulfil several criteria for inclusion into this multisystem since, like hormones or cytokines (see also A. D. Inglot, I. Béládi, M. Tovey and others in this volume):

a) they regulate a variety of normal function of the cell;

b) their action on the cell requires the presence of specific receptors on the cell surface;

c) their action inside of the cell is mediated through the cAMP--GMP-phosphokinase system and, possibly, other second messengers;

d) their activity is influenced by hormones and antagonistic substances (corticosteroid, thymic hormones, platelet-derived growth factor, iterferon antagonists);

e) their level can be raised by a variety of inducers;

f) they appear among the first cytokines after infection ("early" interferons). (Impairment of interferon response may correlate with severity of the disease-29);

g) they are produced by lymphocytes, macrophages and somatic cells and their production is influenced by immunomodulators, monokines, etc. (17);

h) sensitized lymphocytes release specific (gamma) interferon with a non-specific (cytokine-like) effect on the cell;

i) different classes of interferons may differ in their effect on various cell functions (31);

j) their activity is amplified by helper cellular systems (such as macrophages, NK-cells, K-cells, Tc-lymphocytes) and/or secondary effector molecules (such as prostaglandins, adrenergic agents histamin) (15);

k) their activity can be mimiced and potentiated by hormones (ACTH), neutrotransmitters (noradrenalin), immunostimulators (poly I:C), various stimulators of the cAMP-cGMP system, etc. (4, 22, 23);

l) they may exert both physiologic (homeostatic) and pathologic effects (they enhance allergic response through histamine release, provoke tumor appearance in NZB/W mice, may cause immunosuppression, etc.).

In the context of the interconnected activity of the components of this multisystem, a growing role is ascribed to prostaglandins (PG), i.e. cyclic derivatives of C_{20}-oxygenated unsaturated fatty acids:

a) PGs of different classes can be produced by virtually all cells (normal and transformed, immune and somatic – 9);

b) their release can be stimulated by hormones (adrenalin, progesterone), neurotransmitters (noradrenalin), immunomodulators (LPS), interferons and other various manipulations of cells, and antagonized by other hormones and endogenous factors (15, 36, 53);

c) PGs may function both as stimulators of pituitary hormones (particularly growth hormone) and as effector molecules (or antagonists) of neurotransmitters, hormones and immunomodulators (9, 49, 52);

d) PGs influence the immune system by altering the behavior of polymorphonuclears and macrophages, reducing the number of lymphocytes and antibody production, enhancing the serum acute phase protein levels, suppressing the level of hepatoglobins, causing fever, etc. (43);

e) their regulatory, antiviral and other effects in the cell are mediated via the level of cAMP-cGMP-phosphokinase system and Ca^{2+} ions (37);

f) their effect is concentration-dependent and may be antagonized by prostaglandins of other classes (PGE vs. PGA, PGF, leukotrienes, etc.) as well as other concurrently released mediators (histamin, kinins, etc.) (1);

g) recently, a surveillor function in carcinogenesis has been ascribed to them (15, 20).

Prostaglandins resemble IFNs in that: a) they are not stored in any specialized intracellular structures (PGEs), b) are rapidly synthesized and released upon a variety of stimuli, c) as a result

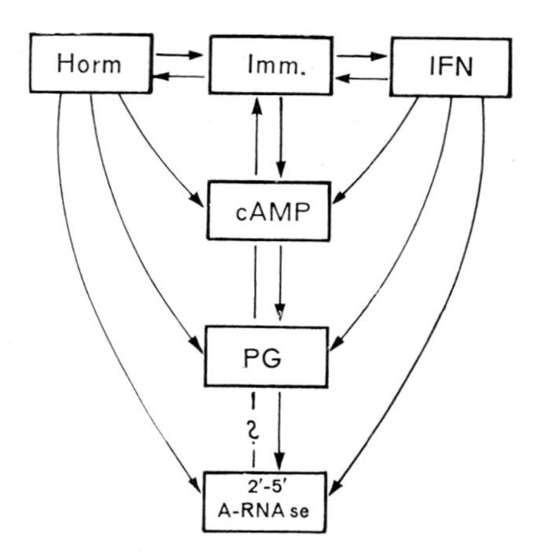

Fig. 1. Interrelations in the HIIP system.

of rapid metabolism in the lung and liver, they act predominantly locally (PGEs), d) different classes of PGs may have different effect on cell multiplication (PGEs often inhibit while PGAs promote cell proliferation), e) their effect on the cell may depend on its functional stage (already activated macrophages may be insensitive to PGs) (fig. 1).

The HIIP Multisystem As Therapeutic Modality

The common properties of hormones, cytokines, interferons, and prostaglandins mentioned above support the emerging concept of the hormone-immune-interferon-prostaglandin (HIIP) multi-system with surveillor and normalizing mission in the organism (24, 46). Since the virus-infected, the transformed and the immunologically deficient cells all represent abnormal cells, the therapeutic use of HIIP agents in viral and cancer diseases may be justified.

While the use of hormones for the therapy of cancer goes back to 1935 (21), attempts to exploit immunomodulators or interferons and their inducers in tumor and/or infectious therapy started around 1968 (34, 45).

In general, these agents can be used in therapy as:

a) direct modulators of transformed or dysabled cells (substitutive or interfering [antagonistic] substances);

b) agents stimulating or modulating various branches of the immune system;

c) as adjuvants for chemo-, radiation- and surgical therapy "last cell killers" operating on a "zero kinetic" basis.

The Hormonal Therapy

The hormones most used in tumor therapy are steroids (oestrogens, androgens, gestagens and corticosteroids) and their natural or sytnhetic antagonists. Although corticosteroids, and to a lesser extent also sex hormones, proved useful in the therapy of various dermatological (allergic), viral (hepatitis B) and autoimmune (SLE) diseases and leukemias — as immunosuppressants or antiinflammatory drugs, only a limited number (types) of cancers and only in certain stages of their development are presently recommended for hormonal treatment (46). They are: cancer of the breast and prostatic gland and, to a lesser extent, uterine cancer, ovarian cancer, renal cancer, malignant melanoma, colonic cancer and acute lymphoblastic leukemia (28). Results of several investigations indicate that, in contradistinction to chemotherapy, the presence of specific receptors on tumor cells is a prerequisite for a successful hormonal therapy. The significance of the presence of receptors on the diseased cells follows from the finding that the beneficial effect of therapy in breast and prostatic cancer was over 50 % in patients with oestrogen receptors but only around 8 % in patients which had no detectable receptors (5).

Further important conclusions drawn predominantly from evaluating the hormone therapy in 2000 prostatic and other hormonally treated cancer patients are as follows (5):

1. There were no significant differences in the curative effect of low (1 mg per day) and high (5 mg per day) pharmacological doses of oestrogen, and enhancement of doses brought no effect in relapses. (However, very small, i.e. "physiological" doses may stimulate the growth of cancer cells) (25). This suggests that the dose in the hormone therapy may be less important than the duration of the therapy (5).

2. The timing of hormonal (oestrogen) treatment is of utmost importance. Stage I and II prostatic and/or breast cancer should not be treated with hormones which is such patients may have an accelerating effect on tumor growth; testosterone has been shown to increase the number of C – particles in cells (26, 50).

3. The hormonal therapy can be applied for a longer period than the chemotherapy which may damage endocrine, immune and other organs and may increase the hormonal imbalance of cancer patients (Vincristine may cause increased antidiuretic hormone secretion, Streptozotocin may be diabetogenic, etc.) (8).

4. In breast cancer patients treated with hormones, the maximum rate of evident tumor regression is constantly reported as about 30 %. A similar therapeutic rate was achieved in uterine (24 %) and other hormonally treated cancers of endocrine organs (27) (tab. I). This may reflect the cellular heterogeneity of tumor tissue (presence of receptor positive and negative cells in different proportion), low proportion of proliferating cells as well as poor vascularization of the tumor tissue which may decrease the effectiveness of the drug therapy, etc. (46).

5. A prolongation of life was observed in more than 50 % of breast and prostatic cancer patients, but the hormonal therapy did not substantially influence the 5-year survival (26).

Table I. Efficacy of hormone therapy in breast cancer (based on data of Stoll, 1981 and Kiang, 1981)

Therapy	Success
Oestrogen	30—40 %
Anti-oestrogen	30—40 %
Hydrocortison + Glutethimide	38 %

6. A beneficial effect of other hormonal treatments (such as adrenalectomy, hypophysectomy, etc.) in these types of tumors suggests a profound hormonal imbalance in these types of cancer. They may involve the following interactions: a) prolactin-oestrogen, b) oestrogen-progesterone, c) androgen-oestrogen and, d) oestriol-oestrone-oestradiol (40).

7. The sensitivity to hormonal therapy of renal or colorectal cancer, of malignant melanoma and acute lymphoid leukemia also shows a correlation with the presence of receptors and the hormonal therapy is moreover justified by some evidence suggesting an inducer role of sex steroids in these diseases (28).

8. In contradistinction to hormone therapy, the acceptable polychemotherapy of breast cancer is presently set at a 50 % tumor regression plateau and has the advantage that it can be effectively used both in oestrogen-receptor positive and negative cases. Nevertheless, the available evidence suggests that the results of chemotherapy can be further improved by combination with hormonal therapy (46).

Immunotherapy

The basic foundation of immunotherapy is to enhance the immune response to tumor antigens, restore the selected immunodeficiency or to break the tolerance (18). It originates probably in observations on dramatic but rare remissions in cancer patients after acute bacterial infection (7). In accordance with this, the first generation of immunomodulators were bacterial vaccines of various origin (BCG or C. parvum). However, the further generation immunomodulators are highly purified bacterial extracts (MER, muramyl dipeptide, Lentinan, etc.), animal cell products such as lymphokines, thymic hormones, etc., and synthetic preparations such as pyran copolymer, polynucleotides, Levamisol, etc. Attempts to stimulate the immune system of pattients with tumor cell vaccination, or substitute their organism with activated macrophages, NK-cells, etc., have been less numerous.

Immunotherapy today resembles the situation which was characteristic of the early days "insulin therapy" when the monitoring of the blood glucose level was not possible. While the balance in the highly complicated immune network of the organism can be easily subverted by a variety of influences, neither the possibility to monitor this balance in detail, nor a selective therapeutic regulation of effector mechanisms is feasible at present. Nevertheless, tumor and viral diseases are regularly accompanied by immunosuppression of a various degree that makes the immune stimulation aimful.

The preliminary conclusions that can be drawn from the therapeutic application of immunostimulants to man are as follows: The antitumor effect was greatest when a) the agent was applied topically (intralesionally) or systematically under conditions that led to localization of the agent in sufficient amount in the tumor deposits (18); b) the tumor load was small; and c) the tumor expressed rejection antigens; d) better results were observed when the disease was accompanied by evident immunodeficiency (34). However, this was not always the case in animal studies (38): 5. The activity of immunostimulating agents may be influenced also by the vehicle used (56).

The best effects with immunotherapy in man have been recorded in (18, 42):

a) Acute leukemia following intensive cytoreduction with chemo- and/or radiotherapy. The immunotherapy resulted in prolongation of disease-free intervals and increased resistance against intercurrent infections;

b) malignant melanoma after intralesional or intralymphatic administration which presumably enabled a better contact between BCG and tumor cells. The immunotherapy resulted in prolongation of diseases-free intervals;

c) lung cancer and colonic cancer after topical (intrapleural or intracolonic) administration; immunotherapy resulted in prolongation of disease-free intervals.

d) recently, in melanoma and breast cancer therapy better results were achieved by using a sequential combination of chemotherapy and immunostimulanst (6, 18, 33), suggesting a heterogeneous sensitivity of tumor cells to agents employed and resembling the experience with hormone therapy.

Inconsistency in the results of immune therapy which led to a decline of this approach can be, at least partially, explained by low selectivity and high individual vulnerability of the immunostimulating effect.

The therapeutic effect of immunomodulators: a) shows a remarkable species, age and sex dependence animal experiments may be misleading in this respect); b) is influenced by nutritional status (Vitamin E, A, C), function of CNS (stress), hormonal status (production of growth factors by tumor cells); c) shows remarkable genetically determined differences in responsiveness; d) the effect of immunotherapy seems to be more evident in chemotherapy resistant patients or in patients treated with highly immunosuppressive drugs (6, 34); e) the effect is often mediated through homeopathic doses (BCG, transfer-factor), while high doses may be detrimental in alergic patients (33) (tab. II).

Table II. Factors that influence the immune therapy

1. Spesies, age, sex
2. Genetic differences
3. Nutritional status
4. Immunological integrity
5. The effect in often mediated by homeopathic doses

The mechanism of the therapeutic effect of immunomodulation is complex. It is mediated through a cascade reaction which involves: a) a cleavage of complement with a subsequent effect of cleavage products on polymorphonuclears, macrophages and other cells; b) direct effect on macrophages leading to release of substances like prostaglandins, interferon, monokines; c) direct effect on various lymphocyte populations (activation of NK-cells, K-cells, T-cell populations); d) direct effect on bone marrow progenitor cells (54); e) the activating effect is under the control of endocrine system (hydrocortitson or thymectomy depress both the number and differentiation of effector lymphoid cells (particularly NK-cells) as well as products of activated cells like PGs and IFN) (30); f) immunomodulators like Levamisol show neurotransmitter (acetylcholin-like) activity (16).

The immunotherapy may be complicated by the following side effects: 1. General (polyclonal) immunostimulation may cause both enhancement and suppression of various immune functions in the same patients. 2. A very weak immune reaction (41) or immunostimulation prior to chemotherapy may result in enhanced tumor growth (34). Per analogy, BCG treatment accelerates the autoimmune disease of NZB/W1F1 mice (10), chronic graft versus host reaction in F1 mice may activate C-type viruses and lead to neoplastic proliferation (19). 3. Immunotherapy (BCG, C. parvum) may inhibit some of metabolizing enzymes in the liver for drugs such as cyclophosphamide, etc. (11). 4. Immunotherapy may cause specific sensitization (51).

Prostaglandins As Antitumor Agents

From several experiments in vitro it might be expected that the proliferation of malignant cells in vitro would be inhibited by PGEs which are considered as "cell-normalizing" agents (20). Conversely, inhibitors of PGE synthesis would be expected to stimulate tumor or normal cell proliferation. In general, this assumption has been confirmed. However, the results of testing the antitumor effect of PGs in vivo have been conflicting and PGs have been described both as metastase facilitating and hindering agents (9, 47).

This controversy may reflect a different degree of activation of various host systems such as the immune and inflammatory systems or alterations in blood supply, etc., by PGs which may modify the end results of experiments. The outcome may be determined also by altering the proportion of PGEs and $PGF_{2}\alpha$-s/or PGAs which usually have opposing action on cell proliferation and function (32).

The interdependence between PG production and the action of other members of the HIIP multisystem was demonstrated by the inhibitory effect of prednisolon, testosteron, or IFN on PG release, as well as an immunosuppressive effect of PGs which included reduced lymphopoiesis, antibody production and immune complex formation in NZB-W mice (12, 15, 39). However, both stimulation and inhibition by PGs was observed when populations of T-cells were examined (2) and it has been found that PG may reactivate IFN production in hyporeactive cells (48). The effect of PGs on the immune system is probably dose and time dependent. This follows from observations that in well ballanced experiments, both poly A:U and PGE can correct immunosuppression in mice

(35, 43). Also, PGs may have both anti- and pro-inflammatory effects.

Preliminary Conclusions from the Therapeutic Use of HIIP Agents and Predictions for IFN Therapy

The hitherto published results of IFN therapy (3, 44) in man are rather contradictory and, on the whole, unsatisfactory. The reasons could be several and include inadequate doses of IFNs used as well as small and unselected groups of patients treated. However, from the results of hormonal and immune therapy in cancer patients, the following preliminary conclusions and predictions regarding the IFN therapy seem to emerge:

1. The therapeutic effects of IFN (and other HIIP agents) in man lay, on the whole, behind the effects registered in laboratory animals. One explanation may be species differences which may be expressed in the deficiency of various components of helper systems or, accelerated catabolism of HIIP agents in man. In addition, it cannot be excluded that the highly IFN (and HIIP) sensitive and worldwide used laboratory strains of viruses or tumor cells represent misleading laboratory artifacts.

2. The experience with hormonal and immune therapy suggests that an approximately 30 % therapeutic success plateau will not be substantially exceeded also in the case of IFN therapy

Table III. Efficacy of alpha-interferon in cancer therapy (based on data of Sikora, 1980)

Disease	No. of treated patients	Response		Total efficacy (%)
		Partial	Clear	
1. Breast carcinoma	38	11 (3)[1]	0	36?
2. Melanoma	4	1	0	25?
3. Myeloma	30	8 (3)	3	36
4. Nodular lymphoma	14	3 (2)	3	37
5. Diffuse lymphoma	7	1	0	14?

[1] In parenthesis: "unclear response"

(tab. III). This makes aimful the attempts to use IFN in combination with chemotherapeutics and/or other HIIP agents (tab. III).

Table IV. Advantages of HIIP-therapy

1. The dose in adjuvant hormonal and/or immunomodulation therapy is less important than the duration of therapy
2. The HIIP therapy is avoid of gross toxicity while chemotherapeutic agents can be carcinogenic or cocarcinogenic.
3. Receptors for HIIP agents are ubiquitous. (although may be altered in the disease.)
4. The HIIP-therapy is multidirectional (may protect cancer patients from infectious complications) and polyvalent (antiviral, antiprotozoal, anticancer etc.)

3. The subversion of the immune and hormonal balance in patients treated with HIIP agents, including IFN and its inducers, may result in unexpected side effects. However, in view of the toxicity of most currently used chemotherapeutic agents, the unsatitsfactory 5-year survival rates of patients treated with chemotherapy as well as physiological considerations, the side effects observed during IFN therapy appear to be of lesser significance and justify further trials. It is highly desirable to use in such trials chemically defined agents with selective biological effects and pay attention also to their vehicle (55).

4. With tumors like papillomas or diseases like chronic hepatitis B, an effective therapy requires prolonged administration of IFN. Its result may then be hindered by specific (antiidiotypic) antibody production as well as appearance of resistant clones of cells and/or viruses (13, 14).

5. Effective IFN therapy would be facilitated by reliable susceptibility tests of diseased cells to IFN and determination of the immunological profile of the patients in different stages of disease.

6. Cells (organs) with profoundly altered cell surface receptors, permeability, blood supply, or 2′—5′A-RNAseF systems etc., will not respond to IFN therapy in a satisfactory way.

7. Due to differences in accessibility of various diseased organs and short half-time of IFN in the body, the topical use of IFN and its inducers seems to be more promising than systemic administra-

tion. Also, the therapeutic success in an early phase of cancer (or before virus dissemination) has higher expectancy than in a late phase. However, the direct (topical) therapeutic effect shall be necesarily transitory due to cytostatic rather than cytotoxic potential of IFN.

8. Paradoxical effects of large and very small IFN doses can be expected.

9. Since several effects of IFN therapy are mediated by helper systems (macrophages, NK-cells, PGs etc.), agents that inhibit these systems (corticosteroids, proteases, etc.) may compromise the resulting therapeutic effect.

Summary

The view that interferon is a component of the hormonal-immune-interferon-prostaglandin (HIIP) multisystem is presented. From a systemic approach to interferon therapy it follows that, like in hormonal or immune therapy: a) a therapeutic success rate substantially higher than 30 % cannot be expected; b) the topical application of interferon should be more effective than the systemic application; c) the effective dose of interferon depends on the reactivity of both diseased cell(s) and helper systems (NK-cells, prostaglandins, etc.).

References

1 Askenase, P. W.; Schwartz, A.; Siegel, J. N.; Gershon, R. K. (1981): Internat. Archs. Allergy Applied Immunol. 66 (Suppl. 1), 225.
2 Bärlin, E.; Leser, H. G.; Deimann, W.; Resch, K.; Gemsa, D. (1981): Internat. Archs. Allergy Applied Immunol. 66 (Suppl. 1), 180.
3 Billiau, A.; De Somer, P. (1980): in "Interferon and Interferon Inducers", ed. by D. A. Stringfellow, Marcell Dekker Inc., N. York, pp. 114.
4 Blalock, J. E.; Stanton, J. D. (1980): Nature 283, 406.
5 Byar, D. P. (1973): Cancer 32, 1126.
6 Carter, S. K. (1976): American Scient. 64, 418.
7 Coley, W. B. (1911): Surg. Gynec. Obstetry 13, 174.
8 Davis, Th. E.; Rose, D. P. (1981): in "Hormonal Management of Endocrine Related Cancer", ed. by B. A. Stoll, Lloyd Luke Ltd., London, pp. 77.

9 Easty, G. C.; Neville, A. M. (1981): in "Hormonal Management of Endocrine Related Cancer", ed. by B. A. Stoll, Lloyd Luke Ltd., London, pp. 41.
10 Engleman, E. G.; Sonnenfeld, G.; Dauphinee, M.; Greenspan, J. S.; Tálal, N.; Mc Devitt, H. O.; Merigan, Th. C. (1981): Arthritis Rheum. 24, 1396.
11 Farquahar, D.; Low, T. L.; Gutterman, J. U.; Hersch, E. M.; Luna, M. A. (1976): Biochem. Pharm. 25, 1529.
12 Fischer, A.; Durandy, A.; Mamas, S.; Mc Call, E.; Dray, F,; Grisielli, C. (1982): Clin. exp. Immunol. 49, 377.
13 Fuchsberger, N., Borecký, L.; Hajnická, V. (1974): Acta Virol. 18, 85.
14 Fuse, A.; Kuwata, T. (1977): J. Natl. Cancer Inst. 58, 891.
15 Fuse, A.; Mahmud, I.; Kuwata, T. (1982): Cancer Res. 42, 3209.
16 Guerrero, J. (1980): Amer. Vetr. Med. Assoc. 176, 1163.
17 Gifford, G. E. (1981—1982): Texas Rep. Biol. Med. 41, 59.
18 Gutterman, J. U. (1977): Cancer Immunol. Imunother. 2, 1.
19 Hirsch, M. S.; Proffitt, M. R.; Black, P. H. (1977): Contemp. Top. Immunobiol. 6, 209.
20 Horrobin, D. F. (1980): Medical Hypothesis 6, 469.
21 Huggins, C.; Hodges, C. F. (1941): Cancer Res. 1, 293.
22 Itkes, A. V.; Turpaev, K. T.; Kartasheva, O. N.; Vagonis, A. I.; Kafiani, C. A.; Severin, E. S. (1982): Biochem. Internat. 5, 15.
23 Itkes, A. V.; Turpaev, K. T.; Kartasheva, O. N.; Kafiani, C. A.; Severin, E. S. (1983): Mol. Cell. Biochem. (in press).
24 Jankovič, B. D.; Isakovič, K.; Mićić, M.; Kneževič, Z. (1981): Clin. Immunol. Immunopathol. 18, 108.
25 Kiang, D. T. (1981): in "Hormonal Management of Endocrine Related Cancer", ed. by B. A. Stoll, Lloyd Luke Ltd., London, pp. 64.
26 Klener, P.; Jakoubková, J.; Marek, J. (1978): in "Chemotherapy of malignant tumors and hemoblastosis" (in Czech). Avicenum, Prague.
27 Kohorn, E. I. (1981): in "Hormonal Management of Endocrine Related Cancer", ed. by B. A. Stoll, Lloyd Luke Ltd., London, pp. 122.
28 Leake, R. (1981): in "Hormonal Management of Endocrine Related Cancer", ed. by B. A. Stoll, Lloyd Luke Ltd., London, pp. 3—12.
29 Levin, S.; Hahn, T. (1981): Clin. exp. Immunol. 46, 475.
30 Low, T. L. K.; Goldstein, A. L. (1981): in "Immunostimulation", ed. by L. Chedid, P. A. Miescher, H. J. Mueller-Eberhard, Springer Vlg, Berlin, pp. 129.
31 Lucero, M. A.; Magdelenat, H.; Friedman, W. H.; Pouillart, P.; Billardon, C.; Billiau, A.; Cantell, K.; Falcoff, E. (1982): Eur. Jour. Cancer. Clin. Oncol. 18, 243.
32 Lynch, N.; Salomon, J. C. (1979): J. natl. Cancer Inst. 62, 117.
33 Mathé, G.; Arniel, J. L.; Halle-Pannenko, O.; Sommler, M. C. (1976): Ann. N. Y. Acad. Sci. 277, 467.
34 Mathé, G. (1978): Cancer Immunol. Immunother. 5, 149.
35 Mozes, E.; Shearer, G. M.; Melmon, K. L.; Bourne, E. H. (1973): Cell. Immunol. 9, 226.

36 Moore, P. K.; Hoult, J. R. S. (1980): Nature 288, 271.
37 Oropeza-Rendon, R. L.; Bauer, H. C.; Fischer, H. (1980): Jour. Immunopharmac. 2, 133.
38 Pavelič, K.; Hršak, I. (1980): Europ. J. Cancer 16, 1297.
39 Plescia, O.; Smith, A.; Grinwich, K. (1975): Proc. natl. Ac. Sci. USA 72, 1848.
40 Poortman, J. (1981): in "Hormonal Management of Endocrine Related Cancer", ed. by B. A. Stoll, Lloyd Luke Ltd., London, pp. 51.
41 Prehn, R. T. (1976): Transplant. Rev. 28, 34.
42 Salmon, S. E. (1977): Cancer Res. 37, 1245.
43 Schultz, R. M.; Stoychkov, J. N.; Pavlidis, N.; Chirigos, M. A.; Olkowski, Z. L. (1979): Jour. RES 26, 93.
44 Sikora, K. (1980): Brit. Med. J. 281, 855.
45 Stewart, W. E. (1979): in "The Interferon System", Springer-Verlag, Wien—N. York.
46 Stoll, B. A. (1981): in "Hormonal Management of Endocrine Related Cancer", ed. by B. A. Stoll, Lloyd Luke Ltd., London, p. 77.
47 Stringfellow, D. A.; Fitzpatrick, F. A. (1977): Nature 282, 76.
48 Stringfellow, D. A. (1981—1982): Texas Rep. Biol. Med. 41, 116.
49 Sundberg, D. K.; Fawcett, C. P.; Hluer, P.; Mc Cann, S. M. (1975): Proc. Soc. exp. Biol. Med. 148, 54.
50 Tihon, C.; Green, M. (1973): Nature New Biology 244, 227.
51 Torisu, M.; Miyahara, T.; Shinohara, N.; Ohsato, K.; Sonozaki, H. (1978): Cancer Immunol. Immunother. 5, 77.
52 Tothill, A.; Rathbone, L.; Willman, E. (1971): Nature 233, 56.
53 Tothill, A. (1974): Jour. IRC 14.
54 Wuest, B.; Wachsmuth, E. D. (1982): Infect. Immun. 37, 452.
55 Yarkoni, E. (1982): Jour. clin. Hmat. Oncol. 12, 105.

L. Borecký, M.D., Institute of Virology, Slovak Academy of Sciences, Bratislava, Mlýnská Dolina (Czechoslovakia)

Contr. Oncol., vol. 20, pp. 333—345 (Karger, Basel 1984)

Clinical Testing of Interferons in Japan

Tsunataro Kishida[a]; *Jiro Imanishi*[a]; *Etsuro Nakajima*[c]; *Tadao Okuno*[c]; *Tatsuro Takino*[c]; *Naoyuki Matsumura*[b]; *Toshikazu Yoshikawa*[b]; *Motoharu Kondo*[b]; *Hiroshi Yoshioka*[d]; *Tadashi Sawada*[d]; *Yoshio Nakagawa*[e]; *Satoshi Ueda*[e]; *Kimiyoshi Hirakawa*[e]; *and Genichiro Hirose*[f]

Kyoto Prefectural University of Medicine, Kyoto 602, Japan
a Department of Microbiology
b 1st Department of Medicine
c 3rd Department of Medicine
d Department of Pediatrics
e Department of Neurosurgery
f Department of Neurology, Kanazawa Medical College, Uchinada 920—02, Japan

Introduction

Interferon (IFN) is recently used for the treatment of many kinds of viral diseases and malignant tumors. However, the therapeutic effects of IFN on viral and malignant diseases have not been elucidated, because sufficient amount of IFN has not been provided for clinical application. The recent progress in the researches of IFN production has made is possible for the authors obtain sufficient amount of IFN, to examine the effects of human leukocyte IFN (HuIFN-α) on some viral and malignant diseases.

IFN Preparations

Human leukocyte interferons (HuIFN-α) were kindly supplied by Dr. H. Nita of the Kyoto Red Cross Blood Center, Kyoto, Japan. It was prepared in human peripheral leukocytes infected with Sendai virus. Crude IFN was purified by the method of Cantell et al. (3). The supernatant was acidified to pH 3.5 by 0.5M KSCN and the precipitate was dissolved.

A High Dosage Therapy

Nine patients with chronic hepatitis B, consisting of 7 males and 2 females, were given a high dosage of HuIFN-α. The liver function, and HBs and HBe Ag-Ab systems were periodically examined in all the patients for 6 months before administration of IFN, and the histological diagnosis was performed within 3 months before the administartion. Especially those patients were selected in whom serum Dane particle-associated DNA polymerase (DNAP) activity was continuously positive, no marked changes were observable in the transaminases for 4 weeks before the administration, and HBeAg was constantly positive over 6 months during the observation period were selected. The histological examination revealed chronic active hepatitis in 5 patients (4 males and 1 female) and chronnic inactive hepatitis in 4 patients (3 males and 1 female).

HuIFN-α was administered intramuscularly at a dose of $10\times \times 10^6$ IU daily for the first 5 days and at a dose of 10×10^6, 5×10^6 IU for 5 to 50 days. The total dose of IFN was 1×10^8 IU in each patients.

The percent negative DNAP increased gradually from 22 to 50 %. In 2 of 7 patients who were observed more than 9 months after the IFN treatment, HBeAg became negative and the seroconversion from HBeAg to anti-HBeAb was found (tab. I).

Table I. Changes of hepatitis B virus associated markers

	Months after ceasing of IFN therapy	1	2	3	4	5	6	7	8	9	10	11	12	13
Incidence (%)	negative DNAP	22 (2/9)	11 (1/9)	22 (2/9)	33 (3/9)	33 (3/9)	38 (3/8)	14 (1/7)	43 (3/7)	43 (3/7)	29 (2/7)	43 (3/7)	43 (3/7)	50 (3/6)
	negative HBeAg (RIA)	0 (0/9)	0	0	0	0	0	0	0	29 (2/7)	29 (2/7)	29 (2/7)	29 (2/7)	17 (1/6)
	positive anti HBe (RIA)	0	0	0	0	0	0	0		0 (0/7)	29 (2/7)	29 (2/7)	29 (2/7)	17 (1/6)

( ) No. of cases

The representative case which showed remarkable effect was indicated in fig. 1. The patient (male, 30 years old) showed the continuously positive HBeAg for 6 years and 7 months. The abnormal liver functions were pointed out 1 year and 9 months ago. HBeAg was positive for 1 year and 9 months. The case was diagnosed as chronic active hepatitis on biopsy under laparoscopy

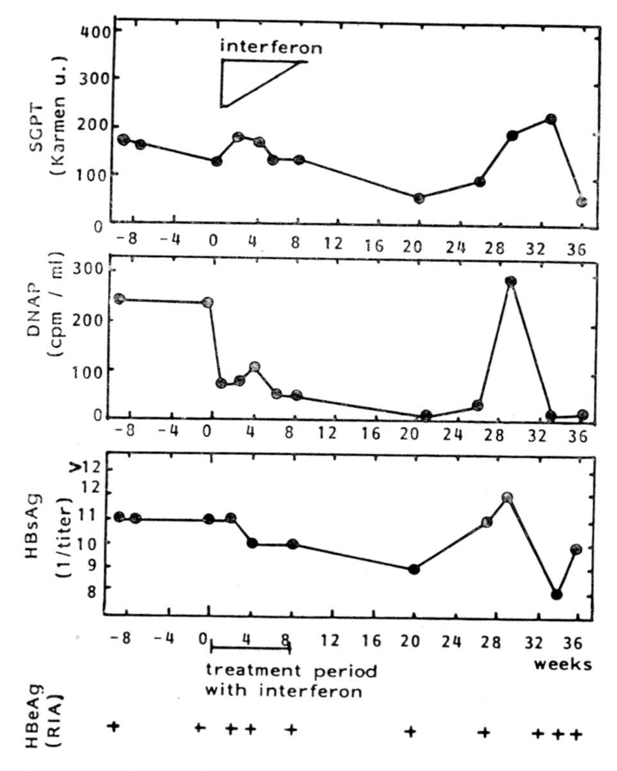

Fig. 1. A typical case of the successful treatment of chronic hepatitis B infection with a large dose of IFN.

2 months before the IFN administration. DNAP activity decreased from 241 to 42 cpm after the completion of IFN administration and became negative 4 months later. The negative DNAP activity was maintained, although the transient elevation of DNAP activity was found 1 month after that. HBeAg changed to negative 9 months after the completion of IFN administration.

Thus it was certain that IFN reduced DNAP activity and caused the seroconversion from HBeAg to anti-HBeAb in some of the patients. It seems that HuIFN-α has a suppressive effect on HBV infection in the chroninc active hepatitis showing a low DNAP activity. Furthermore, in these cases the change of DNAP activity to positive followed the elevation of SGPT and the elevation was coincident with the change of DNAP activity to negative. The seroconversion from HBeAg to anti-HBeAb was found at he same time as the changes of DNAP activity and the elevation of SGPT. It was suggested that the elevation of SGPT was closely related with the suppression of DNAP activity and the seroconversion from HBeAg to anti-HBeAb in the IFN therapy.

A Low Dosage Therapy

A low dosage of HuIFN-α was administered intramuscularly to 47 patients with HBsAg-positive chronic active hepatitis once a week for 4 consecutive weeks (10×10^5, 5×10^5, 2×10^5 and 1×10^5 IU). In 26 of the 47 patients (55 %), HBsAg decreased after the IFN therapy, and in 3 of the patients, it became negative. In only one patient of the three was observed the concomitant appearance of anti-HBsAb (fig. 2). HBsAg titer decreased in most of the patients but not significantly (fig. 3). There was a rapid decrease in DNAP activity in 31 of 34 patients (91 %) investigated. Namely the decrease in DNAP activity was recognized even after the first injection of 1×10^6 IU IFN and reached the maximum level within 4–6 weeks after the start of IFN therapy. In one case, a series of IFN injections did not affect DNAP activity, but the second series of injections caused the reduction and then the elimination of DNAP activity, and the seroconversion from HBeAg to anti-HBeAb was found at he same time. In 10 of 17 patients positive for HBeAg, HBeAg ceased to be detected within 2 months and the seroconversion was noted within 6 months. The levels of SGOT and SGPT were reduced by this series of therapy with IFN. There was a significant decrease in SGOT and SGPT levels in 35 of 38 patients examined. In 10 patients who seroconverted to anti-HBeAb, DNAP activity was reduced, and it became negative in 9 of them. Concurrently SGOT and SGPT levels fell in all of the 10 patients and became normal in 7 patients. However, 7 HBeAg-positive patients remained positive for DNAP activity and abnormal in SGOT and SGPT levels. No changes of HBsAg which was high-titered in all

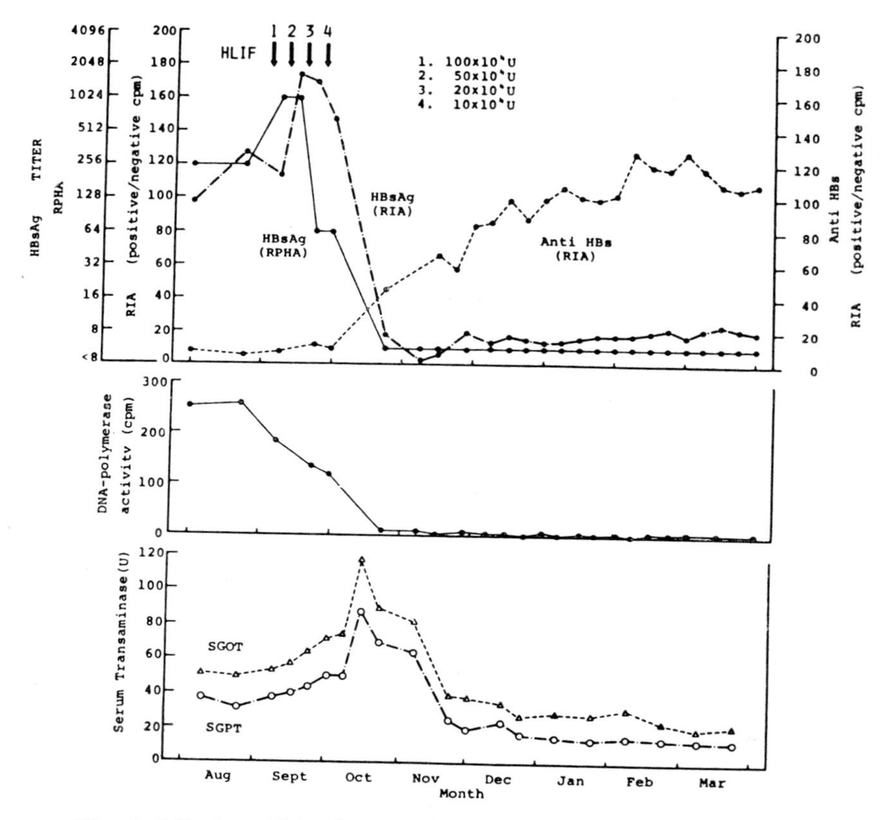

Fig. 2. Effect of HuIFN—α on HBV markers and serum transaminase levels.

the 47 patients were found during or after the IFN therapy. Thus a relatively low dosage of IFN was also effective in HBV infection. The relative merit of the high dosage therapy and the low dosage therapy has not been determined yet. In order to ascertain this, it is necessary to observe the effect of IFN for a long term.

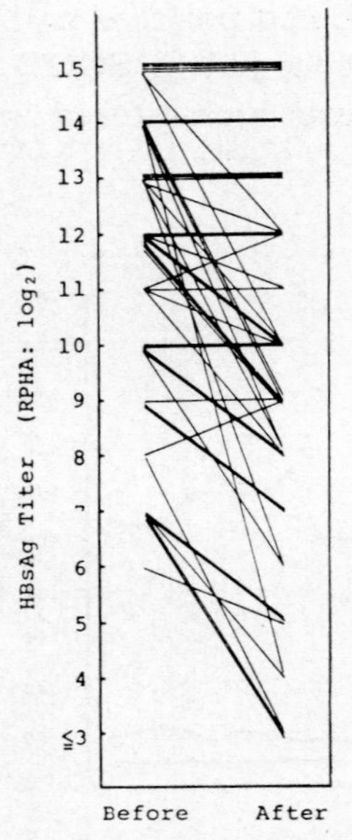

Fig. 3. Changes in HBsAg titers (RPHA) in 47 patients with HBsAg—positive chronic active hepatitis before and after IFN therapy. The net change is insignificant at p 0.1.

Effect of IFN on Chronic Hepatitis B in Children

Eight patients with chronic active hepatitis were treated with HuIFN-α. They were 4 boys and 4 girls, aged from 2 years and 8 months to 11 years and 6 months. All the patients had been observed for 7 to 21 months before the IFN treatment. In all the patients, HBsAg and HBcAb were positive. Five patients showed positive HBeAg, but in 2 of the 5 patients, HBeAg became negative before the initiation of IFN treatment and 1 of the two showed the seroconversion to anti-HBeAb. DNAP activity was positive in 2 of the 3 patients examined. HuIFN-α was administered intramuscularly at doses of 5 to 100×10^4 IU daily or once a week. The total dose of IFN administered per patient was 1,050 to $5{,}395 \times 10^4$ IU.

In 2 of the 5 patients positive for HBeAg, this became negative immediately after the start of IFN treatment. No effect on HBsAg were observed in any of the patients, although a transient rapid decrease in HBsAg in only one case. There was a rapid reduction of DNAP activity after the IFN treatment in 2 patients. SGOT and SGPT increased immediately after the IFN treatment, but they decreased temporarily or continuously to normal levels, and only one case showed normal levels of serum transaminases over 2 years after the IFN therapy.

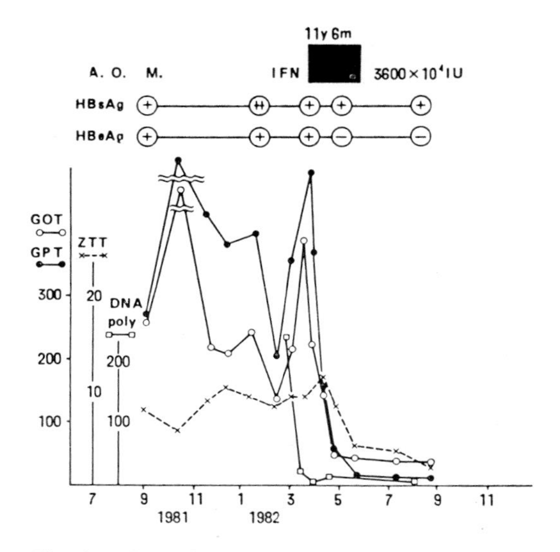

Fig. 4. Effect of HuIFN—α on chronic HBV infection in a child.

The typical case was shown in fig. 4. This case showed the elimination of HBeAg, the rapid decrease in DNAP activity and the reduction of SGOT and SGPT levels by the administration of HuIFN-α. It was certain that HuIFN- could affect he HBV markers and liver functions in some of the patients. There are few reports on the effect of HuIFN on hepatitis B infection in children, although there are many reports for adults (4, 7, 6). It is noteworthy that HuIFN-α may be able to cure chronic hepatitis in children. However it is necessary to examine the effect of IFN in a number of patients. Furthermore, it is necessary to prove that he amelioration of chronic hepatitis was due to the effect of IFN and not to the natural course of the disease, especially in children.

Effect of HuIFN-α on Subacute Sclerosing Panencephalitis (SSPE)

SSPE is a kind of slow virus infection caused by the variant of measles virus. For the treatment of SSPE, some antiviral agents including amantadine, BUdR, IUdR, isoprinosine and transfer factor (TF) are used. However, it is not confirmed that hese agents are effective against SSPE. Interferon is thought to be a candidate of therapeutic agents for SSPE, but here are few data on the effect of IFN on SSPE (1, 5, 2).

The authors tried to examine the effect of IFN in several patients with SSPE. After HuIFN-α was administered by drop infusion at a dose of 3×10^6 IU daily for the first 7 days, it was administered intrathecally at a dose of 1×10^6 IU daily for 7 days and another 4 days 1 or 2 weeks later except for one case. Furthermore, some of the cases received 1×10^6 IU IFN once a week for 2 to 3 months. The results of the typical case were shown in fig. 5. This patient was an 8 year and 11 month-old boy. At he age of 3 years, he suffered from measles. At the age of 8 years, he became forgetful and lost he ability to calculate, especially by subtraction. Then, the tremor, instability in gait, myoclonic jerks were found. The case was diagnosed as SSPE on the basis of electroencephalography (EEG) and the antibody titers against measles in serum and cerebrospinal fluid (CSF). HuIFN-α was administered 9 months after the onset. He had frequent myoclonic jerks and chorea-like movement, and his state corresponded to the stage II of Jabbour. Myoclonic jerks and chorea-like movements were eliminated and the motor activity improved after the IFN treatment (fig. 5). The motor activity was maintained, although myoclonic jerks reappeared. Neurogenic disability index of Dyken changed from 34 % to 10 % 2 months after the IFN treatment.

Another patient was 14-year-old boy. He suffered from measles at the age of 2 years. The somnolence and the reduced activity was noticed by his mother in September of 1981. The neurological symptoms gradually increased, and the disorder was diagnosed as SSPE. HuIFN-α was administered intravenously at a dose of 3×10^6 IU at intervals of 3 days for 27 days and intrathecally once a week for 12 months together with intravenous injection of 1×10^6 IU IFN. The doses of intrathecal injection was increased from 2×10^5 to

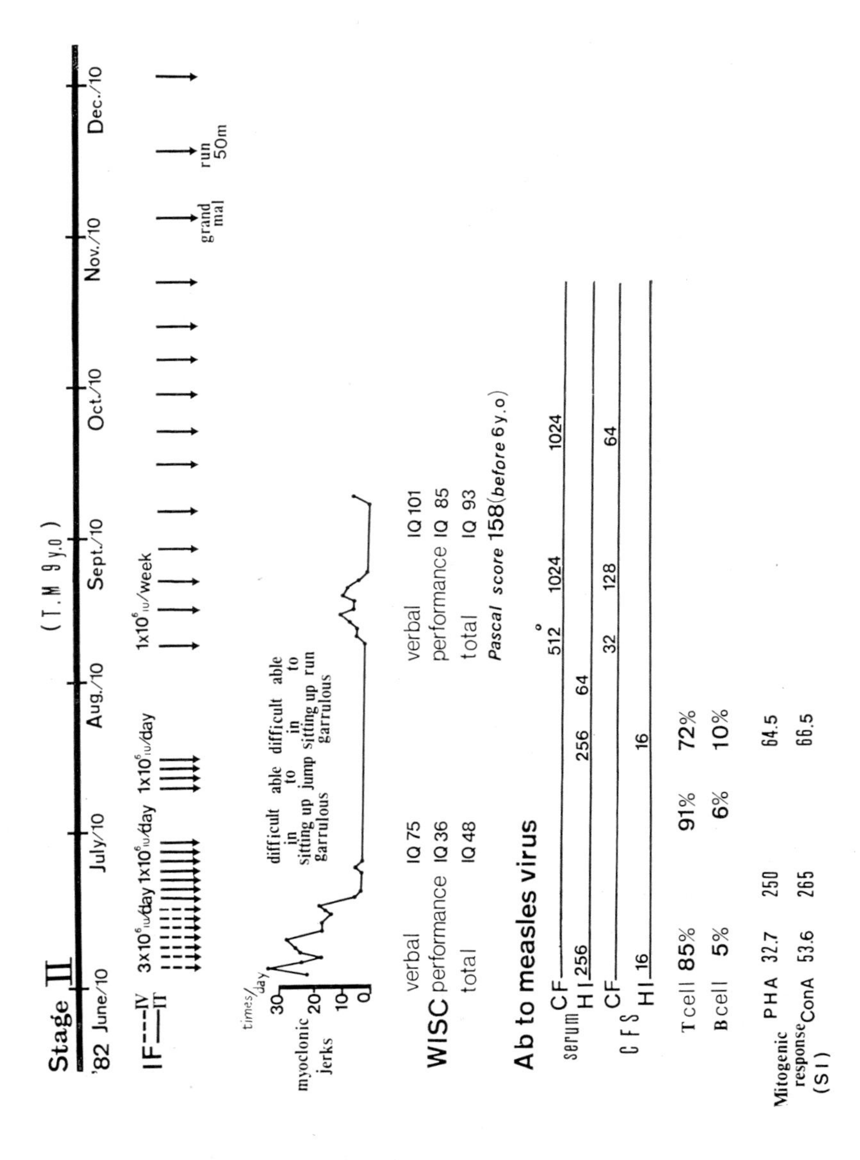

Fig. 5. Effect of HuIFN—α on SSPE.

2×10^6 IU. IFN injection improved the verbal disturbance and the rigidity in the extremities, but the myoclonic jerks increased. The marked depression of IQ was not found over one year. The clinical stage II was maintained. The serum antibody titers against measles decreased from 4,096 to 512 in HI, and those in CSF decreased from 256 to 64 in CF, from 64 to 32 in NT and from 1,024 to 32 in HI. Thus HuIFN-α may not have deteriorated the neurological state.

Similar results were observed in 2 other patients, but it was not certain that IFN really ameliorated SSPE. In the other 7 patients, HuIFN-α had no observable effect on SSPE.

Effect of HuIFN-α of Malignant Brain Tumors

HuIFN was administered, singly, to patient with malignant brain tumor who had been taken off their usual treatment, and its effect on brain tumor was examined.

Twenty-three patients were selected to evaluate the exact clinical effect of IFN. These patients did not receive any other anticancer drugs in combination with IFN. Their surgical specimens were histologically classified as glioblastomas (13), medulloblastomas (3), astrocytomas (2), ependynoma (1), ependymoblastoma (1), potive glioma (1) and metastasis (2). The female to male ratio was 9:14; their ages ranged from 5 to 66 years. All patients were irradiated by ^{60}Co after the surgical removal of tumors. Some patients received additional chemotherapy. However, IFN was administered at least 6 months after the cessation of the usual treatment, in order to exclude any effects of the previous treatment on the tumors. Tumor recurrence was checked from signs and/or CT scan.

General Administration

HuIFN-α was administered intramuscularly into the gluteus of 10 patitents for 4 to 10 months. They were randomly divided into 2 subgroups, the low dosage group and the high dosage group. In the low dosage group, which consisted of 6 patients, HuIFN-α was administered at a dose of 5×10^4 IU once a week, up to a total of 3.4×10^6 IU per patient. In the high dosage group, which consisted

Table III. General Administration of IFN

Case	Age	Sex	Tumor	Single Dose (IU)	Total Dose (IU)	Duration (M)	Side Effect	Response Outcome
1	42	F	Glioblastoma	5×10^4 weekly	2.4×10^6	12	none	PR alive 81M[1]
2	11	F	Medulloblast.	5×10^4 weekly	3.4×10^6	17	none	PR dead 54M
3	55	F	Glioblastoma	5×10^4 weekly	3.4×10^6	17	none	W dead 28M
4	51	M	Glioblastoma	5×10^6 alt. days	129×10^6	4	fever	W dead 13M
5	17	F	Astrocytoma	3×10^6 alt. days	191×10^6	17	fever	U alive 42M
6	9	M	Medulloblast.	3×10^6 alt. days	124×10^6	11	none	U dead 40M
7	33	F	Glioblastoma	3×10^6 alt. days	40×10^6	5	none	W dead 34M
8	56	M	Glioblastoma	5×10^4 weekly	1.2×10^6	6	none	W dead 16M
9	66	M	Metastasis	5×10^4 weekly	3.35×10^6	6	none	W dead 12M
10	63	M	Metastasis	5×10^4 weekly	3.4×10^6	17	none	U alive 32M

[1] M:Month

of 4 patients, HuIFN was administered at a dose of 3×10^6 IU every other day, up to a total of 1.9×10^8 IU. The partial remissions of tumors were noted in 2 of the 10 patients. However, 8 of them showed no response to HuIFN-α therapy (tab. II). It is interesting that a low dosage of HuIFN-α suppressed brain tumor.

A Local Administration

IFN (10^6 IU) was administered through the Ommaya reservoir or by intrathecal injection. The patients were divided into 2 groups according to the schedule of administration. In the first group (7 cases), IFN was administered once or twice a week (weekly group). In the second group (6 cases), IFN was administered everyday for one month after a one month suspension of administration (daily group). Local administration of IFN was combined with general administration in 3 cases and with chemotherapy in 6 cases. HuIFN-α was administered for 2 to 7 months in the weekly group and for 1 to 8 months in the daily group, according to the patient's general condition and the effects of the therapy. The total amount of IFN administered was 12×10^6 IU on an average in the weekly group and 25×10^6 IU in the daily group.

Neurological status and Karnofsky performance scale did not show there was any remarkable improvement. However, 2 patients had a partial remission. The other patients showed no response to HuIFN-α therapy.

There were no serious side effects in the general and local administration. Some of the patients complained of transient fever (38–39° C) after IFN injection, but it was easily controlled with an antifebrile. No remarkable changes in liver or renal functions were found. However, there were 3 cases of meningitis which appeared to have been caused by contamination during and local administration of antibiotics. IFN was used for various malignant tumors, but its effectiveness has not been clarified. This is because the best dosage, the appropriate route of administration and the appropriate time interval have not been established. More clinical data from a number of patients will solve this problem in the near future.

Acknowledgement

The authors are very grateful to Dr. H. Nita of the Kyoto Red Cross Blood Center, Kyoto, for the supply of HuIFN-α.

References

1 Bartram, C. R.; Henke, J.; Treuner, J.; Basler, M.; Esch, A.; Mortier, W. (1982): Eur. J. Pediatr. 138, 187.
2 Behan, P. O. (1981): Lancet, May 9, 1059.
3 Cantel, K.; Hirvonen, S. (1978): J. gen. Virol. 39, 541.
4 Greenberg, H. B.; Pollard, R. B.; Lutwick, L. I.; Gregory, P. B.; Robinson, W. S.; Merigan, T. C. (1976): N. Eng. J. Med. 239, 517.
5 Guggenheim, M. A.; Baron, S. (1977): J. infect. Dis. 136, 50.
6 Hafkin, B.; Pollard, R. B.; Tiku, M. L.; Robinson, W. S.; Mrigan, T. C. (1979): Antimicrob. Agents Chemother. 16, 781.
7 Weimar, W.; Heijtink, R. A.; Schalm, W.; Schllekens, H. (1979): Eur. J. clin. Invest. 9, 151.

T. Kishida, M.D., Departments of Microbiology, Kyoto Prefectural University of Medicine, Kyoto 602 (Japan)

Contr. Oncol., vol. 20, pp. 346—357 (Karger, Basel 1984)

Interferon Inducers: Application As Antiviral and Antineoplastic Agents

D. A. Stringfellow

Cancer Research, Bristol-Myers Pharmaceutical Research & Development, Syracuse, N. Y., USA

Structure Activity Relationships

As summarized in tab. I a variety of substances are capable of inducing an interferon response when introduced into the appropriate animal or cell culture system. Naturally occuring agents ranging all the way from RNA and DNA containing viruses, bacterial extracts, rickettsia, protozoa, endotoxin, mitogens and antigens induce interferon. The type of interferon induced by each of these agents varies depending upon the conditions employed. For example, mitogens and antigens generally induce gamma interferon while the other agents generally induce alpha or beta interferon depending upon the conditions employed. There are a variety of synthetic agents also capable of inducing interferon; polynucleotides, polycarboxalates such as pyran copolymer and a wide spectrum of low molecular weight compounds. Likewise their ability to induce various forms of interferon (alpha, beta and gamma) vary depending upon the conditions employed.

The structures of four classic interferon inducers are illustrated in fig. 1, 2. Chemicals ranging all the way from double stranded RNA such a polyriboinosinic: polyribocytidylic acid (polyI:polyC) to fluorenone compounds such as tilorone and the pyrimidinones induce interferon. These structures illustrate the molecular heterogeneity of chemicals which are active inducing agents. Superficially this would suggest that the molecular induction mechanism is not very specific. This however does not appear to be the case. Various investigators have tried to structurally modify each of these compounds to increase activity but have found that only minor modifi-

Table I. Interferon inducing agents

Class	Example
I. Naturally occurring	
Viruses — RNA	Influenza
Viruses — DNA	Vaccinia
Bacteria	Brucella abortus
Bacterial extracts	B. abortus
Rickettsia	R. tsutsugamushi
Protozoa	Taxoplasma gondii
Endotoxin (LPS)	E. coli
Mitogens	Phytohemagglutinin Poke weed
Antigens	Viral Bacterial (BCG)
II. Synthetic	
Polynucleotides	Poly I : Poly C
Polycarboxylates	Pyran copolymer
Low molecular weight	Tilorone Propanediamine Acridines Pyrimidines

Table II. Factors influencing interferon response in vivo

1. Hyporeactivity due to
   a) virus infection
   b) neoplastic disease
   c) multiple-frequent doses of inducer
2. Physical condition and nutrition
3. Stress
4. Bacterial parasitic infections
5. Age — sex — pregnancy
6. Inducer specificity
7. Cellular destruction (x-irradiation, ALS, silica, etc.)
8. Species variability

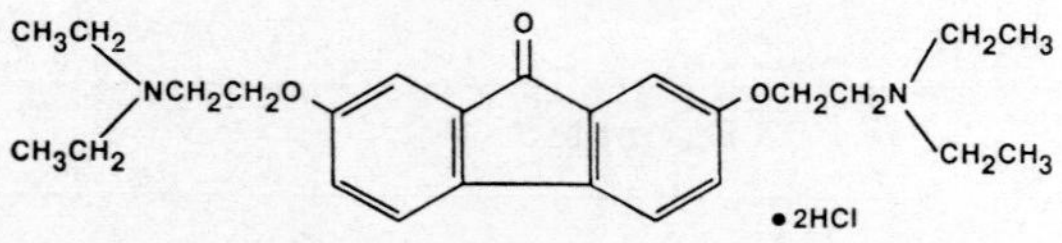

CH_3CH_2 / NCH_2CH_2O / CH_3CH_2 — O — OCH_2CH_2N / CH_2CH_3 / CH_2CH_3 • 2HCl

Bis-diethyl amino ethoxy-Fluorenone
(Tilorone Hydrochloride)

polyriboinosinic: polyribocytidylic acid
(poly I :poly C)

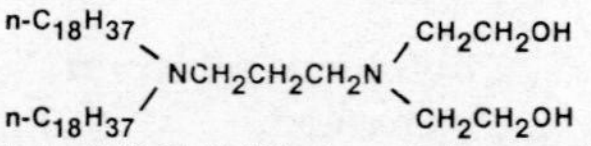

n-$C_{18}H_{37}$ / $NCH_2CH_2CH_2N$ / CH_2CH_2OH / n-$C_{18}H_{37}$ / CH_2CH_2OH

N, N-dioctadecyl-N', N'-bis (2-hydroxyethyl) propanediamine
(CP-20,961)

Fig. 1. Structure of synhetic inducers.

cation of the structure results in major alteration of their ability to induce interferon. Therefore, even though a variety of molecular structures are capable of inducing an interferon response, the molecular characteristic responsible for that ability appears to be fairly specific and cannot be extensively modified without affecting the interferon response.

In Vitro Interferon Response

Tab. II presents a list of factors which influence the interferon response in vivo. These factors range all the way from a hyporeactive condition that can result as a consequences of virus infection, neoplastic disease or multiple doses of an interferon inducing agent to conditions such as nutrition, stress, bacterial and parasitic infections, pregnancy and any other condition which affects the physiologic integrity of cells.

The specifity of the cellular interferon response is more intricate than might be expected. For example, in tab. III the phenomenon of cellular amplification of the interferon response is summarized. Cell cultures in vitro respond by producnig various types

PYRIMIDINONES*

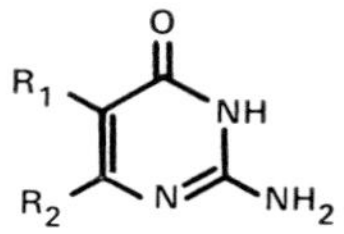

(Structure: pyrimidinone ring with O, NH, N, NH_2, R_1, R_2)

Abbreviation	Substitutions R_1	R_2	IF**	AV***
ABMP	Br	CH_3	++++	+++
ABEP	Br	C_2H_5	++	++
ABPrP	Br	C_3H_7	+	+
ABPrP	Br	$CH(CH_3)_2$	–	–
ABPP	Br	C_6H_5	++++	++++
AIMP	I	CH_3	+++	+++
ACMP	Cl	CH_3	–	–
AIPP	I	C_6H_5	+	++++
ACPP	Cl	C_6H_5	++++	++++

*Compounds were injected i.p. at 1000 mg/kg i.p. Serum was collected at 8 hr for interferon assay. (Mice 20/group) were challenged with Semliki Forest virus 18 hr after compound.

**Serum interferon response: (–) = < 10 units/ml, + = 10-50, ++ = 50-100, +++ = 100-1000, ++++ = 1000-10,000 units/ml.

***Antiviral activity: (–) = < 50% of animals protected, + = 50%, ++ = 50-65%, +++ = 65-80%, and ++++ = 80-100% of animals protected.

Fig. 2. Structure of pyrimidinone interferon inducers.

Table III. Cellular amplification of IFN response

Inducer	Whole Pcs[a]	Glass Adherent	Non-Adherent	Reconstituted PCs
Poly (I : C)	500[b]	450	400	650
ABPP	150	20	< 10	200
NDV	3500	2000	4000	3000
MEM	—	—	—	—

[b]IFN — units/ml. [a]PCs — mouse penitoneal cells

and levels of iterferon based upon the cellular constituency at the time of induction. In tab. III the effect of inducing peritoneal cells from mice with four interferon inducers is summarized. Peritoneal cells were separated based upon their ability to adhere to glass. Whole cell populations as well as those separated by glass-adherent were induced. The interferon response induced by a bromophenol-pyrimidinone (ABPP) illustrates the phenomenon of cellular amplification. When glass-adherent and non-adherent cells were induced in culture together they responded by producing 150 units of interferon/ml of media. However if each was separated, based upon adherence, the non-adherent as well as the adherent cells produced very little interferon when induced separately. Together the peritoneal cells amplified one another's interferon response. This phenomenon suggests that minor modification of the host's cellular components might dramatically influence the levels as well as the types of interferon induced in vivo.

Hyporeactivity to Interferon Induction

As previously indicated, animals develop a state of hyporeactivity, that is a reduced ability to produce interferon, as a consequence of virus infection, neoplastic disease or previous exposure

Table IV. Serum interferon response in mice to daily administration of compound

	Daily dose number			
Compound	1	2	3	4
Poly I : C	4200[a]	3600	250	460
tilorone	6500	150	>10	>10
ABPP	3824	1175	36	>10
AIPP[b]	91	818	292	56
ABmFPP[b]	654	152	31	>10
AlmFPP[b]	50	1025	126	37

a: Units of IFN/ml of serum.
b: Pyrimidinone derivatives

to an interferon inducing agent. This is illustrated in tab. IV where the ability of mice to respond to daily injections of one of several interferon inducing agents is summarized. With each sequential injection of inducer the animals became less responsive to interferon induction. The rate of development of hyporeactivity was inducer-dependent. With tilorone hydrochloride animals were hyporeactive after only a single injection of the compound. With other compounds two to three injections were required before the animals lost their ability to respond. Previous studies have shown that this hyporeactive state persists for 5 to 7 days after discontinuation of inducer therapy.

It became of interest to determine if hyporeactivity to one inducer was specific for that agent or if animals were cross-hyporeactive with other inducers. To evaluate this, animals first received a single injection of one inducer and then received a second dose of the same or another agent as illustrated in fig. 3. With some inducing agents the state of hyporeactivity was found to be general. For example, once animals received a single injection of the bromophenol-pyrimidinone they had a suppressed ability to respond to any of the other inducing agents used. However if animals first receiv-

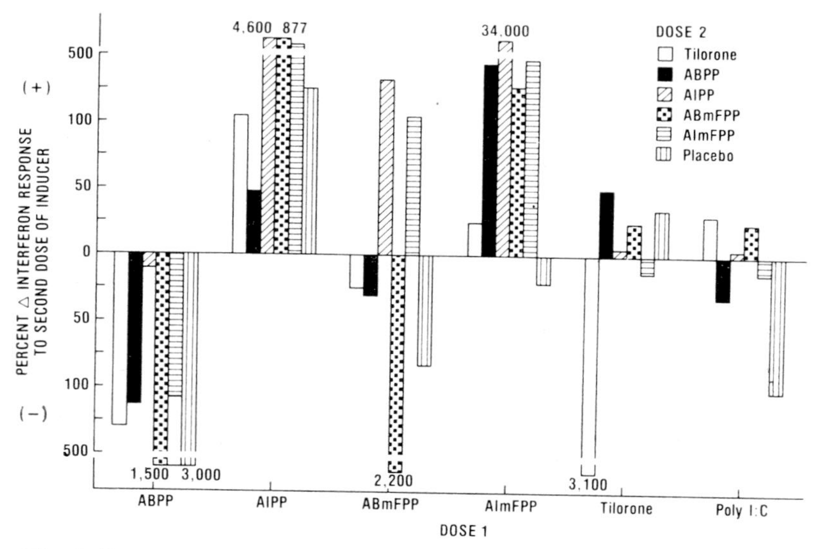

Fig. 3. Hypeoreactivity to second dose of inducers given 24 hrs following first dose.

ed a single injection of the iodophenol-pyrimidinone (AIPP) they had an enhanced ability to respond to other interferon inducing agents. Animals that had become hyporesponsive to a particular interferon inducing agent were still responsive to other selected inducers and the cross reactivity of the hyporeactive response was a function of the cell population induced by the compound. That is, a compound that induced interferon primarily in lymphocytes would make animals generally hyporeactive to other lymphoid interferon inducing agents but an agent which induced interferon in macrophages would still stimulate a response in vivo. These results suggest that one means of maintaining high levels of interferon in vivo would be to alternate between various interferon inducing agents.

Immune Modulation

Interferon inducing agents in general have a capacity to modulate the immune response. Agents which induce an interferon response can enhance natural killer cell, macrophage, T cell and the antibody response of animals depending upon the conditions employed. This has previously been primarily attributed to the interferon induced by the agent; the theory being that the interferon induced by these compounds modulated the subsequent immune effect. Using the pyrimidinone interferon inducers we found that

Table V. Summary of immune modulating activity of pyrimidinones

Increased:	Murine natural killer cells in vivo.
Increased:	Murine macrophage-mediated cytotoxicity after in vivo or in vitro treatment.
Increased:	In vivo antibody formation in unimmunized and immunized mice.
Decreased:	In vitro spleen cell response to SRBCs.
Decreased:	T killer cell cytotoxicity in vitro.
Increased:	Bone marrow colony-forming units.
No effect:	In vitro compounds were not mitogenic nor did they affect mitogen response

the ability of these agents to modulate the immune response was independent of their ability to induce interferon as summarized in tab. V. We found no clear correlation between the ability of several pyrimidinones to induce interferon and increase macrophage, natural killer cell and B responsiveness suggesting separate mechanisms of action.

Antiviral Activity

Interferon inducers in general have broad spectrum antiviral activity. Most investigators have attributed the antiviral activity of inducers to their ability to stimulate a circulating interferon response. In a series of studies design to explore the antiviral activity of the pyrimidinones, Renis et al. (2) and Stringfellow et al. (3) explored the anti-herpetic HSV and anti-Semliki forest virus SFV activity of a variety of pyrimidinone interferon inducers. They found that the ability to induce interferon could be directly correlated with the anti-Semliki forest virus activity but the activity against Herpes simplex virus was independent of the ability of the compound to induce circulating interferon (tab. VI). In fact, Renis et

Table VI. Effect of anti-thymocyte serum on protection by ABPP and AIPP of mice infected with HSV-1, EMC and SFV

Virus	Saline			ABPP[b]			AIPP[b]		
	Sal.	ATS[c]	NRS[d]	Sal.	ATS	NRS	Sal.	ATS	NRS
	% S[e]	% S	% S	% S	% S	% S	% S	% S	% S
HSV-1 (i.p.)[a]	13	6	0	56	7	64	81	25	80
EMC (i.p.)	0	0	0	100	100	100	100	100	100
SFV	19	7	7	100	100	94	100	100	

[a] Mice were inoculated at 0 hr with i.p. 0.1 ml HSV-1 (7.4×10^4 PFU) or 0.1 ml of SFV (5×10^1 PFU). EMC (250 PFU in 0.05 ml) was inoculated intranasally after light anesthesia with ether.

[b] Mice were treated i.p. with 200 mg/kg/day of each drug.

[c] Rabbit anti-mouse thymocyte serum.

[d] Normal rabbit serum.

[e] % S = percent survivors.

Table VII. Duration of antiviral effect

Days: —16 → —1 (—3 hr) 0 + 1 day ↑ SFV

Compound	Route	Dose (mg/kg)	Time of treatment (days)[a]													
			—16	—14	—12	—10	—8	—7	—6	—5	—4	—3	—2	—1	0	+1
AlmFPP	IP	500	12[b]	50	50	75	83	83	78	79	83	75	88	96	46	33
ABmFPP			—	—	—	—	—	19	54	32	47	46	26	85	94	27
ABPP	IP	500	—	—	—	—	—	—	4	17	15	24	42	92	82	15
AIPP	IP	500	30	62	57	67	65	64	75	77	80	83	89	94	42	4
Tilorone	PO	250	—	—	—	—	—	19	25	25	56	88	94	100	100	25
Poly (I : C)	IP	5	—	—	—	—	—	19	38	31	50	75	88	100	100	19

[a] SFV (10 LD_{50}, a 100 % lethal dose) was injected IP at 0 hr; a single dose of drug was administered between 16 days before to 1 day after infection.

[b] Values are % survivors (30 mice/group).

al. (2) found that the anti-herpetic activity but not the anti-Semliki forest virus or interferon inducing activity of the pyrimidinones could be inhibited by injection of antithymocyte serum. This data suggested that thymocyte activation by the pyrimidinones mediated anti-Herpes virus activity independent of the interferon response. This was strengthened in studies by Stringfellow et al. (3) where the duration of the antiviral effect of the pyrimidinones in Semliki forest and Herpes simplex virus infected mice was studied. They found (tab. VII) that animals were protected up to 14 days after a single injection of one of the pyrimidinones independent of the levels or type of interferon induced. Since exogenous transfer of interferon only protected mice for a few hours, the data suggested that host defense mechanisms independent of interferon were involved which persisted for up to two weeks.

Antitumor Activity

Both interferon and interferon inducers have had activity against transplated murine tumors. With each of these agents however the antitumor activity has been tumor load dependent as summarized in tab. VI. By increasing the number of tumor cells injected (in this case B16 malignant melanoma cells), the antitumor activity of inducers could be overwhelmed. These data suggest that the antitumor activity of biologic response modifiers such as inducers will be limited by the tumor load the host has at the time of

Table VIII. Combination of inducers with cyclophosphamide

B 16 Injected IP	Cyclophosphamide alone	Cyclophosphamide + ABPP	ABPP alone	Placebo
2×10^7	30[1]	70	0	0
2×10^6	60	75	0	0
2×10^5	65	100	10	0
2×10^4	80	100	45	10

30 mice/group
[1] percent survivors

initiation of chemotherapy. The antitumor activity of these types of compounds may therefore be limited clinically unless used in combination with agents such as irradiation, surgery or cytotoxic chemotherapy that can reduce the tumor load. To explore this possibility we treated animals with cyclophosphamide in conjunction with inducers, tab. VIII. Cyclophosphamide alone was superior to the inducers in increasing survival time. When the two were combined together however a synergistic effects was observed. In fact, 100 % of animals were completely protected from B16 malignant melanoma when treated with cyclophosphamide and the inducer. Also, Milas et al. (1) found that a combination of surgery with inducer treatment significantly enhanced the antitumor activity over either treatment given alone.

Summary

The data presented indicate that: 1. A wide range of structures are capable of inducing interferon but only modest modification of these structures can influence the levels of iterferon induced. 2. The serum interferon response is not a true reflection of the interferon levels the cells are exposed to at the organ level. This may make interpretation of clinical studies with interferon inducers difficult since the most meaningful data may only be obtained by monitoring organ not circulating interferon levels. 3. The immune modulating activity of iterferon inducers can be significantly modified through only slight modification of the chemical structure which interestingly suggested an independence from the circulating interferon levels induced. 4. The antiviral activity of interferon inducers can also be modified independent of their ability to induce circulating levels of interferon. 5. The antitumor activity of inducers is tumor load dependent but through combination of inducer therapy with methods of reducing tumor load such as cytotoxic therapy or surgery a synergistic enhancement of antitumor activity was observed. 6. Clinical evaluation of interferon inducers should be aimed at monitoring interferon induction as well as immune modulation.

One of the obvious questions confronting interferon research at the present time is, are interferon inducers really necessary now that interferon has been cloned, modulated and in some cases mutilated? The answer to this question is obvious since interferon inducers are not only capable of inducing a circulating interferon response but independent of that ability are capable of modulating various immune functions. Also by specific modulation of chemical structure it is possible to selectively make these agents more specific in their ability to modulate specific immune processes. This would suggest that one mechanism to specific modulation of the immune response is through utilization of compounds such as "inducers" which have been specifically modified to enhance specific cellular functions. And lastly, even though passive administration of interferon can achieve fairly high circulating levels of interferon, the studies with interferon inducers suggest that organ interferon levels, not circulating interferon levels are of primary concern and modulate the biologic activity observed. Interferon inducers can specifically modulate organ interferon levels which are not necessarily reflected in the circulation of the animal, thereby giving an avenue of achieving some selectivity and avoiding the systemic – undesirable side effects and toxicity observed with systemic administration of interferon.

References

1 Milas, L.; Hersch, E. M.; Stringfellow, D. A.; Hunter, N. (1982): J. natl. Cancer Inst. 68, 139—145.

2 Renis, H. E.; Eidson, E. E. (1980): Current Chemonth. Inf. Dis. 11, 1411—1412.

3 Stringfellow, D. A.; Vanderberg, H. C.; Weed, S. D. (1980): J. Inf. Res. 1: 1—13.

D. A. Stringfellow, M.D., Cancer Research, Bristol-Myers Pharmaceutical Research & Development, P. O. Box 657, Syracuse, NY 13201 (USA)

Contr. Oncol., vol. 20, pp. 358—374 (Karger, Basel 1984)

Studies with the IFN Inducer and Immune Modulator, Poly ICLC

H. B. Levy; M. Chirigos

NIAID and NCI, Ft. Detrick, Frederick, Maryland, USA

Introduction

The double stranded RNA, polyinosinic-polycytidylic acid (Poly I . Poly C), is a good IFN inducer, antiviral agent, and anti tumor agent in rodents, (14, 19). However in primates including man, it is none of these (16). Primate serum has a higher level of hydrolytic activity against Poly I . Poly C than does that of rodents (13) and it is thought likely that this is the explanation for the poor activity of Poly I . Poly C in primates. Poly I . Poly C forms a stable complex

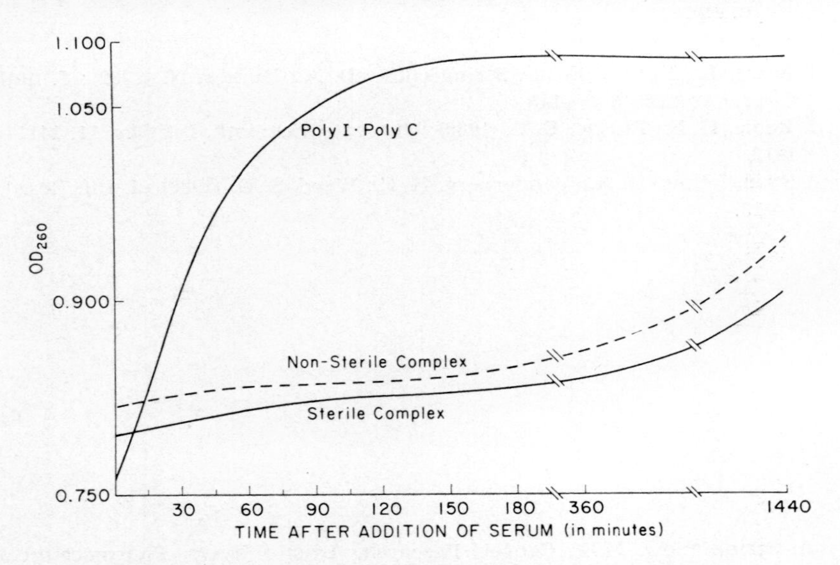

Fig. 1. Hydrolysis of Poly I . Poly C complexes by human serum.

with a complex of poly lysine and carboxymethyl cellulose. The final material is referred to as Poly ICLC. Poly ICLC is relatively resistant to hydrolysis by serum or RNase as shown in fig. 1 (10).

This report will deal with some aspects of studies with Poly ICLC, emphasizing clinical effects in animals and man, immune modifying activities, and toxic manifestations.

Antiviral activities

Poly ICLC has been tested against a number of viruses in several species of animals. A few studies will be reviewed in slight detail, and all the studies will be listed.

Rabies

Classical treatment of rabies consists of two phases. In the first, antibody to rabies is given to cover the infecting virus with passive immunity. In the second phase, vaccine is given to stimulate the host to develop his own active immunity to neutralize the virus at later stages. These two treatments tend to negate each other, and repeated injections of both are needed. In highly advanced industrial countries purified antibody and vaccine are available, but in those countries where rabies is a major human problem, the antibody and vaccine are much less purified and potent, and larger numbers of injections of both products are needed.

Table I. Effects of poly (ICLC) treatment in post exposure prophylaxis of rabies in monkeys

Treatment	Dead/Treated
Poly ICLC + Vaccine 24 and 72 hours post infection	1/8
Vaccine 24, 48, and 72 hours post infection	7/8
Controls-untreated	8/8

As a result, perhaps as much as 50 % of the patients in some countries do not complete the course of treatment. The use of Poly ICLC to control the infecting virus, together with vaccine, has resulted in a drastically reduced treatment in an experimental monkey model tab. I. One dose of the inducer followed by two doses of vaccine has resulted in control of lethal infection in monkeys

There are many problems involved in mounting a competent

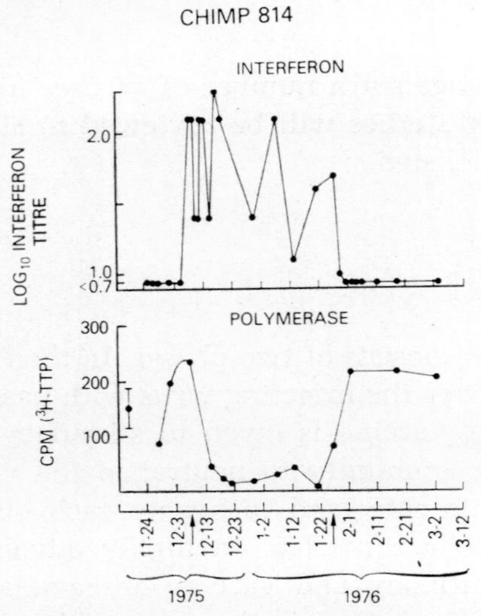

Fig. 2. **Treatment of chronic hepatitis in young chimpanzees with poly ICLC: The IFN titer is shown above and the virus-associated polymerase** activity is shown below. The beginning and the end of the treatment period are indicated by arrows on the abscissa. The dot and bar on the graph of DNA polymerase response indicate the mean ($\pm$ 1 SD) of polymerase activity detected in 6 serum samples obtained during the 5 weeks immediately preceding the experiment.

trial in humans. For example, as will be mentioned later, Poly ICLC is a potent immune adjuvant with some antigens. There was concern that the use of Poly ICLC might increase the incidence of allergic encephalitis in patients receiving a vaccine made from suckling mouse brain. However, in tests in rats, this was found not to be the case. (Baer and Levy, unpublished)

Chronic hepatitis B is an infection for which there is no proven therapy. It is carried by many millions of people in the world, particulary in the Orient. There is a model of this disease in chimpanzees which bears some resemblance to the human disease. Fig. 2 shows the results of treating hepatitis B-carrying chimps with Poly ICLC. (15)

It will be seen that while chimps are being treated with Poly ICLC they produce IFN (and presumably enhanced immune reactivity). During this time, evidence of the disease, as measured by the level of DNA polymerase in serum, declines to a vanishing low level. When treatment was stopped after 16 weeks, the IFN levels disappeared and evidence of disease returned. These data resemble

Table II. Virus diseases of animals that have been treated with poly ICLC

Disease	Animal	Results
Simian hemorrhagic fever	Monkey	Complete protection if given before virus, none if given after virus
Venezuelan equine encephalitis	Monkey	No animals with light virus challenge died; Poly ICLC reduced viremia by 50 %
Yellow fever	Monkey	75 % protection up to 8 hr post-challenge
Japanese encephalitis	Monkey	50 % protection up to 24 hr post-challenge
Tacaribe virus	Mouse	No effect by Poly ICLC
Rabies	Monkey and mouse	See text
Hepatitis	Chimpanzee	Virus controlled whille on drug. Control ends when treatment stopped
Bolivian hemorrhagic fever	Monkey	Possible worsening of disease
Tick-borne encephalitis	Monkey	Strong protection
Vaccinia	Monkey	Strong protection
Vaccinia skin lesions	Rabbit (topical treatment)	Spread of lesions stopped

those seen by Merigan in humans treated with IFN. (5) It was only after treatment in man was continued for a year, that some apparent cures were noted. It is not known what would be the result if chimpanzees would be treated for a year.

Tab. II summarizes the data for a variety of diseases in animals treated with Poly ICLC. In most instances there was some beneficial effect. However, in the case of Bolivian Hemorrhagic fever, the disease may have been worsened by treatment. (11)

Immune Modulatory Effects in Animals

As with a number of IFN unducers, Poly ICLC can modify immune responses, both humoral and cell mediated.

Humoral Antibody

Poly ICLC has been found to increase the amount of circulating Ab induced in response to the following vaccines; Swine flu, Venezuelan equine encephalitis, Japanese B encephalitis, Hemophilus influenzae, Herpes virus envelope and Rift valley fever. Low levels of the drug are capable of inducing this enhancement (12).

Vaccine against Swine flu was relatively ineffective in young people. It is also poorly effective in young monkeys. However, when the vaccine is augmented with Poly ICLC, the level of Ab is increased (18), fig. 3.

As can be seen, in the absence of Poly ICLC, the levels of Ab produced are considered to be ineffective, but with addition of Poly ICLC, Ab levels of 40 units were found, which are considered capable of protecting against disease. The levels of Poly ICLC used in this study, 10 μg/kg, did not cause any toxic manifestation and did not induce detectable serum interferon.

The enhancement of Ab production can be much greater than that with Swine flu. With Venezuelan equine encephalitis, for example, the degree of antibody enhancement can be several hundred fold, as seen in fig. 4 (6).

Vaccine to VEE, with or without Poly ICLC was given to monkeys. After 28 days, another injection of vaccine without Poly ICLC was given. After a total elapsed time of 42 days, antibody levels in monkeys that received the adjuvant was 250 times as high as in those monkeys that received vaccine alone. Poly ICLC has no effect on the sequential development of IgM and IgG. At the end of the experiment, after about 60 days, all the Ab was IgG.

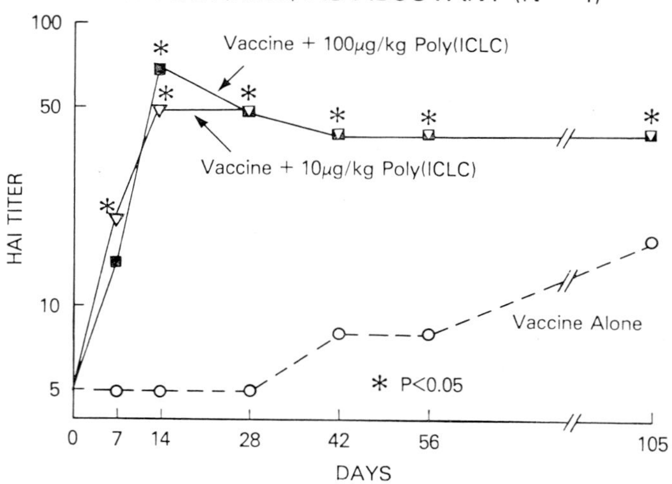

Fig. 3. Effect of one injection of poly (ICLC) on antibody production by rhesus monkeys in response to a subunit vaccine to Swine flu (4 monkeys per group).

Protection axainst death by subsequent challenge with the lethal pathogen was the criterion used to determine effectiveness of Poly ICLC in Herpes virus infection. Mice were immunized with or without Poly ICLC, with a vaccine prepared from Herpes virus envelope. Fig. 5 shows that he addition of Poly ICLC to the vaccine significantly increased the survival of mice subsequently challenged intracerebrally with live virus (16).

Analogous results have been observed with the vaccines listed earlier.

It should be noted that Poly ICLC is not a universal adjuvant. Of those antigens tested, those that were relatively weak induced

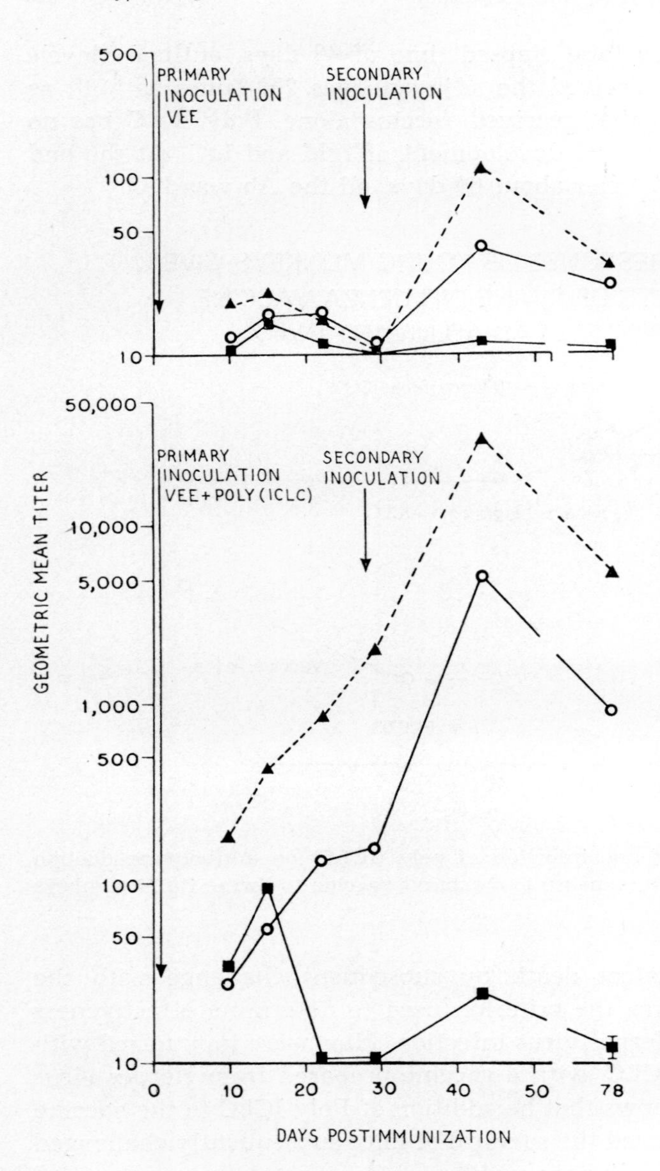

Fig. 4. Serum-neutralizing antibody response by immunoglobulin class of rhesus monkeys innoculated on days 0 and 28 with (A) inactivated VEE virus vaccine ($n = 4$), or (B) vaccine combined with 200 μg of Poly ICLC/kg ($n = 4$). △: whole serum antibody titers; ○: antibody for IgG fractions; ■: antibody from IgM fractions.

more antibody when Poly ICLC was also given; those that were strong antigens, like albumin and pneumococcal polysaccharide type III, induced less antibody when Poly ICLC was also given. With rabies infection, Poly ICLC has no effect on Ab production. However, not enough antigens have been tested to justify any general conclusions.

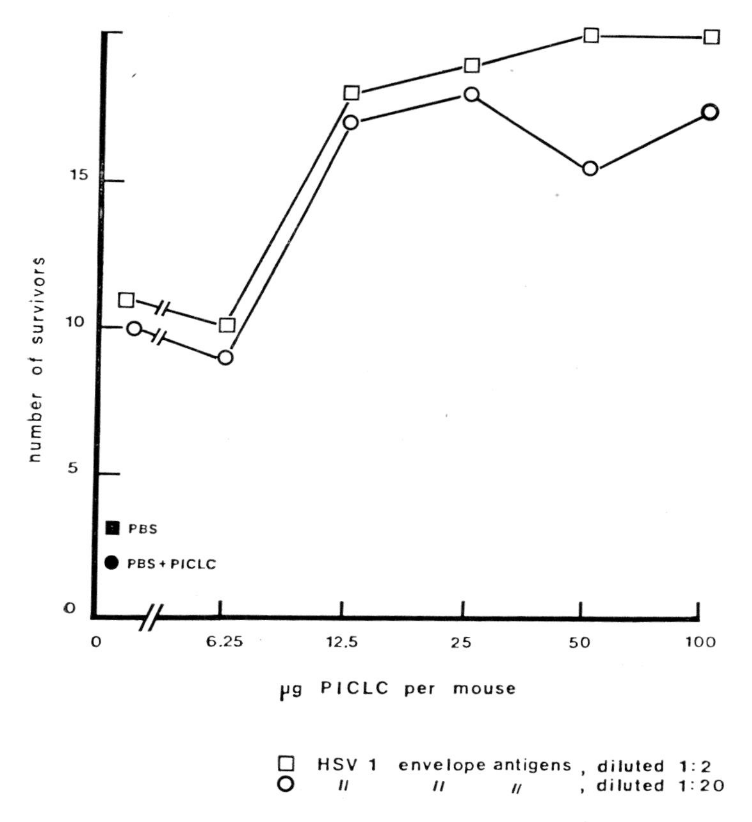

Fig. 5. Effects of Poly ICLC given with envelope antigen from herpes virus on survival of mice against subsequent challenge with live virus.

Cell Mediated Immunity

Both interferon and Poly ICLC modulate cell mediated immunity in a variety of ways, not always identically. Poly ICLC in mice enhances: a) delayed type hypersensitivity, as measured by food pad swelling, b) natural killer (NK) cell activity vs. several tumors, both in vivo and in vitro, c) macrophage activity vs. tumors, d) transformation of T lymphocytes by phytohemagluttinin but not transformation of B lymphocytes by lipopolysaccharide, e) graft vs. host reaction, f) number of pluripotential stem cells (CFC). These effects are still present 48—72 hours after a single injection.

Treatment in mice with exogenous interferon also activates NK cells and macrophages. On the other hand, IFN mostly inhibits GVH reaction, delayed type hypersensitivity, and CFC development. Tab. III shows a comparison between the effects of treating macrophages in vitro with interferon vs. treating them with Poly ICLC (3).

Tab. III shows that both IFN and Poly ICLC stimulate macrophages to produce CSF and prostaglandin, but that here are differences. Treatment with ten μg/ml of Poly ICLC leads to the formation of 50 U of IFN and 1,400 pg of PGE, while treatment with 500 U of IFN leads to the production of only 225 pg of PGE, suggesting that PGE stimulation by Poly ICLC is not attributable solely to IFN. This concept is substantiated by the observation that addition of sufficient antibody to IFN to the mixture abolishes the CSF production by exogenous IFN, but has no effect on the stimulation by Poly ICLC.

In initial trials in cancer patients (H. Levy, R. Herberman, and J. Reid), i.m. injection of Poly ICLC at 1 mg/m^2 caused a significant and prolonged augmentation of NK cell activity in about 70 % of the patients.

Activation of NK cells and tumoricidal macrophages probably contribute to the antitumor activity of the drug. Fig. 6 demonstrates

Table III. Effect of IFN and poly ICLC on production by macrophage of soluble products

Agent	IFN units	CSF units	Prostaglandin E pg
Control	0	30	70
Poly ICLC:			
10 μg/ml	50	N.D.	1400
50 μg/ml	1000	120	2000
IFN 500 U/ml	N.D.	100	225

a point that is likely to prove relevant to the antitumor activity of biological response modifiers in general. When Poly ICLC is given shortly after inoculation of the tumor, there is a pro-

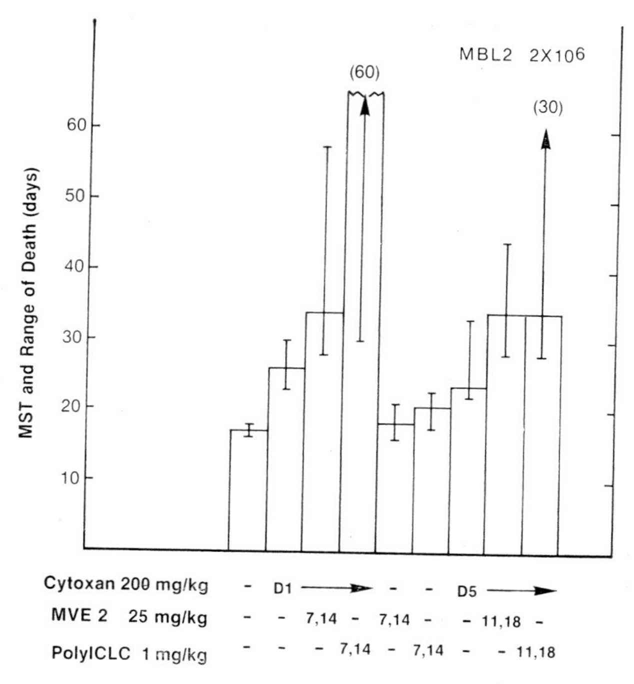

Fig. 6. Effect of Poly ICLC and MVE_2 (pyran copolymer) + cytoxan (to decrease tumor burden) on mean survival time (MST) and percent survival (numbers on top of column) of mice given MBL_2 tumor.

nounced antitumor effect, but if the tumor is allowed to grow to a large size, there is much less effect. Hower, if the tumor is allowed to grow and Cytoxan is given to reduce the tumor burden the Poly ICLC again has a pronounced effect on survival time and percent survival.

Human Studies

Prior to initial human studies, preclinical toxicity studies were done.

Animal Toxicity Studies

Mice treated with one i.v. dose of Poly ICLC have an LD_{10} of 31 mg/m^2 and LD_{50} of approximately 46 mg/m^2 in a large number of trials. For rats, however, the LD_{10} occurs with one i.v. dose of 100 mg/m^2. The LD_{50} is 140 mg/m^2. Lethal doses in rodents were associated with bloody nares, pulmonary edema, seizures and hepatosplenomegaly. Mice treated i.v. with approximately 10 mg/m^2 twice (4 days apart) had no apparent toxicity. Specifically, there were no significant changes in hematologic indices or clotting times, the bone marrow remained normal, and there was no gross pathology on sacrifice. In cats treated i.v. with approximately 66 mg/m^2 twice (4 days apart), the following was observed: emesis, decreased food consumption, mildly diminished erythropoiesis, and a slight left shift in the WBC series. There were no other changes in hematology, chemistry or urinalysis. Rhesus monkeys were treated intravenously with 60 mg/m^2 of Poly ICLC for two consecutive days with the following findings: BSP retention occurred in all, and a slight hyperbilirubinemia in one, but there was no gross hepatic or gallbladder pathology. There were no other significant changes in hematology or blood chemistry, except for a mild erythoid hyperplasia in one monkey. There was no other gross or microscopic pathology attributable to the drug upon sacrifice. Rhesus monkeys were also given a dose of 36 mg/m^2 of Poly (ICLC) daily for six days and then three times weekly for the following 14 weeks with no apparent toxicity nor gross pathology. Chimpanzees received 60–90 mg/m^2 in one intravenous dose with a transient elevation in SGOT, and slight thrombocytopenia and leukopenia, but no other chemical abnormalities. Four of four chimpanzees given this dose every other day for 6 weeks developed transient hepatotoxicity thought to be due to the carboxymethyl cellulose component.

In all monkey, and chimpanzee studies, the Poly ICLC preparation has been pyrogenic. A maximum temperature elevation up to 5° F has occurred at about 4–8 hours post treatment (first dose) with defervescence occurring 6–12 hours later. With subsequent doses, fever occurred at an earlier time in relation to the treatment, and defervescence continued to occur in 6–12 hours.

Clinical Use of Poly ICLC

The first clinical use of Poly ICLC was in a phase I trial in advanced cancer patients (9). The protocol consisted of treating at least 3 patitents at each level of escalating doses. Patients were given 1 dose at a low lewel, a week was allowed to pass, then 14 consecutive daily doses at that same level were given. If toxicity was not excessive, 3 different patients went into the next level of drug, at the same schedule and so on until unacceptable levels of toxicity were encountered. The mean levels of serum IFN are shown in tab. IV, along with those achieved by monkeys of comparable doses.

In a study done by the Childrens Cancer Testing Group in null type acute lymphocytic leukemia very ill children were treated with Poly ICLC at 12 mg/m^2. This level of drug proved too high for these children, and severe toxicity was noted. As will be mentioned later, current dosage is less than 12 mg/m^2 and less frequent than daily. When the dosage was reduced to 6 or 9 mg/m^2 toxicity

Table IV. Interferon induction with Poly ICLC

Dose (mg/m^2)	Mean peak serum titers (IU/ml)	
	Rhesus monkeys	Man
2.5		15 (0—25)
7.5		200 (25—250)
12	1500 (600—2000)	2000 (200—5000)
18		4500 (600—15,000) (toxic)
27		6000 (200—10,000) (toxic)
36	10,000 (3000—15,000)	
240	100,000 (20,000—200,000) (lethal)	

The highest acceptable level in man in these studies was 12 mg/m^2, and corresponding mean peak serum IFN levels were about 2,000 IU/ml. The toxicity seen in these patients is summarized in table V.

was less. IFN levels ranged widely, depending on dose, frequency, and status of patient, ranging from 50 to 1,000 units/ml. There were no clinical responses in these resistant and far advanced patients. There were a number of antileukemia responses in the marrow, a few of which are shown in tab. VI. It can be seen that the percentage of blasts decreased, but the patients died too soon to know whether there would have been replacement with normal cells.

Table V. Poly ICLC toxicity

Manifestation	No./total
Fever	25/25 (100)[a]
Nausea	11/25 (44)
Thrombocytopenia and leukopenia	17/25 (68)
Hypotension[b]	7/25 (28)
Syndrome of erythema, polyarthralgia, and polymyalgia[b]	4/25 (16)
Renal failure[b]	1/25 (4)
Trial aborted[c]	5/25 (20)

[a] Numbers in parentheses, percentage of total.
[b] Related to dose level and/or magnitude of interferon induction.
[c] 3, hypotension; 1, renal failure (27 mg/m^2); 1, serum sickness (?). Maximum tolerated dose, 12 mg/m^2.

Table VI. Poly ICLC in treatment of leukemia

Type of Leukemia	Dosage	Days given	Before		After	
			WBC	% Blasts	WBC	% Blasts
T Cell	9 mg/m^2	14	91,600	95	100	0 %
Null ALL	9 mg/m^2	14	18,850	99	25	60 %
Null ALL	9 mg/m^2	14	140,000	94	900	0 %
Null ALL	9 mg/m^2	14	77,900	96	100	24 %

Slightly more encouraging results were seen in 6 patients with multiple myeloma, studied at the University of Arizona. A summary of the result with these patients is given in tab. VII (1).

It can be seen that there were several partial remissions, using

Table VII. Poly ICLC in patients with multiple myeloma

M-component type	Comments
Kappa light chains	67 % decrease in B-J excretion, plus correction of hypercalcemia and disease stabilization with first period of treatment. 44 % decrease in B-J excretion when Poly ICLC restarted 2 months later. Normocalcemic for 5 months.
IgG kappa	M-component decrease from 5.2 gm to 3.9 gm, Improved bone pain and performance status (became ambulatory) for 2—3 months.
IgG lambda	Trial stopped because of toxicity (malignant hypertension).
IgM kappa	Plasmapheresis requirement decreased from q 14 days to q 28 days.
IgM kappa	Stable disease parameters for 2 months.
Lambda light chains	Died one week after initiation of Poly ICLC.
IgG kappa	50 % decrease in serum IgG M-component, plus symptomatic benefit.

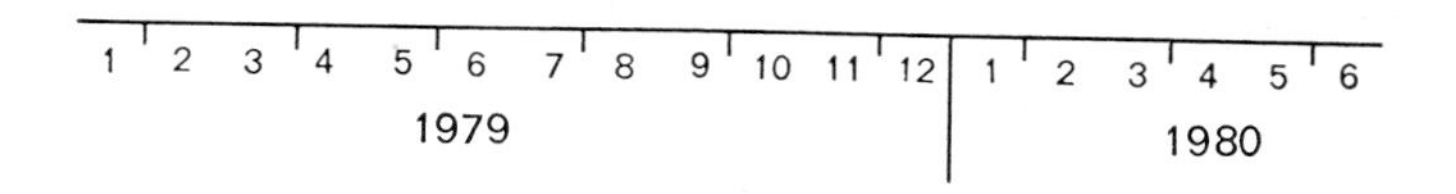

= endoscopic procedure

= Poly ICLC 12 mg/m² t.i.w.

= Poly ICLC 12 mg/m² weekly

Fig. 7. Effect of Poly ICLC of frequency of surgical intervention in juvenile laryngopapilloma in a child. Each dot represents surgical intervention.

the indicated criteria. These were far advanced patients who had become resistant to other forms of therapy.

It may be that, as suggested by the data in fig. 6 that prior chemotherapy to reduce the tumor load would enable the biological response modifier Poly ICLC to have a more pronounced effect.

Juvenile laryngopapilloma can occur in a fulminating form that requires surgical intervention at biweekly intervals. In studies on seven young patients at Johns Hopkins Medical School and the University of North Carolina Medical School, five patients showed significant improvement. Fig. 7 shows results in one of these patients. It will be seen that surgery was very frequent prior to onset of therapy, and it greatly decreased when Poly ICLC treatment was given. In the year subsequent to the completion of this figure only one surgical intervention was necessary. Tab. VIII summarizes the results with 5 other patients (8).

Table VIII. Effect of Poly ICLC on 6 patients with juvenile laryngo papilloma

Patients	Age (Years)		Frequency of Surgery	
	at Dx	at Rx	During 9 months before Rx	During 6 months of Rx
1	1.5	4	18	0
2	1.5	3	23	4
3	2	3	8	0
4	2	6	5	1
5	54	61	4	0
6	41	47	3	∅

Dx: diagnosis Rx: treatment

Several neurologic diseases, whose etiology is presumed to be associated with immune dystrophy are under study.

Twelve cases of peripheral paralytic neuropathies, have been treated with Poly ICLC (4, 17, Levy, Engel, and Salazar, unpublished). The types are indicated in fig. 8. Most were classified as Landre-Gullian-Barré syndrome. Patients were treated i.v. once a week, beginning at a dose of 50 μg/kg and increasing weekly up to 100 μg/kg. Eight of the twelve have shown moderate to marked

Poly — ICLC Treatment, 20—70 μg/kg i. v.

	improved
NEUROPATHY, DYSIMMUNE	4 of 8
NEUROPATHY, Dysimmune (?) Dysneural	2 of 2

Benefit lasts 2—6 weeks; **Respasivity** lasts > 3 years

Acute Effects:

- flu-like syndrome (fever, chills, malaise), 4 – 18 hrs.
- lymphocytopenia to 1—20% of baseline, 4—5 days
- granulocytosis 2—3× baseline, 24—48 hrs.
- hypercorticoidemia, 48 hrs.
- hyperinterferonemia minimal, < 24 hours.

Fig. 8. Effect of Poly ICLC on neurologic diseases.

improvement, in which the patients started as severely restricted to paraplegic and went to being able to resume a more or less normal life (2—3 months of treatment were needed). It is characteristic of patients with this type of disease that fever causes increased weakness. Poly ICLC always induces fever, and always caused increased weakness in measurable and persistent increase in strength. This cycle repeated with each treatment. In some cases, after 4—5 weeks they began to feel weak again, and required maintanance injections about once every month.

Six patients with chronic multiple sclerosis have been entered in a protocol to be treated with Poly ICLC. While on treatment, disease in these patients did not progress. Since M. S. can show spontaneous transient stabilization, one cannot at this time attribute this effect to Poly ICLC. In several patients there was increase in polymorphonuclear leucocytes that persisted for about 24 hours. There was also a pronounced transient lymphocytopenia, and transient weakness associated with the induced fever.

Studies with malignant melanoma and renal cell carcinoma are in too early a stage to be evaluable.

In summary, a derivative of Poly I. Poly C that resists hydrolysis by primate serum has been shown to have therapeutic value in viral diseases in mice and monkeys, and causes modification in a number of immune parameters, both humoral and cell associated. It is undergoing trial in several human diseases. It induces formation of IFN in people, and also demonstrates immune modulatory effects, but it is too early to make any real clinical evaluation.

References

1 Alexanian, R.; Guttermann, J.; Levy, H. B. (1982): in "Clinics in Haematology". Ed. S. E. Salmon, W. B. Saunders Co., Philadelphia, pp. 211—220.
2 Bar, G. M.; Shaddock, J. H.; Moore, S. A.; Yager, P. A.; Baron, S.; Levy, H. B. (1977): J. infect. Dis. 136, 58—62.
3 Chirigos, M. A.; Papademetriou, V.; Bartocci, A.; Read, E.; Levy, H. B. (1981): Int. J. Immunopharmac. 3, 329—337.
4 Engel, W. K.; Cuneo, R.; Levy, H. B. (1978): Lancet 503—504.
5 Greenberg, H. B.; Pallard, R. B.; Intevick, L. I.; Gregory, P. B.; Robinson, W. J.; Merigan, T. C. (1976): New Engl. J. Med. 295, 519—522.
6 Harrington, D. L.; Crabbs, C. L.; Hilmas, D. E.; Braun, J. R.; Higbee, C. A.; Cole, F. E.; Levy, H. B. (1979): Infect. Immun. 24, 160—166.
7 Klein, R. J.; Klein-Bulmovice, E.; Moser, U.; Mouch, R.; Hilfenhaus, J. (1981): Arch. Virology 68, 73—80.
8 Leventhal, B. G.; Kashiman, H.; Levine, A. S.; Levy, H. B. (1981): J. Pediat. 99, 614—616.
9 Levine, A. S.; Sivulich, M.; Wiernik, P. H.; Levy, H. B. (1979): Cancer Research 39, 1645—1650.
10 Levy, H. B.; Baer, G.; Baron, S. Buckler, C. E.; Gibbs, C. J.; Iadarola, M. J.; London. W. T.; Rice, J. (1975): J. infect. Dis. 132, 434—439.
11 Levy, H. B.; Riley, L. F. (1983): in Interferons and their Application. Eds. W. Carter and P. Came, Springer.
12 Levy, H. B.; Riley, F. L. (1983): in The Lymphokines", vol. 8, Ed. E. Pick, Academic Press, New York, pp. 303—322.
13 Nordlund, J. J.; Wolff, S. M.; Levy, H. B. (1970): Proc. Exp. Med. 133, 439.
14 Park, J. H.; Baron, S. (1968): Science 162, 811.
15 Purcell, R. H.; London, W. T.; McAliffe, V. J.; Palmer, A. E.; Kaplan, P. M.; Levy, H. B.; Gerin, J. L.; Popper, H.; Lvovsky, E.; Wong, D. C. (1976): Lancet, Oct. 9, 757.
16 Robinson, R. A.; De Vita, V. T.; Levy, H. B.; Baron, S.; Hubbard, S. P.; Levine, A. S. (1975): J. natl. Cancer. Inst. 57, 599.
17 Salatzar, A. M.; Engel, W. K.; Levy, H. B. (1981): Arch. Neurol. 38, 382—384.
18 Stephen, E. L.; Hilmas, D. E.; Mangiafico, J. A.; Levy, H. B. (1977): Science 197, 1289—1290.
19 Worthington, M.; Baron, S. (1971): Proc. Soc. exp. Biol. Med. 136, 323.

H. B. Levy, NIAID and NCI, Ft. Detrick, Frederick, M.D. 21 701 (USA)

Contr. Oncol., vol. 20, pp. 375—386 (Karger, Basel 1984)

Poly (G). Poly (C) As an Inducer of Interferon

E. De Clercq[a]; *P. F. Torrence*[b]

[a] Rega Institute for Medical Research, Katholieke Universiteit Leuven, Leuven, Belgium, and [b] Laboratory of Chemistry, National Institute of Arthritis, Diabetes, and Digestive and Kidney Disease, National Institutes of Health, Bethesda, Maryland, USA

Although interest in interferon inducers have somewhat waned during the last years, they may be considered as an attractive alternative to exogenous interferon therapy, particularly in those conditions where it may be difficult to sort out the most appropriate interferon species (or subspecies). A staggering variety of compounds belonging to the most diverse chemical classes have been described as inducers of interferon, but only two classes of inducers yield promise for clinical use. These are the double-stranded RNAs, such as poly(I) . poly(C) (polyriboinosinic acid . polyribocytidylic acid), and the low-molecular-weight compounds, such as the 5-halo--6-(2,3-difluorophenyl)pyrimidinones. The double-stranded RNAs are by far the most potent, on a weight basis, and also the most widely active of all interferon inducers. As prototype of the interferon inducers, poly(I) . poly(C) has been routinely used to produce interferon in various cell culture and animal systems. However, the clinical use of poly(I) . poly(C), i.e. in the prophylaxis and therapy of virus infections and cancer, is fraught with a number of problems: (a) a short half-life, due to a rapid degradation by plasma nucleases, resulting in (b) a rather poor and short-term interferon response, (c) toxic side effects, such as fever, leukopenia, thrombocytopenia, nausea, vomiting, diarrhea and hypotension, and (d) hyporeactivity of the host to repeated administration of the inducer.

In attempts to develop more active or less toxic inducers of interferon, various chemical modifications have been carried out at several positions of the poly(I) . poly(C) molecule. Within the interior of the double helix, both the pyrimidine ring (at C-2 and

C-5) and the purine ring (at N-1, C-2, N-3, N-7 and N-9) have been the subject of a series of substitutions. The sole modifications that appeared compatible with interferon inducing activity were substitution of CH for N-7 at the purine ring (30, 9), substitution of sulfur for oxygen at C-2 of the pyrimidine ring (24), and substitution of bromine for hydrogen at C-5 of the pyrimidine ring (30, 9) (Fig. 1). At the lower concentration range, poly(I) . poly(br^5C) proved even more potent as an inducer of interferon than poly(I) . poly(C) (9).

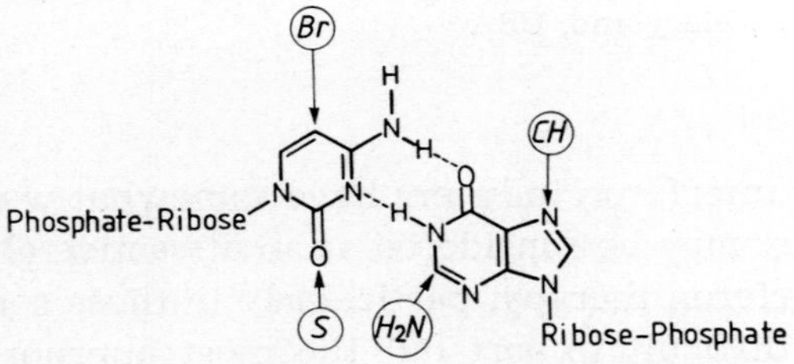

Fig. 1. Substitutions at the base pairs of poly(I) . poly(C) which are compatible with a significant interferon inducing activity:
— Br at C-5 of cytosine, as in poly(I) . poly(br^5C)
— S for O at C-2 of cytosine, as in poly(I) . poly(s^2C)
— CH for N-7 of hypoxanthine, as in poly(c^7I) . poly(C)
— NH_2 at C-2 of hypoxanthine, as in poly(G) . poly(C).

At the ribose-phosphate moieties of poly(I) . poly(C), the phosphate groups have been replaced by thiophosphate, and the C′-2 hydroxyl groups have been substituted by either hydrogen, fluorine, chlorine, methoxy, ethoxy, amino or azido groups. Some of these modifications were compatible with interferon inducing activity, i.e. substitution of thiophosphate for phosphate in either the poly(I) or poly(C) strand, or both, and substitution of fluorine, chlorine or azido for hydroxyl at C′-2 of the poly(I) strand (14, 12) (fig. 2). Of these poly(I) . poly(C) analogues, there was one, namely poly-(DIFL) . poly(C), which proved even more potent as an interferon inducer than the parent compound, at least in primary rabbit kidney cells (19).

Base-mismatching, as in poly(I_x,U) . poly(C), poly(I) . poly(C_x,U) and poly(I) . poly(C_x,G) is also compatible with interferon induction, provided $x \geqq 10$ (10, 31, 5). Thus, the polynucleotide does not have to be double-stranded over its whole length. A double-helical stretch of 10 base pairs may suffice to trigger the interferon response. In fact, the interferon induction process may occur in a biphasic manner involving first recognition of rather large segment of the RNA to allow for proper binding to the putative cellular receptor, followed by recognition of a much smaller region of the RNA corresponding to no more than one double-helical turn, which would then serve as the actual trigger of the induction process (18). This biphasic concept of interferon induction is supported by a series of studies with partially 2′-0-methylated poly(I) . poly(C) derivatives (18), mismatched poly (I_x,U) . poly(C) analogues (10) and spin (nitroxide radical)-labelled poly(I) . poly(C_x,ls^{4}U) analogues (4).

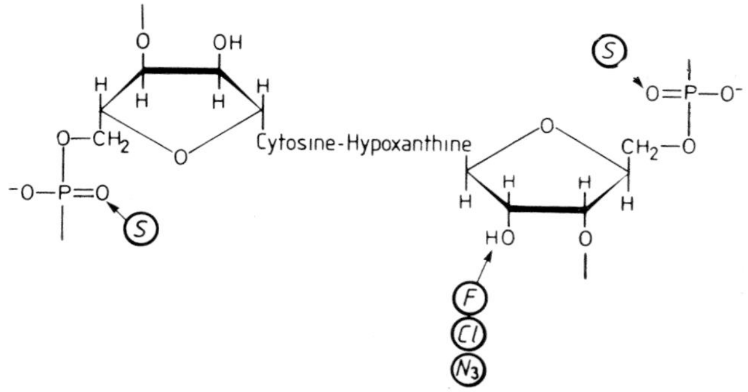

Fig. 2. Substitutions at the ribose-phosphate groups of poly(I) . poly(C) which are compatible with a significant interferon inducing activity:
— F, Cl or N_3 for OH at C'-2 of inosine, as in poly(dIfl) . poly(C), poly(dIcl) . poly(C) or poly(dIz) . poly(C)
— S for O at the phosphates of either poly(I) or poly(C) or both, as in poly(sI) . poly (C), poly(I) . poly(sC) or poly(sI) . poly(sC).

In the design of more active or less toxic analogues of poly(I) . . poly(C) two divergent approaches can be considered. The first is based on the development of analogues which are more resistant to degradation by nucleases and which may, therefore, persist for longer time in biological fluids and induce higher levels of interferon. The second approach is based upon the premise that poly(I) . . poly(C) analogues which are more sensitive to nuclease destruction may be equally active as the parent compound but less prone to toxic side reactions. Typical examples of poly(I) . poly(C) derivatives that follow the first approach are poly(I) . poly(br^{5}C), poly(I) . poly-

-(s^2C) and poly(ICLC). The latter represents poly(I) . poly(C) stabilized with poly-L-lysine and carboxymethylcellulose and induces much higher levels of interferon in both human and nonhuman primates than poly(I) . poly(C) itself (21, 20). Surprisingly, poly-(ICLC) is not more toxic than poly(I) . poly(C) (Stringfellow and Weed, 1980). Proceeding in the opposite direction are the mismatched analogues poly(I_x,U) . poly(C), poly(I) . poly(C_x,U) and poly(I) . poly(C_x,G), which might be less toxic by virtue of their increased sensitivity to nuclease degradation. However, this is a premise that remains to be borne out (27).

The double-stranded RNA complex poly(G) . poly(C) [poly(riboguanylic acid) . poly(ribocytidylic acid)] is another example of a poly(I) . poly(C) analogue, which follows the first approach (see supra). Poly(G) . poly(C) differs from poly(I) . poly(C) by the presence of an amino group at C-2 of the purine ring (fig. 1). This amino group is responsible for a third hydrogen bond between the purine and pyrimidine base pairs and thereby stabilizes this double-helical complex. Apparently, poly(G) . poly(C) fulfils all requirements which are generally deemed essential for the polynucleotide inducers of interferon, viz. a stable double-helical structure of sufficiently high molecular weight with properly paired purine and pyrimidine bases within the interior and intact ribose-phosphate moieties at the outside of the double helix. Yet, the interferon inducing ability of poly(G) . poly(C) has been a matter of conjecture. Thus, Colby and Chamberlin (6), Aksenov et al. (1), Timkovsky et al. (28), Vilner et al. (32), and Novokhatsky et al. (23) reported poly(G) . poly(C) to be an active interferon inducer, whereas Field et al. (17), Matsuda et al. (22), De Clercq et al. (8) and De Clercq and Torrence (13) found no interferon inducing activity with their poly(G) . poly(C) preparations.

After several attempts we have finally succeeded in obtaining an interferon response with poly(G) . poly(C) and also delineated the conditions that are required for the induction of interferon by poly(G) . poly(C) (tab. I). First, the constituent homopolymers should possess a high molecular size ($s_{20,w} \geqq 13$ S). If poly(G) is of medium molecular weight (9 S), the interferon inducing activity is completely lost, but if poly(C) is of medium molecular weight (8 S), the interferon inducing activity is partially retained (29). Thus, the interferon inducing activity of poly(G) . poly(C) is critic-

Table I. Requirements for an active poly(G) . poly (C) preparation

1. Composed of homopolymers of sufficiently high molecular size ($s_{20,w}$ 13 S).
2. Annealed by heating at 95° C in the presence of 8 M urea.
3. Exposed to the cells for 24 hrs (or, alternatively, for 1 hr in the presence of DEAE-dextran or $CaCl_2$).

ally dependent on the molecular size of the homopolymers and more dependent on maintaining a high molecular size of poly(G) than of poly(C). This confirms the findings of Vilner et al. (32).

Even more crucial are the conditions at which poly(G) and poly(C) are annealed. Because of the tendency of poly(G) to form self-aggregated complexes, it is impossible to obtain duplex formation with poly(C) under the conditions which are normally applied

Table II. Interferon induction by poly(G) . poly(C) in primary rabbit kidney cells, as evaluated under different assay conditions

Procedure	IFN titer (log_{10} units/ml)	
	poly(G) . poly(C) at 10^{-5} Mp	poly(I) . poly(C) at 10^{-5} Mp
Complex exposed to the cells for 1 hr; then cells superinduced with cycloheximide and actinomycin D.	<0.8	4.7
Complex exposed to the cells for 24 hrs; then cells superinduced with cycloheximide and actinomycin D.	2.5	2.0
Complex exposed to the cells for 1 hr in the presence of DEAE-dextran (100 μg/ml)	3.0	3.5
Complex exposed to the cells for 1 hr in the presence of $CaCl_2$ (5 mM).	3.5	4.5

The conditions for measuring interferon induction in primary rabbit kidney cells have been described previously (9, 14, 30)

for the preparation of poly(I) . poly(C). However, a genuine poly(G) . . poly(C) complex with 1 : 1 stoichiometry was obtained when the homopolymers were annealed in 8 M urea at a temperature > 90° (tab. I). The annealing method we have employed (29) was modification of the method originally proposed by Englander et al. (16).

Another critical determinant for the interferon inducing ability of poly(G) . poly(C) is the time of exposure to the cells (tab. I). If it is exposed to the cells for 1 hr, as is routinely done with poly(I) . . poly(C), no interferon is induced; but, if left in contact with the cells for 24 hr, poly(G) . poly(C) is as efficient, if not more efficient, than poly(I) . poly(C) in inducing interferon in primary rabbit kidney cells (tab. II). The incubation period can be shortened from 24 hr to 1 hr if poly(G) . poly(C) is incubated on the cells in the presence of DEAE-dextran or $CaCl_2$. Under these, conditions, poly(G) . poly(C) induces even higher interferon levels than after a 24 hrs contact period, and the cells do not have to be "superinduced" (treated with cycloheximide and actinomycin D) to further enhance the interferon response (tab. I).

Table III. Interferon induction by poly(G) . poly(C) in different human cell lines

Cell line	IFN titer ($\log_{10}$ units/ml)			
	Poly(G) . poly(C) at 10^{-4} Mp		Poly(I) . poly(C) at 10^{-4} Mp	
	Procedure A	Procedure B	Procedure A	Procedure B
VGS	...	<0.5	...	2.6
E_6SM	<0.2	<0.7	2.0	2.0
RL (T-21)	...	<0.7	...	1.5
MG-63	2.0	<1.0	2.0	2.3
Primary rabbit kidney	3.3	3.2	2.0	4.1

Procedure A: complex exposed to the cells for 24 hrs; then cells superinduced with cycloheximide and actinomycin D.

Procedure B: complex exposed to the cells for 1 hr in the presence of DEAE-dextran (100 μg/ml); cells not superinduced with cycloheximide and actinomycin D.

While poly(G) . poly(C) proved very effective in inducing interferon in primary rabbit kidney cells following a 1 hr incubation period with the cells in the presence of DEAE-dextran, a similar procedure failed to elicit any interferon production in human cells (tab. III). A number of human cell lines were used, i.e. the diploid fibroblast cell lines VGS and E_6SM, the 21-trisomic cell line RL and the aneuploid fibroblastoid cell line MG-63, but none produced interferon in response to a 1 hr contact with poly(G) . poly(C) in the presence of DEAE-dextran. However, the MG-63 cells consistently yielded interferon levels of 10^2 units/ml when exposed to poly(G) . . poly(C) for 24 hrs and then superinduced (tab. III).

The antiviral principle induced by poly(G) . poly(C) in cell culture behaved in all aspects like interferon: it was destroyed by trypsin, stable at pH 2 and inactive on heterologous cells. The interferon induced by poly(G) . poly(C) in MG-63 cells was characterized as HuIFN-β, based on its neutralization by antiserum to HuIFN-β (tab. IV). This anti-(HuIFN-β) serum neutralized the poly(G) .

Table IV. Neutralization of interferon induced by poly(G) . poly(C) in MG-63 cells with antiserum to HuIFN-β

Sample	Actual IFN titer, as used in assay (log_{10} units/ml)	Neutralization titer (log_{10} units/ml)
Poly(G) . poly(C) at 10^{-4} Mp	1.2	5.0
Poly(G) . poly(C) at 10^{-5} Mp	1.2	5.0
Poly(I) . poly(C) at 10^{-4} Mp	1.2	5.0
Poly(I) . poly(C) at 10^{-5} Mp	1.2	5.0
Authentic HuIFN-β	1.0	5.1

The IFN samples were prepared by procedure A (complex exposed to the cells for 24 hrs; then cells superinduced with cycloheximide and actinomycin D). Mixtures of the IFN samples, diluted so as to contain 10 IFN units per ml, with serial dilutions of goat anti-(HuIFN-β) serum were incubated at 37° C for 1 hr, after which the residual antiviral activity was determined in RL (T-21) cells challenged with VSV. The neutralization titers correspond to the highest dilution of antiserum required to neutralize the protective activity of IFN by 50 %, multiplied by the interferon titer of the sample assayed.

. poly(C)-induced MG-63 interferon to the same extent as authentic HuIFN-β.

The interferon inducing capacity of poly(G) . poly(C) was not limited to cell cultures. It was also effective in inducing interferon in rabbits and mice (fig. 3). The kinetics of the interferon response

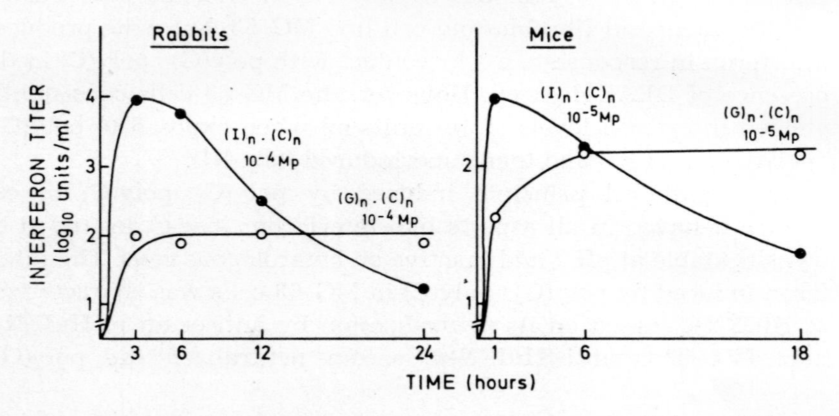

Fig. 3. Interferon inducing activity of poly(G) . poly(C) in rabbits and mice. Albino-white rabbits (weighing ∼1 kg) were injected intravenously with poly(G) . poly(C) or poly(I) . poly(C) at 10^{-4} Mp in 1 ml Dulbecco's phosphate buffered saline (PBS). Female NMRI mice (weighing 20—25 g) were injected intravenously with poly(G) . poly(C) or poly(I) . poly(C) at 10^{-5} Mp in 0.2 ml Dulbecco's PBS (3 mice/group). Blood samples were taken at the indicated times.

to poly(G) . poly(C) was clearly different from that of poly(I) . . poly(C). While poly(I) . poly(C) gave higher peak levels of interferon than poly(G) . poly(C), the interferon response to poly(G) . . poly(C) lasted longer; even 24 hours after poly(G) . poly(C) had been injected, no decrease in interferon titer was noted. A similar prolonged interferon production pattern has been previously noted by Aksenov et al. (1) with poly(G) . poly(C) in mice.

The protracted interferon response generated by poly(G) . . poly(C) in vivo may be related to a longer persistance of the polynucleotide in biological fluids. Poly(I) . poly(C) is rapidly degraded by human plasma nuclease(s), and these nuclase(s) resemble pancreatic ribonuclease A in substrate specificity in that they specifically hydrolyze the poly(C) strand of the poly(I) . poly(C) complex (10). In contrast with poly(I) . poly(C), poly(G) . poly(C) was completely resistant to degradation by ribonuclease A and human serum (fig. 4). Even when exposed to ribonuclease A at 100 μg/ml or human serum at 50 % v/v, poly(G) . poly(C) did not lose any of its interferon inducing potential. These observations are in agreement with those of Vilner et al. (34) showing lack of inactivation of poly(G) . poly(C) after 3 hr incubation with monkey serum.

According to Smorodintsev (25) and Smorodintsev et al. (26), poly(G) . poly(C) would be considerably less toxic than poly(I) .

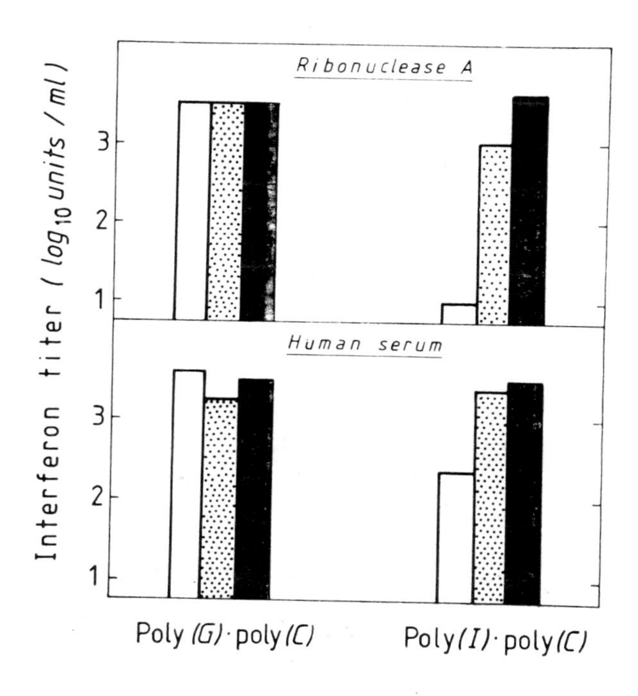

Fig. 4. Sensitivity of poly(G) . poly(C) to degradation by pancreatic ribonuclease A and human serum, as monitored by residual interferon induction in primary rabbit kidney cells. Poly(G) . poly(C) and poly(I) . poly(C) were incubated at a final concentration of 10^{-5} Mp with either ribonuclease A or human serum for 1 hr at 37 °C. DEAE-dextran was then added at a final concentration of 100 μg/ml and the reaction mixtures were incubated with the cells for 1 hr. The cells were further monitored for IFN production. Concentrations of ribonuclease A: □ 1, 100 μg/ml; ▒ 2, 10 μg/ ml; ■ 3, control. Concentrations of human serum: □ 1, 50 % (v/v); ▒ 2, 10 % (v/v); ■ 3, control.

. poly(C) in vivo. Its acute LD_{50} would be greater than 200 mg/kg in mice and greater than 10 mg/kg in rabbits (tab. V). Unlike poly(I) . poly(C) which causes an increased death rate when injected shortly after the mice have been infected with Newcastle disease virus or Vesicular stomatitis virus (11), poly(G) . poly(C) would be devoid of such increased toxicity for virus-infected animals (33).

Table V. Toxicity of poly(G) . poly(C) in vivo[1]

Animal species	Acute LD_{50} (mg/kg)	
	Poly(I) . poly(C)	Poly(G) . poly(C)
Mice	15.8	>200
Rabbits	0.22	>10

[1] Following intraperitoneal administration according to Smorodintsev (25) or intravenous administration according to Smorodintsev et al. (26).

It appears, therefore, that poly(G) . poly(C) has several advantages over poly(I) . poly(C) (tab. VI). It is longer acting as an interferon inducer in vivo, more resistant to degradation by nucleases, and, apparently, less toxic. This also means that, if the lack of toxicity of poly(G) . poly(C) were to be confirmed, greater resistance to degradation not necessarily leads to increased toxicity, as was originally postulated by Ts'o et al. (31) and Carter et al. (5).

Table VI. Advantages of poly(G) . poly(C) over poly(I) . poly(C)

1. Longer acting as an interferon inducer in vivo (mice, rabbits).
2. More resistant to degradation by nucleases.
3. Less toxic in vivo (mice, rabbits).

The clinical usefulness of poly(G) . poly(C) in the prophylaxis and treatment of virus infections in humans remains to be defined. In a clinical study reported by Smorodintsev (25), poly(G) . poly(C) was administered intranasally as an aerosol (6 mg/2 ml) to volunteers infected with Influenza virus A_2 (H_3N_2) Hong Kong. Optimal protection was achieved if poly(G) . poly(C) was given 24 hr prior to virus infection. Under these conditions poly(G) . poly(C) reduced the incidence of virus shedding from 70 % to 27 %.

Considering the distinct advantages of poly(G) . poly(C) over poly(I) . poly(C) (tab. VI), it would now seem mandatory to further assess the therapeutic and prophylactic potential of poly(G) . poly(C) in vivo. These studies should be focussed on various aspects, such as the biodegradation of poly(G) . poly(C), its protective activity against virus infections, its inhibitory effects (if any) on tumor growth, its short- and long-term toxicity and, finally, the host's responsiveness to repeated administration of the polynucleotide. Depending on the results of these studies, one may then envisage the possibility of clinical trials in which the efficacy of poly(G) . . poly(C) as an antiviral or antitumor agent would be evaluated, either as such or in comparison with poly(I) . poly(C).

References

1 Aksenov, O. A.; Bresler, S. E.; Zakabunin, A. I.; Kogan, E. M.; Rait, V. K.; Salganik, R. I.; Sidorova, N. S.; Smorodintsev, Al. A.; Timkovsky, A. L. (1978): Voprosy Virusologii 23, 466.

2 Aksenov, O. A.; Timkovsky, A. L.; Ageeva, O. N.; Kogan, E. M.; Bresler, S. E.; Smorodintsev, A. A.; Tikhomirova-Sidorova, N. S. (1973): Voprosy Virusologii 18, 345.

3 Black, D. R.; Eckstein, F.; De Clercq, E.; Merigan, T. C. (1973). Antimicrob. Agents Chemother. 3, 198.

4 Bobst, A. M.; Langemeier, P. W.; Torrence, P. F.; De Clercq, E. (1981): Biochemistry 20, 4798.

5 Carter, W. A.; O'Malley, J.; Beeson, M.; Cunningham, P.; Kelvin, A.; Vere-Hodge, A.; Alderfer, J. L.; Ts'o, P. O. P. (1976): Mol. Pharmacol. 12, 440.

6 Colby, C.; Chamberlin, M. J. (1969): Proc natl. Acad. Sci. USA 63, 160.

7 De Clercq, E. (1979): Eur. J. Biochem. 93, 165.

8 De Clerq, E.; Eckstein, F.; Merigan, T. C. (1970): Ann. N. Y. Acad. Sci. 173, 444.

9 De Clercq, E.; Edy, V. G.; Torrence, P. F.; Waters, J. A.; Witkop, B. (1976): Mol. Pharmacol. 12, 1045.

10 De Clercq, E.; Huang, G.-F.; Bhooshan, B.; Ledley, G.; Torrence, P. F. (1979): Nucleic Acids Res. 7, 2003.

11 De Clercq, E.; Stewart II, W. E.; De Somer, P. (1973): Infect. Immun. 7, 167.

12 De Clercq, E.; Stollar, B. D.; Hobbs, J.; Fukui, T.; Kakiuchi, N.; Ikehara, M. (1980): Eur. J. Biochem. 107, 279.

13 De Clercq, E.; Torrence, P. F. (1977): J. Gen. Virol. 37, 619.

14 De Clercq, E.; Torrence, P. F.; Stollar, B. D.; Hobbs, J.; Fukui, T.; Kakiuchi, N.; Ikehara, M. (1978): Eur. J. Biochem. 88, 341.

15 De Clercq, E.; Torrence, P. F.; Witkop, B. (1974): Proc. natl. Acad. Sci. USA 71, 182.
16 Englander, J. J.; Kallenbach, N. R.; Englander, S. W. (1972): J. mol. Biol. 63, 153.
17 Field, A. K.; Tytell, A. A.; Lampson, G. P.; Hilleman, M. R. (1967): Proc. natl. Acad. Sci. USA 58, 1004.
18 Greene, J. J.; Alderfer, J. L.; Tazawa, I.; Tazawa, S.; Ts'o, P. O. P.; O'Malley, J. A.; Carter, W. A. (1978): Biochemistry 17, 4214.
19 Kakiuchi, N. C.; Marck, C.; Rousseau, N.; Leng, M.; De Clercq, E.; Guschlbauer, W. (1982): J. Biol. Chem. 257, 1924.
20 Levine, A. S.; Sivulich, M.; Wiernik, P. H.; Levy, H. B. (1979): Cancer Res. 39, 1645.
21 Levy, H. B.; Baer, G.; Baron, S.; Buckler, C. E.; Gibbs, C. J.; Iadarola, M. J.; London, W. T.; Rice, J. A. (1975): J. infect. Dis. 132, 434.
22 Matsuda, S.; Kida, M.; Shirafuji, H.; Yoneda, M.; Yaoi, H. (1971): Arch. Ges. Virusforsch. 34, 105.
23 Novokhatsky, A. S,; Ershov, F. I.; Timkovsky, A. L.; Bresler, S. E.; Kogan, E. M.; Tikhomirova-Sidorova, N. S. (1975): Acta virol. 19, 121.
24 Reuss, K.; Scheit, K.-H.; Saiko, O. (1976): Nucleic Acids Res. 3, 2861.
25 Smorodintsev, A. A. (1979): Acta biol. med. Germ. 38, 867.
26 Smorodintsev, A. A.; Aksenov, O. A.; Konstantinova, I. K.; Vilner, L. M.; Tyufanov, A. V.; Timkovský, A. L.; Bresler, S. E.; Kogan, E. M.; Tikhomirova-Sidorova, N. S. (1978): Voprosy Virusologii 23, 201.
27 Stringfellow, D. A.; Weed, S. D. (1980): Antimicrob. Agents Chemother. 17, 988.
28 Timkovsky, A. L.; Aksenov, O. A.; Bresler, S. E.; Kogan, E. M.; Smorodintsev, A. A.; Tikhomirova-Sidorova, N. S. (1973): Voprosy virusologii 18, 350.
29 Torrence, P. F.; De Clercq, E. (1983): in "Antiviral Drugs and Interferon: The Molecular Basis of their Activity", ed. by Y. Becker, Martinus Nijhoff Publishers, The Hague (The Netherlands) (in press).
30 Torrence, P. F.; De Clercq, E.; Waters, J. A.; Witkop, B. (1974): Biochemistry 13, 4400.
31 Ts'o, P. O. P.; Alderfer, J. L.; Levy, J.; Marshall, L. W.; O'Malley, J.; Horoszewicz, J. S.; Carter, W. A. (1976): Mol. Pharmacol. 12, 299.
32 Vilner, L. M.; Brodskaya, L. M.; Kogan, E. M.; Timkovsky, A. L.; Tikhomirova-Sidorova, N. S.; Bresler, S. E. (1974): Voprosy Virusologii 19, 45.
33 Vilner, L. M.; Kogan, E. M.; Timkovsky, A. L.; Tyufanov, A. V.; Finogenova, E. V.; Platonova, G. A.; Tikhomirova-Sidorova, N. S. (1980): Voprosy Virusologii 25, 67.
34 Vilner, L. M.; Timkovsky, A. L.; Kogan, E. M.; Naumovich, N. G.; Brodskaya, L. M.; Chernyshov, V. I.; Tikhomirova-Sidorova, N. S. (1979): Voprosy Virusologii 24, 181.

E. de Clercq, M.D., Riga Institute for Medial Research, Katholieke Universiteit Leuven, Minderbroedersstraat 10, B-3000 Leuven (Belgium)